Biological Diversity and Function in Soils

Soil has generally been regarded as something of a 'black box' by ecologists. The importance of soil is obvious: it provides physical support for plants, and both the living and non-living components contribute to a variety of important environmental functions. Soil is a species-rich habitat, but many questions about the ecological significance of the soil's biological diversity, and in particular how it affects ecosystem function, have never been asked. The linkages between above-ground ecology, which is rich in ecological theory, and below-ground ecology, where investigation has been restricted by methodological difficulties, have not been made.

Recent technical developments, including isotopic and molecular methods as well as new experimental and modelling approaches, have led to a renaissance in soil biodiversity research. The key areas are reflected in this exciting new volume, which brings together many leading contributors to the new understanding of the role and importance of soil biota.

RICHARD D. BARDGETT is Professor of Ecology at Lancaster University and has published widely on plant–soil interactions, nutrient cycling and soil ecology. He is especially interested in studying linkages between plant and soil biological communities, and examining how these links are affected by herbivores.

MICHAEL B. USHER is retired. For more than 40 years he has had a research interest in the soil biota, especially the Collembola (springtails), Isoptera (termites) and Mesostigmata (mostly predatory mites). He also has a strong interest in nature conservation, and is currently working on a number of aspects of the conservation of biodiversity.

DAVID W. HOPKINS is Professor of Environmental Science at the University of Stirling. His interests cover the microbiology and biochemistry of carbon and nitrogen cycling in soils, particularly processes of plant residue decomposition and nutrient cycling in soils at high latitudes and altitudes.

Ecological Reviews

Ecological Reviews will publish books at the cutting edge of modern ecology, providing a forum for volumes that discuss topics that are focal points of current activity and of likely long-term importance to the progress of the field. The series will be an invaluable source of ideas and inspiration for ecologists at all levels from graduate students to more established researchers and professionals. The series will be developed jointly by the British Ecological Society and Cambridge University Press and will encompass the Society's Symposia as appropriate.

Biological Diversity and Function in Soils
Edited by Richard D. Bardgett, Michael B. Usher and David W. Hopkins

Biotic Interactions in the Tropics: Their Role in the Maintenance of Species Diversity
Edited by David F. R. P. Burslem, Michelle A. Pinard and Sue E. Hartley

Biological Diversity and Function in Soils

Edited by

RICHARD D. BARDGETT
Institute of Environmental and Natural Sciences, Lancaster University, UK

MICHAEL B. USHER
School of Biological and Environmental Sciences, University of Stirling, UK

DAVID W. HOPKINS
School of Biological and Environmental Sciences, University of Stirling, UK

CAMBRIDGE
UNIVERSITY PRESS

CAMBRIDGE UNIVERSITY PRESS
Cambridge, New York, Melbourne, Madrid, Cape Town, Singapore, São Paulo

Cambridge University Press
The Edinburgh Building, Cambridge CB2 8RU, UK

Published in the United States of America by Cambridge University Press, New York

www.cambridge.org
Information on this title: www.cambridge.org/9780521847094

© British Ecological Society 2005

First published 2005

A catalogue record for this publication is available from the British Library

ISBN 978-0-521-84709-4 hardback
ISBN 978-0-521-60987-6 paperback

Transferred to digital printing 2007

Contents

Contributors

JONATHAN M. ANDERSON School of
Biological Sciences, University of Exeter,
Exeter EX4 4PS, UK. Email:
J.M.Anderson@exeter.ac.uk

RICHARD D. BARDGETT
Institute of Environmental and Natural
Sciences, Lancaster University, Lancaster LA1
4YQ, UK. Email: r.bardgett@lancaster.ac.uk

ISABELLE BARIOS
Instituto de Ecología A. C., Xalapa, Veracruz,
Mexico. Email: Isabelle@ecologia.edu.mx

EDMUNDO BARRIOS
Centro Internacional de Agriculture Tropical
(CIAT), Cali, Colombia. Email:
E.Barrios@cgian.org

JENNIFER BENNETT
Forest Nutrition Coop, North Carolina State
University, Raleigh, NC 27695-8008, USA.
Email: Jbennett@ncsfnc.cfr.ncsu.edu

MATTY P. BERG
Department of Animal Ecology, Institute of
Ecological Science, Vrije Universiteit, De
Boelelaan 1085, 1081 HV Amsterdam, The
Netherlands. Email:
matty.berg@ecology.falw.vu.nl

DAVID BIGNELL
Queen Mary College, University of London,
London, UK. Email: D.Bignell@qmw.ac.uk

L. BODDY
Cardiff School of Biosciences, Cardiff
University, Cardiff CF10 3TL, UK. Email:
BoddyL@cardiff.ac.uk

MICHAEL BROCKHURST
Department of Plant Sciences, University of
Oxford, South Parks Road, Oxford, OX1 3RB,
UK. Email:
michael.brockhurst@plant-sciences.
oxford.ac.uk

LIJBERT BRUSSAARD
Soil Quality Section, Wageningen University,
PO Box 8005, 6700 AA Wageningen, The
Netherlands. Email: lijbert.brussaard@wur.nl

ANGUS BUCKLING
Department of Biology and Biochemistry,
University of Bath, Bath BA2 7AY, UK. Email:
bssagjb@bath.ac.uk

S. R. COLVAN
Institute for Research in Environment and
Sustainability, King George VI Building,
University of Newcastle upon Tyne, Newcastle
upon Tyne NE1 7RU, UK. Email:
stephanie.colvan@ncl.ac.uk

THOMAS P. CURTIS
School of Civil Engineering and Geosciences
and Centre for Molecular Ecology, University
of Newcastle upon Tyne, Newcastle upon
Tyne NE1 7RU, UK. Email:
Tom.Curtis@newcastle.ac.uk

LEWIS J. DEACON
Centre for Ecology and Hydrology Lancaster, Library Avenue, Bailrigg, Lancaster, LA1 4AP, UK. Email: lewis.deacon@kcl.ac.uk

RON G. M. DE GOEDE
Soil Quality Section, Wageningen University, PO Box 8005, 6700 AA Wageningen, The Netherlands. Email: ron.degoede@wur.nl

PETER C. DE RUITER
Department of Environmental Sciences, Copernicus Research Institute for Sustainable Development and Innovation, Utrecht University, PO Box 80115, 3508 TC Utrecht, The Netherlands. Email: P.deRuiter@geog.uu.nl

D. P. DONNELLY
Cardiff School of Biosciences, Cardiff University, Cardiff CF10 3TL, UK. Email: damian.donnelly@btinternet.com

NOAH FIERER
Department of Ecology, Evolution and Marine Biology, University of California Santa Barbara, Santa Barbara, CA 93106, USA. Email: Fierer@lifesci.ucsb.edu

ALASTAIR H. FITTER
Department of Biology, University of York, York YO1 5DD, UK. Email: ahf1@york.ac.uk

KEN GILLER
Department of Plant Sciences, Wageningen University, P. O. Box 430, 6700 AK, Wageningen, The Netherlands. Email: ken.giller@wur.nl

NEIL D. GRAY
School of Civil Engineering and Geosciences and Centre for Molecular Ecology, University of Newcastle upon Tyne, Newcastle upon Tyne NE1 7RU, UK

E. G. GREGORICH
Agriculture and Agri-Food Canada, Central Experimental Farm, Ottawa, Ontario K1A OC6, Canada. Email: gregoriche@agr.gc.ca

P. GROGAN
Biology Department, Queen's University, Kingston, Ontario K7L 3N6, Canada. Email: groganp@biology.queensu.ca

J. A. HARRIS
Institute of Water and Environment, Cranfield University, Silsoe, Bedfordshire MK45 4DT, UK. Email: j.a.harris@cranfield.ac.uk

IAN M. HEAD
School of Civil Engineering and Geosciences and Centre for Molecular Ecology, University of Newcastle upon Tyne, Newcastle upon Tyne NE1 7RU, UK

LIA HEMERIK
Soil Quality Section, Wageningen University, PO Box 8005, 6700 AA Wageningen, The Netherlands, and Biometris, Department of Mathematics and Statistical Methods, Wageningen University, PO Box 100, 6700 AC Wageningen, The Netherlands. Email: lia.hemerik@wur.nl

R. J. HOBBS
School of Environmental Science, Murdoch University, Murdoch, WA 6150, Australia. Email: rhobbs@essun1.murdoch.edu.au

DAVID J. HODGSON
School of Biological Sciences, Hatherly Laboratories, University of Exeter, Prince of Wales Road, Exeter EX4 4PS, UK. Email: D.J.Hodgson@exeter.ac.uk

D. W. HOPKINS
School of Biological and Environmental Sciences, University of Stirling, Stirling FK9 4LA, UK. Email: d.w.hopkins@stir.ac.uk

JEROEN HUISING
Tropical Soil Biology and Fertility Institute, Centro Internacional de Agriculture Tropical (CIAT), Nairobi, Kenya. Email: J.Huising@cgiar.org

D. JOHNSON
School of Biological Sciences, University of
Aberdeen, Cruickshank Building, St Machar
Drive, Aberdeen AB24 3UU, UK. Email:
d.Johnson@abdn.ac.uk

T. HEFIN JONES
Cardiff School of Biosciences, Cardiff
University, PO Box 915, Cardiff CF10 3TL, UK.
Email: JonesTH@cardiff.ac.uk

NANCY KARANJA
Department of Soil Science, University of
Nairobi, Nairobi, Kenya. Email:
biofix@arcc.or.ke

REES KASSEN
Department of Plant Sciences, University of
Oxford, South Parks Road, Oxford OX1 3RB,
UK. Email: rkassen@uottawa.ca

K. KILLHAM
School of Biological Sciences, University of
Aberdeen, Aberdeen AB34 3UU, UK. Email:
k.killham@abdn.ac.uk

PATRICK LAVELLE
UMR 137 BIOSOL, Centre 1RD Ile de France,
Université de Paris VI, 32 rue Henri Varagnat,
93143 Bondy Cedex, France. Email:
Patrick.Lavelle@bondy.ird.fr

J. R. LEAKE
Department of Animal and Plant Sciences,
University of Sheffield, Sheffield S10 2TN, UK.
Email: j.r.leake@sheffield.ac.uk

E. MALOSSO
Institute for Research in Environment and
Sustainability, King George VI Building,
University of Newcastle upon Tyne, Newcastle
upon Tyne NE1 7RU, UK. Email:
elaine_malosso@hotmail.com

ALAN J. McCARTHY
School of Biological Sciences, The Biosciences
Building, University of Liverpool, Liverpool
L69 7ZB, UK. Email: aj55m@liverpool.ac.uk

A. MEHARG
School of Biological Sciences, University of
Aberdeen, Aberdeen AB24 3UU, UK. Email:
a.meharg@abdn.ac.uk

JOHN MOORE
Department of Biology, University of
Northern Colorado, Greeley, CO 80523, USA.
Email: jcmoore@bentley.unco.edu

FATIMA MOREIRA
Universidade Federal de Lavras, Minas Gerais,
Brazil. Email: fmoreira@ufla.br

ANJE-MARGRIET NEUTEL
Department of Environmental Sciences,
Copernicus Research Institute for Sustainable
Development and Innovation, Utrecht
University, PO Box 80115, 3508 TC Utrecht,
The Netherlands. Email: neutel@geog.uu.nl

A. G. O'DONNELL
Institute for Research in Environment and
Sustainability, King George VI Building,
University of Newcastle upon Tyne, Newcastle
upon Tyne NE1 7RU, UK. Email:
tony.o'donnell@ncl.ac.uk

ELDOR A. PAUL
Natural Resource Ecology Laboratory,
Colorado State University, Fort Collins, CO
80523, USA. Email: eldor@nrel.colostate.edu

J. I. PROSSER
Department of Molecular and Cell Biology,
Institute of Medical Sciences, University of
Aberdeen, Aberdeen AB25 2ZD, UK. Email:
j.prosser@abdn.ac.uk

E. JANIE PRYCE MILLER
Department of Life Sciences, King's College,
University of London, Franklin-Wilkins
Building, 150 Stamford Street, London SE1
9NN, UK, and Centre for Ecology and
Hydrology Lancaster, Library Avenue,
Bailrigg, Lancaster, LA1 4AP, UK. Email:
elizabeth.pryce_miller@kcl.ac.uk

PAUL B. RAINEY
School of Biological Sciences, University of
Auckland, Private Bag 92019, Auckland, New
Zealand, and Department of Plant Sciences,
University of Oxford, South Parks Road,
Oxford OX1 3RB, UK. Email:
p.rainey@auckland.ac.nz,
paul.rainey@plants.ox.ac.uk

J. I. RANGEL CASTRO
School of Biological Sciences, University of
Aberdeen, Aberdeen AB24 3UU, UK, and
Department of Molecular and Cell Biology,
Institute of Medical Sciences, University of
Aberdeen, Aberdeen AB25 2ZD, UK. Email:
i.rangel@abdn.ac.uk

D. J. READ
Department of Animal and Plant Sciences,
University of Sheffield, Sheffield S10 2TN, UK.
Email: d.j.read@sheffield.ac.uk

KARL RITZ
National Soil Resources Institute, Cranfield
University, Silsoe, MK45 4DT, UK. Email:
k.ritz@cranfield.ac.uk

CLARE H. ROBINSON
Department of Life Sciences, King's College,
University of London, Franklin–Wilkins
Building, 150 Stamford Street, London SE1
9NN, UK. Email: clare.robinson@kcl.ac.uk

JOSHUA P. SCHIMEL
Department of Ecology, Evolution and
Marine Biology, University of California
Santa Barbara, Santa Barbara, CA 93106, USA.
Email: schimel@lifesci.ucsb.edu

HEIKKI SETÄLÄ
Department of Ecological and Environmental
Sciences, University of Helsinki, FIN-15140
Lahti, Finland. Email:
heikki.setala@helsinki.fi

D. B. STANDING
School of Biological Sciences, University of

Aberdeen, Aberdeen AB24 3UU, UK. Email:
d.standing@abdn.ac.uk

S. SUPAPHOL
Department of Soil Science, Faculty of
Agriculture, Kasetsart University, Bangkok
10900, Thailand. Email: orn478@yahoo.com

MIKE SWIFT
Tropical Soil Biology and Fertility Institute
(TSBF), Centro Internacional de Agriculture
Tropical (CIAT), Nairobi, Kenya. Email:
Mike.Swift@mpl.ird.fr

MICHAEL B. USHER
School of Biological and Environmental
Sciences, University of Stirling, Stirling FK9
4LA, UK. Email: m.b.usher@stir.ac.uk

WIM H. VAN DER PUTTEN
Netherlands Institute of Ecology, PO Box 40,
6666 ZG Heteren, The Netherlands. Email:
putten@nioo.knaw.nl

MEINE VAN NOORDWIJK
World Agroforestry Centre, ICRAF Southeast
Asia, Bogor, Indonesia. Email:
M.Van-Noordwijk@cgiar.org

BART C. VERSCHOOR
Soil Quality Section, Wageningen University,
PO Box 8005, 6700 AA Wageningen, The
Netherlands, and De Groene Vlieg BV,
Houtwijk 75, 8251 GD Dronten, The
Netherlands. Email:
verschoor@degroenevlieg.nl

DIANA H. WALL
Natural Resource Ecology Laboratory,
Colorado State University, Fort Collins, CO
80523, USA. Email: Diana@nrel.colostate.edu

DAVID A. WARDLE
Department of Forest Vegetation Ecology,
Faculty of Forestry, Swedish University of
Agricultural Sciences, S901-83 Umeå, and
Landcare Research, PO Box 69, Lincoln, New
Zealand. Email: david.wardle@svek.slu.se

GREGOR W. YEATES
Landcare Research, Private Bag 11-052,
Palmerston North, New Zealand. Email:
yeatesg@landcareresearch.co.nz

IAIN M. YOUNG
SIMBIOS Centre, University of Abertay
Dundee, Bell Street, Dundee DD1 1HG, UK.
Email: i.young@tay.ac.uk

Editors' Preface

Soil has generally been treated as something of a 'black box' by ecologists. It provides the physical support for plants, and both the living and non-living components contribute to a variety of important environmental functions. These include decomposition and the recycling of nutrients, which are both key functions in terrestrial ecosystems. Other roles, such as the breakdown of pollutants and the storage of bioelements, have immense applied significance in a changing environment. Soil provides a habitat for many species of bacteria, fungi, protists and animals; it is generally recognised as a habitat that is species rich. But many questions about the ecological significance of the soil's biological diversity, and in particular how it affects ecosystem function, have never been asked. Until fairly recently this has been because the linkages between above-ground ecology, which is rich in ecological theory, and below-ground ecology, where investigation has been restricted by methodological difficulties, have not been made. It is now time to open the 'black box' and to start to understand how it works.

At the end of the twentieth century and with the start of the twenty-first century, efforts have been going on around the world to gain a greater understanding of the diversity of life in the soil and of the functions that these many species perform. In the UK there have been two major programmes of research on biological diversity and the function of soil ecosystems. As a result of these, and research programmes in other parts of the world, and many other activities by soil ecologists, there have been numerous recent technical advances, including isotopic and molecular methods, as well as new experimental and modelling approaches to understanding the functions of life in the soil. Concurrent with these technical developments, there has been an increasing recognition within the wider ecological community of the importance of soil organisms. This has largely come from the increasing knowledge about the functions of the soil biota. For a long while it has been known that the soil biota regulates major ecosystem processes such as organic matter turn-over and nutrient mineralisation. But now it is suspected that it has a key role in feedbacks between above-ground and below-ground communities, and hence that it influences how the whole of the ecosystem functions.

It was, therefore, topical for the British Ecological Society to hold its 2003 symposium with the theme 'Biological diversity and function in soils', presenting a synthesis of what we know about soil biodiversity and its role in the soil ecosystem. Indeed, in 2003, under the heading 'Areas to watch for in 2004', *Science* (Vol. 302, p. 2040) highlighted soil biodiversity and ecosystem function as one of the research themes set for a big change. This is because of the happy coincidence of technique development, the advancement of a theoretical framework and concentration of effort. To put this into perspective, soil biodiversity featured alongside human genome research, exploration of Mars, subatomic physics and open-access scientific publishing. Soils and soil ecologists have joined the big league!

It seems likely that no soil is 'non-functional', but that all soils are either more or less favourable from a human point of view. We have known for a long time that soils host a huge variety of organisms, which jointly provide a range of 'ecosystem services', but we know very little about which organism is doing what and when it is doing it. Recently, molecular tools, clone libraries and a range of other taxonomic techniques have allowed researchers to gain new insights into biological diversity and the processes that these various organisms perform. But are all of these new techniques outstripping our theoretical framework and our ability to design experiments that can fully exploit and test their potential?

We are sure that the symposium in 2003, and these proceedings, will go a long way in demonstrating the recent advances in soil biology, the developments in soil ecology, and the research basis for both policy development and practical management of soil. We are equally sure that readers of this book will appreciate that at last we are beginning to open the soil's 'black box', and what we find inside is both wonderful and exciting.

Acknowledgements

We should like to thank the many scientists, around the world, who have acted as reviewers of each of the chapters, and Jo Newman, who has assisted us in undertaking all of the minutiae of editing this series of 20 chapters. We are also extremely grateful to Hazel Norman and Richard English of the British Ecological Society, both of whom assisted greatly with the organisation of the symposium, to the people who chaired the symposium sessions, and to the staff at Lancaster University who ensured that all went well on the days of the meeting.

PART I

Introduction

CHAPTER ONE

Developing new perspectives from advances in soil biodiversity research

DIANA H. WALL
Colorado State University

ALASTAIR H. FITTER
University of York

ELDOR A. PAUL
Colorado State University

SUMMARY

1. We use a historical context to examine the accomplishments of soil biodiversity and ecosystem research. These accomplishments provide a framework for future research, for enhancing and driving ecological theory, and for incorporating knowledge into sustainable management of soils and ecosystems.
2. A soil ecologist's view of the world differs from that of a terrestrial ecologist who focuses research primarily on above-ground organisms. We offer 'ten tenets of soil ecology' that illustrate the perspectives of a soil ecologist.
3. Challenges for the future are many and never has research in soil ecology been more exciting or more relevant. We highlight our view of 'challenges in soil ecology', in the hope of intensifying interactions among ecologists and other scientists, and stimulating the integration of soils research into the science of terrestrial ecology.
4. We conclude with the vision that healthy soils are the basis of global sustainability. As scientists, we cannot achieve our future goals of ecological sustainability without placing emphasis on the role of soil in terrestrial ecology.

Introduction

Despite the visionary appeals of an earlier generation of soil scientists, soil biologists and others (Jacks & Whyte 1939; Hyams 1952), above-ground ecologists have hitherto shown insufficient awareness of the significance and fragility of soils and the need to understand how life in soils relates to sustaining our global environment. However, many scientists, including microbial ecologists, atmospheric scientists, biogeochemists and agronomists, as well as economists and policy makers, are now starting to take heed of the multiple issues involving

Biological Diversity and Function in Soils, eds. Richard D. Bardgett, Michael B. Usher and David W. Hopkins.
Published by Cambridge University Press. © British Ecological Society 2005.

soils and their biota, on both local and global scales. Phenomena that affect global sustainability via their impact on soil biota include wind and water erosion, pollution, the use of genetically engineered plants and microorganisms, invasive species, atmospheric deposition, land use change, and changes in soil structure. Such changes have rippling effects on the hydrologic cycle, loss of carbon and loss of fertile soils for cropping as well as societal needs. These issues pinpoint how little scientists knew about the relationship between soil biodiversity and ecosystem functioning, and whether scientists could, based on rigorous experiments, predict how future changes might impact human interactions with soil and their biota.

Papers throughout this volume summarise many of the recent results and how these might apply to future sustainability of soils, their biota and both ecosystem functions and processes. The advances and research priorities that they highlight need to be evaluated against short- and long-term needs for determining solutions to environmental problems. Recent data together with the immediacy of environmental changes that affect soil and soil biota create a mandate for

- assessing the present state of knowledge,
- developing and recording new challenges, and
- prioritising a new research agenda.

Past successes in research are the platform for new inquiry and preparation for future challenges. When combined with discussions and a synthesis and assessment of how the world works below-ground, they can serve to forge new ecological theories and directions. Future challenges for soil biodiversity and soil ecology research are a part of the broader global sustainability agenda that involves all academic disciplines and policy makers. Communication beyond the scientific community to the public, managers and government agencies about options for maintenance of soils and soil biodiversity must contribute to increasing public awareness of their dependence on soils. Such communication will highlight society's important role in decisions on sustaining soils for the functioning of the Earth's biosphere.

The definition of biodiversity we use here comes from the 1992 Convention on Biodiversity (CBD), the 'variability among living organisms, within species, between species, and of ecosystems'. Since the term was first coined (Takacs 1996), and as the global loss of biodiversity has escalated, there has been a dramatic increase in research on soil biodiversity. The literature survey by Morris *et al.* (2002) showed a ten-fold increase in publications from about 1985 to 2000 on such subjects as the rhizosphere, microbial habitats in soil and microbe–plant systems such as mycorrhiza. International conferences on mycorrhiza and soil ecology now routinely attract gatherings of around 500 scientists, which demonstrates the interest in this field. Inherent in research discussions is the additional consideration of the role of soil biodiversity in ecosystem functioning across temporal and spatial scales.

Soil ecology is the study of soil organisms and their interactions with their environment, and should therefore encompass the study of soil biodiversity in the broadest sense. We examine how a history of scientific accomplishments in soil biodiversity will contribute to new ecological perspectives and to global sustainability. Using past and present developments in soil biodiversity and soil ecological research as a foundation, we offer a list of 'ten tenets of soil ecology' and conclude with a discussion of our perspectives on six 'challenges in soil ecology'.

A brief history of advancements in soil biodiversity research

Many (Usher et al. 1979; Fitter 1985; Usher 1985; Wardle 2002) have proposed that a better understanding of soil ecology would present new ecological theories. As outlined by these authors and others (Coleman 1985; Coleman & Crossley 1996; Giller 1996), it is clear that above-ground ecological theories dominate present concepts in soil ecology. This is partially because we approach ecological questions based on differences in education and disciplinary training and from familiarity with the habitats and organisms we study. Rapid technical advances in molecular biology have enhanced research in all fields of soil biodiversity, and have transformed the study of microbial ecology both above and below-ground (Tiedje & Stein 1999; Tiedje et al. 2001). In some cases, soil biodiversity and microbial ecology are practised by scientists with little training in soils or ecology, using very different techniques and asking very different questions. Scientists trained in soil biodiversity gain a perspective that depends on an integrative, holistic and systems approach. For example, consider the question that permeates recent ecological research: does species diversity affect ecosystem functioning? Above-ground ecologists design experiments with many plants (species, traits, functional groups), across large spatial scales, and consider soil as a medium of physical and chemical properties needed for plant growth. They measure ecosystem process rates with rare consideration of soil biodiversity (Naeem et al. 1994, 2000; Tilman et al. 1997a, b). Soil ecologists, in contrast, must design field experiments with attention to both larger (m^2) scales for above-ground parameters and smaller (mm, cm) scales for soil organisms, and accordingly reduce the number of plant species to quantify effects on soil biodiversity (Bradford et al. 2002). This latter approach is integrated above and below-ground, but has a flip side: soil microbes and soil invertebrates are so abundant that researchers are limited to studying one or a few biotic groups at the species level. Identifying invertebrates to species level is labour intensive, often demanding molecular technologies. In short, above- and below-ground ecologists approach questions differently, and much less is known about below-ground biodiversity compared to that above-ground. Soil species have been minor players in past ecological studies due to the greater number of above-ground ecologists, the charismatic nature of many above-ground animals and plants, as well as human consumption and use of primarily plant shoots, leaves, flowers and other products.

Soil ecology, biodiversity and microbial process analyses have contributed and will continue to contribute to general ecological concepts and analysis of ecosystem functioning (Virginia & Wall 2000; Wardle 2002). Developments from numerous disciplines comprising soil ecology are crucial for studies of soils in managed ecosystems. The evolution of progress in soil biodiversity and ecosystem functioning has raised awareness in the terrestrial ecological community, as a whole, of the importance of inserting the interactions of soil organisms and ecosystem processes in what has traditionally been 'above-ground' ecology (Wardle 2002). The diversity and abundance of species (however defined) and operational taxonomic units (OTUs) in soil may be greater than the number above-ground (Virginia & Wall 2000; Wardle 2002). Thus, the inclusion of soil biodiversity in ecological research, as we include above-ground biodiversity, is critical if we are to manage soils as a renewable natural resource and in an environmentally sustainable manner.

The period 1900 to 1950 may be described as the era of 'soil biodiversity natural history' (Fig. 1.1). Soil ecologists compared extraction and culture techniques to enable quantification of major groups of organisms including soil invertebrates. However, in soil microbiology, ecology was regarded as a second-order problem, a view that is now being challenged. Fenchel (2003) credits Beijerinck with the ubiquitously quoted and (in retrospect) pernicious statement that any bacterial species can be found wherever its environmental needs are met ('the environment selects . . .'). The era was dominated largely by systematics and agriculture (soil science, plant pathogens). Quantification of soil pest and pathogen population dynamics, and establishing and modelling thresholds for plant damage, were integral to management of plant disease and increasing crop yields.

In natural, less-managed systems, experts continued to explore and measure the abundance and taxonomic diversity of groups of microbes and invertebrates. They documented geographic origin and dispersal mechanisms of pathogen species as a basis for plant quarantine regulations. During the early 1900s, general ecological theories were proposed and tested, based on larger above-ground and aquatic biota. For example, there were advances in plant and animal ecology in understanding the ecosystem (Tansley 1935; Clements 1936; see also Worster 1994), plant species community associations (Gleason 1926) and niche theory (Grinnell 1917; Elton 1927; Gause 1934). Soil ecology also was developing in Scandinavia (Bornebusch 1930) and elsewhere, and investigations in soil microbial ecology showed a succession of fungi occurring on various types of organic resources (Waksman 1932). Experiments on individual species, their food sources, microbial–faunal interactions and predator–prey relationships established the basis for soil food web research (Hunt *et al.* 1987; Killham 1994).

Field experiments with both radioactive and stable tracers in the 1950s to 1970s measured the processes involved in soil organic matter dynamics, nitrogen

Influences on research in soil biodiversity

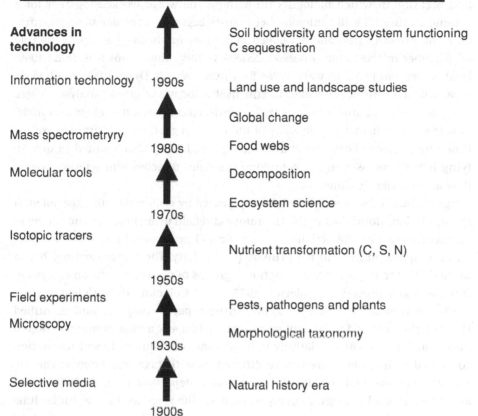

Advances in
technology

Information technology   1990s

Mass spectrometryry
                         1980s

Molecular tools

                         1970s

Isotopic tracers

                         1950s

Field experiments

Microscopy
                         1930s

Selective media

                         1900s

Soil biodiversity and ecosystem functioning
C sequestration

Land use and landscape studies

Global change

Food webs

Decomposition

Ecosystem science

Nutrient transformation (C, S, N)

Pests, pathogens and plants

Morphological taxonomy

Natural history era

Figure 1.1. A brief history of recent accomplishments in soil biodiversity research.

mineralisation, nitrification, denitrification and nitrogen fixation, as well as the transformations and availability of sulphur and phosphorus. This period also saw the advent of chromatographic and other instrumental measurements, which were often associated with the use of metabolic inhibitors or the use of alternative substrates such as specific antibiotics, acetylene reduction and acetylene inhibition, to measure the same soil processes (Paul & Clark 1996). Selective media were often employed to determine the soil biota associated with these processes.

The late 1960s to mid 1970s saw the development of ecosystem science and further development of the concept of species diversity or 'biodiversity'. Investigations on species diversity accelerated after Hutchinson's (1959) article, asking 'why are there so many kinds of animals?' and, in the soil, Wallwork's (1976) *The Distribution and Diversity of Soil Fauna*. Some ecosystem highlights included Odum's (1969) ecosystem article, Swift *et al.*'s (1979) *Decomposition in Terrestrial Ecosystems*, and van Dyne's (1969) edited volume, *The Ecosystem Concept in Natural*

Resource Management. The International Biological Programme (IBP) marked the first occasion in which biologists throughout the world worked together for a common cause. The IBP microbial ecologists began to consider biomass, turnover and growth rates, and developed many new methods to establish the role of microbes in ecosystem processes. Extensive IBP publications (van Dyne 1969; Heal & French 1974; Stewart 1976; Breymeyer & van Dyne 1980) resulted in a wealth of scientific information still useful today for global analyses. These ecosystem–science studies stimulated scientists (ecologists, modellers and meteorologists) to consider the diversity of life through grouping of 'similar taxa', or functional groups, both above and below-ground. They also resulted in quantifying biotic processes in the common currencies of ecosystem science, energy flow and nutrient cycling.

Agricultural ecosystems were less emphasised by ecologists, but experimental manipulations flourished in the laboratory and field using biocides and soil management to reduce the activities of specific soil groups and measure the effect on decomposition rates or net primary productivity. These investigations began to establish the importance of functional groups of soil organisms on ecosystem functions and processes (Anderson 1975, 1978; Coleman 1976; Coleman *et al.* 1983). Petersen and Luxton's (1982) IBP synthesis paper compared and quantified the contribution of faunal groups in carbon budgets across biomes; although most soil fauna seemed relatively unimportant contributors to soil respiration compared to microbes, there were differences with taxa and biomes. The IBP was an extensive global network and a radical departure from traditional soil and above-ground ecology. In many ways, it set the stage for future studies linking soil organisms and ecosystem processes. The Swedish ecosystem project on arable land (Andrén *et al.* 1988), which synthesised the role of soil fauna and microorganisms in carbon and nitrogen cycling, is an example of a later initiative, as is Fox and MacDonald's (2003) project for soil biodiversity in Canadian agroecosystems.

There was increasing awareness that mycorrhizae and rhizobia played key roles in nutrient (phosphorus, nitrogen) transfer, a concept that had major implications for plant root competition and plant community development in managed ecosystems. The attention to the detritus food web and organic matter transformation became a major driver for soil ecology and ecosystem research during the 1970s. Today it is used as a foundation for understanding soil organic matter dynamics and nutrient availability in no-till agricultural systems (Groffman *et al.* 1986; Hendrix *et al.* 1986; Hunt *et al.* 1987; Frey *et al.* 1999). Research on soil biodiversity as a direct linkage to plants, e.g. the rhizosphere–soil food web, had less emphasis. These important research contributions from numerous ecosystems were major steps towards integrating soil research and extending awareness of soil food webs to ecology and ecosystem science. Today's publications (Wardle 2002; Fox & MacDonald 2003) synthesising soil biodiversity in several ecosystems continue this successful approach.

Since the 1980s, research in soil ecology has exploded as a result of collective, individual and international scientific efforts, affirming the urgency of soil bio-diversity research. Local concerns became global as we tackled issues such as the number of species on Earth (May 1988), the increased rate of species loss, changes in land use, environmental indicators, invasive species and atmospheric change. The Convention on Biological Diversity, Montreal Treaty, Kyoto Protocol and Desertification Treaty all had aspects that resulted in more attention to soil biodiversity and ecosystem processes. For example, global models were miss-ing data on the allocation of carbon to plant roots and soils under elevated atmospheric CO_2 concentration. This fostered additional research on soil food webs.

A major question dominating ecology has been the role of biodiversity in ecosystem functioning. Results from a final Scientific Committee on Problems in the Environment (SCOPE) workshop (Schulze & Mooney 1994) recommended a synthesis of the relationship of biodiversity in soils and sediments to criti-cal ecosystem processes. This led to another SCOPE committee which produced many synthetic papers that identified new areas of research (Wall Freckman et al. 1997; Groffman & Bohlen 1999; Adams & Wall 2000; Hooper et al. 2000; Bardgett et al. 2001; Wall et al. 2001b, c).

Experimentally, scientists asked how they could rigorously determine if species richness in soils directly affects the rate of an ecosystem process, given the large abundance and largely unknown diversity of soil microbes and invertebrates. To address this, a meeting was held in 1994 at the UK's Natural History Museum with disciplinary representatives from ecosystem sci-ence, ecology, systematics and soil science (Freckman 1994). The varying views resulted in a collective agreement on research priorities that became a basis for the UK's NERC Soil Biodiversity and Ecosystem Functioning (Soil Biodiver-sity) Programme (http://mwnta.nmw.ac.uk/soilbio/index.html), and, in the USA, a National Science Foundation US/UK collaborative soil biodiversity and ecosystem functioning grant (http://www.nrel.colostate.edu/projects/soil/us_uk/index.html). Major experimental networks were also leading and contributing to this focus in soils, including the Tropical Soil Biology and Fertility Program, the Macro Faunal Network, the Global Litter Invertebrate Decomposition Experiment, and the Long Term Ecological Research network (Symstad et al. 2003). Additionally, the European Union and European Science Foundation funded cross-EU experi-ments targeting, or including, soil biodiversity such as CLUE (Changing Land Use Experiment). These and other research projects in soil ecology are evolving to integrate soil biodiversity and ecosystem processes into research topics such as above-ground/below-ground coupling, multi-trophic interactions, biogeography, soil carbon sequestration, land abandonment, invasive species and atmospheric change.

In general, most experiments on biodiversity and ecosystem functioning have been resolved for a group of soil organisms at the functional level or lower taxonomic resolution. It is widely accepted by ecologists that earthworms and termites are ecosystem 'engineers' (Lavelle & Spain 2001) and that microfauna such as mites, nematodes and protozoa affect rates of mineralisation (Coleman *et al.* 1983; Coleman 1985; Ingham *et al.* 1985). We know that groups of organisms, and in some cases species below-ground, can influence plant community composition and contribute to succession of plants (Read 1991; van der Heijden *et al.* 1998; de Deyn *et al.* 2003). We also have learned that disturbances, such as land fragmentation and agricultural intensification, decrease species diversity of earthworms with resulting changes in soil porosity and soil structure (Hooper *et al.* 2000; Bignell *et al.* in press; Giller *et al.* this volume). Evidence at the species level in microcosms and field experiments shows that there is considerable redundancy in soil species; this suggests little effect on rates of general decomposition processes with loss of soil species (Hunt & Wall 2002). As long as there is a functional group available to perform a particular role in a given ecosystem function, it may not matter whether there are many or few species within the functional group (see Setälä *et al.* this volume). However, there are situations where the abundance of a particular species can have a disproportionate impact on a process, such as with invasive species. The earthworm species *Aporrectodea tuberculata* and *Lumbricus terrestris* (which are two of only 45 introduced earthworm species in North America) homogenise upper soil horizons and increase erosion and runoff (Burtelow *et al.* 1998; Groffman & Bohlen 1999; Hendrix & Bohlen 2002). The nematode species *Bursaphelenchus xylophilus*, introduced from North America to Japan and Portugal (Mota *et al.* 1999), kills pine trees in plantations and forests within a couple of months (Mamiya 1983; Rutherford *et al.* 1990). These situations indicate that key species may sometimes strongly affect a range of soil-based ecosystem processes, including decomposition pathways, carbon and nitrogen cycling, hydrologic pathways and the maintenance of soil structure.

The multiple activities and experiments that have contributed to the present state of knowledge mark a new era of research in soil ecology. That is a recognition by ecologists and other disciplines that soil biodiversity at any taxonomic level is worthy of study for its own sake and as a major component of ecological and ecosystem research. There is a new fascination from terrestrial, atmospheric, aquatic and marine ecologists about these mostly unknown soil organisms and how they interact in ecosystems, where they are, whether general principles exist for below-ground microbes and fauna, and if these generalities might extend to microbes and invertebrate fauna elsewhere. Soils and their processes are a natural meeting place for fostering interdisciplinary studies, and it is interdisciplinary studies that are going to be the key in solving environmental problems. Soil biodiversity, global changes and both nutrient transformations and movement, involve ecologists, agronomists, foresters, biogeochemists, biochemists, pedologists and geologists. Interdisciplinary studies tie together many terrestrial and aquatic processes, and are an excellent way of integrating otherwise difficult to integrate scientific disciplines. Thus, identifying general belowground ecological principles from interdisciplinary research holds promise for broadening our understanding of ecosystems.

Ten tenets of soil ecology

Based on advances in soil biodiversity research over the past decades, we offer a short, unprioritised list that highlights the perspectives soil ecologists have when studying terrestrial ecosystems. These are the supporting groundwork for the 'challenges in soil ecology', which we discuss later.

1. *The terrestrial world is brown and black, not green.* Soils are brown to red and humus black. Despite our perception of the world as driven by photosynthesis, virtually all net primary production ends up as soil organic matter and, because of its relatively long residence time, soils contain twice as much carbon as vegetation (Schimel 1995). Consequently, more ecology occurs below than above-ground.

2. *The world seems primarily microscopic.* In soils, microbes and most groups of invertebrates (as adults or juveniles) are microscopic, or even at times at the electron microscope level, complicating studies of phylogeny, population and community ecology (Wilson 2002). However, some clones can be immense: fungal mycelia may be the largest organisms on Earth. Research is yet to determine whether ecological theory for larger organisms will apply to microbes (fungi and bacteria). However, microscopic is not a synonym for prokaryotic. Although life and all vitally associated processes could probably continue (as it did for two billion years) in the absence of vascular plants and vertebrate animals (Nabonne 2003), and despite Knoll's (2003) assertion that 'eukaryotic food webs form a crown – intricate and unnecessary – atop ecosystems maintained primarily by prokaryotic metabolism', modern ecosystems depend on a much broader range of organisms, still principally microscopic, but, in the case of several key groups (fungi, nematodes, arthropods), eukaryotic.

3. *We do not know their names or what they do.* Estimates indicate that less than 5% of species or less than 1% of operational taxonomic units (OTUs) in soils are described. One estimate based on DNA similarity was that there were 10^4 bacterial species per gram of soil (Torsvik et al. 1990). Hawksworth (2001) and Hawksworth and Rossman (1997) estimate 1.5×10^6 fungal species globally. Although molecular approaches enable the recognition of extreme

diversity in many soil-inhabiting taxa, we are only beginning to connect identity to function and consequently know little about what each species does. Today, using microarrays, we can determine hundreds of thousands of an organism's genes, but have only vague ideas of what most of the genes do. Similarly, we may discover millions of species in soils, but have little idea of what each does. Thus, field soil biodiversity experiments are rarely at the species level, even for invertebrates (Brussaard *et al.* 1997; Hooper *et al.* 2000). There are no soils where all species/OTUs of microbes or even all invertebrates have been described or quantified. Studies of natural history to determine food source and functional role are based on relatively few examples.

4. *Food webs do not follow traditional rules.* Omnivores are prevalent in soils (Moore *et al.* 1988; Moore & de Ruiter 1997). Many fungi are generalists and their extensive mycelial structure means that they may operate on a different spatial scale than their predators. Extreme environments may have no predators and few invertebrate species (Wall & Virginia 1999; Stevens & Hogg 2002). For example, in Antarctic Dry Valley soils, Stevens and Hogg (2002) noted only one mite species, *Stereotydeus mollis* (family Penthalodidae – not the predaceous mite family Tydeidae) and one Collembolan, *Gomphiocephalus hodgsoni.* Wall Freckman and Virginia (1998) found only one species of a nematode, *Eudorylaimus antarcticus,* capable of being a predator (Freckman & Virginia 1989). However, there is no evidence that these three species, which rarely co-occur, are predators (Fitzsimmons 1971; Block 1985; Davidson & Broady 1996; Sinclair & Sjursen 2001; Sjursen & Sinclair 2002). Soil food webs, unlike above-ground food webs, include a strong element of recycling within the decomposer component. For example, no-till agriculture favours a fungal-dominated food web on the soil surface (Beare *et al.* 1992; Frey *et al.* 1999), which leads to increased aggregation of soil particles, increased decomposition rates but overall greater soil carbon storage.

5. *Indirect effects can dominate, and are hard to quantify.* An individual species may have an identifiable effect on function in the field and laboratory, but the species richness in soils results in multiple species interactions that are indirect and difficult to measure. Facilitative interactions between species may be important to a process, but experiments are few. Thus, our knowledge of whether two, three or a succession of species are important in an ecosystem function is largely unknown. Based on a number of experiments to determine direct relationships between a single soil species and an ecosystem process, most species appear to be interchangeable (Setälä *et al.* this volume).

6. *Scale is a dividing issue between above- and below-ground ecology.* Spatial and temporal scales of most organisms above and below-ground are not in synchrony. The small spatial heterogeneity in soil allows for multiple complex

habitats and diversity. For example, the rhizosphere scale (mm to cm) can have extremely large diversity compared to the same scale above-ground. Consider the major differences in biodiversity and a process (nitrogen transformation) occurring vertically, from 0 to 30 cm soil depth and compare them to the same differences above-ground. The magnitude of change in a process is similar to that from a forest floor to a tree canopy. In soil, the species composition of the rhizosphere can differ from that of bulk soil organic matter a few centimetres away (Blackwood & Paul 2003). Organisms vary in their ability to cross scales (e.g. roots may spread for tens of metres; mycorrhizal fungi operate at the millimetre to metre scale between plants, whereas nematode species may move only a few centimetres, and some bacteria will be limited to spaces within aggregate microhabitats). Soil biota (fungi, invertebrates, vertebrates and roots) connects these otherwise isolated habitats through movement, mycelial structure and process reactions. Soils are also less temporally dynamic than above-ground systems, being buffered somewhat in nutrient dynamics from sudden change and human impact. Symptoms of damage to soils and their biota may take longer to observe than damage above-ground, and may be irreversible even before being identified (Amundson et al. 2003).

7. *Soil legacy imprints soil biodiversity (and can override plant effects).* Soils take hundreds to thousands of years to form, and their structure and biodiversity in an ecosystem today reflects past geologic, climatic and vegetative history. Soil horizons change slowly over time until disturbed. Soil organic matter, even at only 50 cm depth in the field, can be 1400 years older than that at the surface (Paul et al. 2001). Yet, when brought into the laboratory it decomposes at the same rate as that from the surface (Collins et al. 2000). The age and amount of carbon held in soils is greater than above-ground, and as it becomes available, provides an energy base for biodiversity, above and below-ground. This evolution of physical and chemical patterning over eons, as well as consequences of biological evolution, produces many characteristic and complex biotic habitats within and across ecosystems, and may have a greater impact on patterning of soil biodiversity than plants in some ecosystems (Virginia & Wall 1999; Kaufmann 2001; Williamson & Harrison 2002).

8. *Soils and their biota are not isolated in terrestrial ecosystems: they have multiple landscape connections.* Soils and biota are often separately examined, yet there are many important soil interactions including the mediation of plant productivity by soil-dwelling animals (Warnock et al. 1982; Masters et al. 1993). Soils are linked physically and biologically to sediments of freshwater, estuaries and marine systems (Wall Freckman et al. 1997; Wall et al. 2001c), and the atmosphere. They also are an intermediary habitat for above-ground animals. For example, the mountain plover (*Charadrius montanus*) nests in

bare soil of the US shortgrass prairie, but due to decreasing habitat, began
nesting in agricultural soils where spring ploughing contributed to loss of
nests and its decline (Knopf & Rupert 1999; US Fish and Wildlife Service
1999). A species of amphibian, *Oscaecilia ochrocephala* (a Caecilian), moves to
10 m depth in tropical soils (Wake 1983, 1993), and various soil invertebrates, as juveniles and adults, serve as food for above-ground vertebrates.
Soils and biota are subjected to fluctuations and inputs from erosion and
flooding, which change the biotic interactions. The lateral connections of
biodiversity from soils to sediments in aquatic systems are studied less than
in clearly defined ecosystems (e.g. soils vs. aquatic systems) but are often
interdependent (Polis *et al.* 1997). Detrital food webs and process controls
are similar in soils and sediments, as are major taxonomic groups, but
species differ.

9. *Small creatures have biogeography too.* Our knowledge of latitudinal and land-
    scape patterns of soil biodiversity is sketchy (Bardgett *et al.* this volume).
    Most process studies are conducted on relatively flat land in northern tem-
    perate ecosystems with very little information from the large part of our
    Earth that is hilly or tropical. We know little about how diversity and pro-
    cesses differ at depth (Jackson *et al.* 2002) because the majority of reports
    on soil species diversity and processes are from within the top 30 cm of the
    soil profile (Schenk & Jackson 2002) where, globally, 50% of root biomass
    occurs. However, many roots occur below 1 m (Schenk & Jackson 2002) and
    diversity and biological processes occur wherever conditions and energy can
    support life, which, in aquifers and buried deposits, is often well below the
    rooting layer.

10. *Decomposition is one of the two major life-generating processes.* Decomposition
    pairs with photosynthesis to maintain the function of ecosystems and the
    balance of the atmosphere. It differs from photosynthesis in the enor-
    mous diversity of organisms involved. As a tenet of soil ecology, we believe
    it is important to emphasise that one process does not dominate: these
    two processes must be considered of parallel importance for research in
    soil ecology and the consideration of our sustainable future. A small frac-
    tion of photosynthetic products and biotic degradation products is pro-
    tected from decomposition in soil, and this forms the core of soil organic
    matter.

Challenges in soil ecology

Recent advances in technologies that can be applied to studying soil biota have
been exceptionally rapid, and they will accelerate. We are virtually unlimited
in the types of questions we can ask and the hypotheses we can test. Our confidence in extending findings from many taxonomic groups at higher resolution

will be tested at multiple scales. However, we have to take advantage of many developments occurring now. For example, model organisms (e.g. the nematode *Caenorhabiditis elegans*) might extend our knowledge, in soil, of bacterial-feeding nematode physiological ecology (survival rates, life stages, periodicity, mechanisms) and life history (life cycle length, death), as well as the effects of agronomic management and sustainable agriculture. Availability of annotated gene sequences for soil organisms or their close relatives will enable investigations of links between diversity and function. Use of stable isotopes to track carbon and nitrogen transformations is revolutionising our ideas of nutrient sources through species and food webs (Johnson *et al.* 2002; Manefield *et al.* 2002; Radajewski *et al.* 2002). Multiple technologies, and concurrent use of technologies (e.g. NMR and chemical fractionation techniques), will open new frontiers for soil biodiversity. Global research networks such as the Below-ground Biodiversity (BGBD) Network (Tropical Soil Biology and Fertility (TSBF) Institute of the International Center for Tropical Agriculture (CIAT); see Giller *et al.* this volume) magnify our understanding of sustainable soil management.

While the issue of scale, both temporal and spatial, will continue to be an overriding challenge, in the following list of six questions we have not highlighted it, nor have we highlighted technologies such as global positioning system (GPS) and remote sensing. These challenges are not unique to soil ecology, and overlap with many issues, questions and priorities listed elsewhere (Cracraft 2002). We see the following as challenges that will radically change our view of life in soils and how soils are intertwined with the Earth system.

What are the organisms?

Species as we once knew them, even for invertebrates, no longer really exist. We are now dealing with a new 'tree of life'. 'What is a species?' is considered by Cracraft (2002) as one of the seven great questions for systematic biology, and is integral to many experiments we do. The species concept has been continually evolving and many of the recently tested ecological hypotheses and resulting conclusions are based on concepts of a species, derived from studies of organisms in which reproduction is normally sexual and individuals are easily defined (Cohan 2002). We must have an idea of our organism or OTU if we are to accept the soil as a living habitat and if we are to propose hypotheses at broader scales. How will we test what is considered 'a defined species' above-ground (e.g. plants and large vertebrates and invertebrates) when for 'soil species' (e.g. invertebrates, protozoa and microbes) we have OTUs or even different resolution of DNA sequence data? This uncertainty about the nature of ecological and evolutionary units in soil ecosystems has profound operational consequences. Simply quantifying biodiversity in soil (how many species and how abundant are they?) is challenging.

Building on that quantification, we need to be able to link diversity – now typically measured by some attribute of DNA sequence variation – with function. Yet most of the organisms appear to be unculturable, so that without new and imaginative experimental protocols, technologies such as stable isotope probing (Radajewski *et al.* 2002) and accelerator mass spectrometry (Staddon *et al.* 2003b), we will not be able to say which organisms perform which processes. A further problem in addressing soil biodiversity and ecosystem functioning at the species level is the difficulty of experimentally teasing apart the massive number of species interactions that occur in soil.

Where are they?

There is a missing field of study in soil ecology: biogeography (Bardgett *et al.* this volume). We do not know what geographical patterns of soil biodiversity exist. Can we predict where most soil organisms will be found? Are there endemic species and hot spots of diversity that are related to ecosystem functions at different scales (Usher this volume)? Are these based on soil physical and chemical coupling, climate, geology or plant characteristics? Can such knowledge be used to predict soil fertility and ecosystem productivity in a changing, anthropogenically stressed environment? The words of Whitaker *et al.* (2003), on bacteria in an isolated hot spring, challenge the original model of Beijerinck (see Fenchel 2003) in which unrestricted dispersal constrains species richness. C. B. Blackwood (personal communication), using DNA analysis and multi-variate spatial statistics, showed significant differences in types of bacteria at the scale of 1 mg to 1 g of soil. Another break occurred at 10^3 to 10^5 g of soil where spatial organisation was hypothesised to be controlled by plant type and density.

What do the organisms do?

It has become unfashionable (and unfundable) simply to study an organism to discover how it operates in the wild (although funding sometimes is available to do the same for a gene). Consequently, we extrapolate understanding of functional groups in an ecosystem process based on life histories and physiologies of a few hopefully representative species and sites. Using stable isotopes to track sources of carbon and nitrogen in species of enchytraeids previously placed within the same feeding group indicates problems with lumping together species of similar morphology (Cole *et al.* 2002). Somehow, we need to compile more information on more taxa to determine generalities (for example, what are the food sources and who eats what?) if we are to build predictive models of many soil taxa and processes. How does periodicity in feeding affect transfer of nutrients? What aspects of physiological ecology make organisms in the same location either vulnerable or successful? Under what conditions does a facultative organism become a pathogen? How do multiple species interact? How do multiple element interactions change through the soil food web? And, how do multiple species affect root architecture, rooting depth, plant community succession and feedbacks to organic matter dynamics? How do rhizosphere and root symbionts and parasites affect nutrient and carbon flux when interacting with species of detrital food webs? Can we extend knowledge of interactions (facultative, competitive) horizontally across soils and sediments, and vertically to above-surface processes?

What are commonalities in survival and dispersal mechanisms at the species level that occur under different environmental stresses? How do multiple species and microbes respond to simple changes in soil moisture? We know different organisms enter a survival state under water stress at different moisture potentials. We also know that in soils a large proportion of soil organisms are in the resting state (Paul & Clark 1996). When do the organisms become active? The resting nature of much of the microbiota represented by isolated DNA makes it difficult to interpret activity and process studies based on DNA analysis of gene frequency and abundance. This problem of a comprehensive lack of natural history of our organisms is acute for the generally uncharismatic soil biota. For birds, mammals, plants and many insects, there is a great wealth of enthusiastic amateurs, often extremely knowledgeable, whose commitment to the gathering of information can be put to productive use. Who will do the same for collembola, nematodes, ascomycetes or archaea?

Do soil species matter in ecosystem processes?

One of the most frequently recurring questions in soil ecology is whether there is a link between biodiversity and the way in which an ecosystem operates. A first problem is how to pose this question in an interesting and yet experimentally tractable form. The UK Soil Biodiversity Programme focused on carbon fluxes within soil as a measure of ecosystem function. With stable isotopes, these can now be quantified in the field (e.g. Johnson et al. 2002; Staddon et al. 2003a), but we still have only the crudest techniques for manipulating biodiversity. As part of the programme, Bradford et al. (2002) recreated a field system in a controlled environment and varied maximum body size of the soil fauna – the most remarkable result was how robust the system was to such a major perturbation.

In most soils, there is a surprising balance between carbon inputs and losses. This is one reason why soils take so long to form. We have little idea as to why this should be, nor the extent to which soil organic matter and soil biodiversity are related. The accumulation of organic carbon in soil is closely linked to nitrogen dynamics. In soils of high carbon : nitrogen ratio, partially decomposed organic matter tends to accumulate because the decomposers are limited by substrate quality, pH and oxygen. This accumulation, however, locks up potentially huge amounts of nitrogen, exacerbating the problem. Similar relationships must exist with phosphorous and probably other elements.

How will soil species respond to a changing world?

How predictable are species changes? Can we use evolutionary and geologic history to predict the impact of future changes on soil communities, and on their survival or dispersal? Soils are the geologic legacy that produced heterogeneity in soil biodiversity. Soil parent material and climate contributed to individual habitats (and species that evolved in those habitats). Recent estimates indicate 4.5% of US soil series are in danger of serious loss or extinction (Amundson *et al.* 2003). We need to examine how soils and geologic history of plant–animal–microbe associations have or have not imprinted today's soil communities. Additionally, we know little about the environmental ranges of soil biota so we cannot easily predict how it will respond to change. Above-ground ecologists and soil systematists have long recognised the power of preserved specimens to provide clarification of geographic patterns and local diversity patterns. Many are now aggressively intensifying analyses of museum collection data with GIS, climate data and overlying major groups of plants and animals in a region. These museum collections provide a phylogenetic context that allows systematists to place past ecological and natural history collections in a historical and global framework and broaden the database to predict how species distributions will react in the future. These and ongoing global experiments when synthesised provide modellers with the ability to answer questions such as what happens to ecosystem functions when soil warms, or how effectively can soil microbes disperse (Finlay 2002).

Should we care?

While the first five questions have relevance to global environmental problems, there must also be a targeted research effort, and an effort to extend and communicate our findings, and to make them relevant to society as a whole. Management of soil biota and its functions is critical for carbon sequestration, for agriculture, for forestry and for watersheds. In the past 50 years we have advanced considerably in indirect management of decomposers and soil engineers (termites, earthworms), and in direct management of symbionts and pathogens/pests (the latter while reducing pesticide use, such as methyl bromide; Swift & Anderson 1994; Giller *et al.* this volume). The molecular revolution may offer the opportunity to influence the soil community through genetic manipulation of the plant, by altering the biochemical signals sent out to the soil. Society, at least in some countries outside Europe, is starting to accept the introduction of genetically engineered plants (or genetically modified organisms, GMOs), but not GM microorganisms. Yet one of the most widely used GMO plants, incorporating resistance to insects with Bt, which is derived from soil bacteria. Its effects on soil invertebrates and the soil food web are only beginning to be known (Stotzky & Bollag 1992; Hopkins & Gregorich 2003). There is great need for compiling and synthesising the overlapping data from soils, climate and soil biodiversity to use in a predictable manner for problems such as desertification and abandoned land (and other land use changes).

How comfortable are we at extending the evidence we have now gathered towards pressing global problems? Can we extend our knowledge to explain, for example, how we manage soil ecologically to decrease plant, animal and human disease in the light of global change? Our knowledge of soil organisms as vectors or pathogens under varying environmental or soil conditions is, at present, scattered (Jamieson 1988; Hendrix & Bohlen 2002).

Ecosystem services are now an accepted part of our socio-economic system. These include a broad variety of functions from soil and water decontamination, prevention of the spread of plant and animal diseases, and responses to stresses such as global and climatic change to visual landscape diversity (Wall *et al.* 2001a). The soil biota is an integral part in all the above. For example, can associated mycorrhiza move rapidly enough to allow whole forests to adapt to changing temperatures? Will society be more willing to pay for desirable habitats and landscapes?

Conclusions

The question arises, how can we possibly tackle all these challenges? It can be done. The British Ecological Society (BES) conference in 2003 on Soil Biodiversity and Ecosystem Functioning was one of the largest conferences held by the BES. Repercussions of this meeting are multiple and very promising. Delegates of all disciplines were energised with recent experimental progress, and with their cutting edge technologies, and are ready to work towards meeting these challenges and to test and develop below-ground theory.

There are several ways to proceed. These include terrestrial ecology institutes that integrate all aspects of soil ecology from organism to global models and cross disciplines (e.g. the Tropical Soil Biology and Fertility Institute of the International Center for Tropical Agriculture, http://www.ciat.cgiar.org/tsbf_institute/index_tsbf.htm). Others include integrated international experimental networks focused on long-term or short-term projects, such as the Integrated Tundra Experimental Network, ITEX (Marion *et al.* 1997; Welker *et al.* 1997); Global Litter Decomposition Experiment, GLIDE, http://www.nrel.colostate.edu/projects/glide/index.html; the EU project, Conservation of Soil Biodiversity under Global Change, CONSIDER, http://www.nioo.knaw.nl/CTE/MTI/Index.htm; and International Long Term Ecological Research, ILTER, http://www.ilternet.edu.

Biodiversity research needs by definition to be extended in time. Will long-term, well-organised global research sites and networks, such as agricultural and forest experimental stations (e.g. Fluxnet and Precipnet), highlight soil biodiversity and help to synthesise the vastly different disciplines needed to manage soils? Projects need to emphasise informatics and utilise statistical analysis to synthesise information hidden in taxonomic collections and relate it to soil characteristics. There are many ongoing efforts to link scientists and managers; an example is the FAO Portal on Soil Biodiversity (http://www.fao.org/landandwater/agll/soilbiod/activity.stm), which also has links to soil biodiversity projects. It is perhaps time to link knowledge on soil biodiversity, from these various disciplines and networks in natural and managed ecosystems, into a global soil assessment that could provide information useful for the future management and sustainability of soils. A synthesis of the ecosystem goods and services provided by the functioning of soils and soil biodiversity, and an evaluation of their vulnerability under future scenarios for global change (land use, atmospheric, climatic and biotic), are both critically needed, but must involve all disciplines involved in soil research as well as economists and social scientists.

There is now a great international interest in the subject of soil sustainability. The International Technical Workshop on Biological Management of Soil Ecosystems for Sustainable Agriculture (Embrapa 2002) states 'Sustainable agriculture including forestry involves the successful management of agricultural resources to satisfy human needs while maintaining or enhancing environmental quality and conserving natural resources for future generations'. The effective management of crop species, soil fertility, soil physical factors and water availability as well as pests and diseases are all involved in soil sustainability. Soil biodiversity and biological processes are both impacted by the above and in turn have major roles to play in understanding and management of the processes involved. The general indicators of soil sustainability involve soil structure maintained by the interaction of physical and biotic forces, soil organic matter levels and soil biotic populations as well as a healthy, economically successful crop. We hope that the tenets of soil biodiversity and the challenges in this field can be used to enhance our understanding further, and even future improvement, of the sustainability of soils. The present melding of the various fields of soil science with ecosystem ecology and soil biodiversity research, the great advances in molecular biology and the close relationship of soil biodiversity research to present programmes in global change carbon scenarios, as well as in organic agriculture, make us believe there is great hope for the future.

This chapter echoes May's (1997) statement that implicitly emphasises the need for research in soil biodiversity, 'A full understanding of the causes and consequences of biological diversity, in all its richness, probably cannot be had until the contribution made by decomposers to the structure and functioning of ecosystems is fully understood'. The urgency to provide information and concepts that can be a general theory for soil biodiversity, and contribute to sustainability of ecosystems at local, regional and global levels, is a goal for all of us.

Acknowledgements

The authors thank Byron Adams, Emma Broos, Johnson Nkem, Mike Swift, two anonymous referees and the editors for their critical and helpful comments. D. Wall acknowledges Holley Zadeh, Lily Huddleson and US National Science Foundation Grants, DEB 98 06437 and OPP #9810219.

References

Adams, G. A. & Wall, D. H. (2000). Above and below the surface of soils and sediments: linkages and implications for global change. *BioScience*, **50**, 1043–1048.

Amundson, R., Guo, Y. & Gong, P. (2003). Soil diversity and land use in the United States. *Ecosystems*, **6**, 470–482.

Anderson, J. M. (1975). Succession, diversity and trophic relationships of some soil animals in decomposing leaf litter. *Journal of Animal Ecology*, **44**, 475–495.

Anderson, J. M. (1978). Inter- and intra-habitat relationships between woodland Cryptostigmata species diversity and the diversity of soil and litter microhabitats. *Oecologia*, **32**, 341–348.

Andrén, O., Paustian, K. & Rosswall, T. (1988). Soil biotic interactions in the functioning of agroecosystems. *Agriculture, Ecosystems and Environment*, **24**, 57–67.

Bardgett, R. D., Anderson, J. M., Behan-Pelletier, V., et al. (2001). The influence of soil biodiversity on hydrological pathways and transfer of materials between terrestrial and aquatic ecosystems. *Ecosystems*, **4**, 421–429.

Beare, M. H., Parmelee, R. W., Hendrix, P. F. & Cheng, W. (1992). Microbial and faunal interactions and effects on litter nitrogen and decomposition in agroecosystems. *Ecological Monographs*, **62**, 569–591.

Bignell, D. E., Tondoh, J., Dibog, L., et al. (in press). Below-ground biodiversity assessment: the ASB rapid, functional group approach. *Alternatives to Slash-and-Burn: A Global Synthesis* (Ed. by P. J. Ericksen, P. A. Sanches & A. Juo), pp. XXX–YYY. Madison, WI: American Society for Agronomy.

Blackwood, C. B. & Paul, E. A. (2003). Eubacterial community structure and population size within the soil light fraction, rhizosphere and heavy fraction of several agricultural systems. *Soil Biology and Biochemistry*, **35**, 1245–1255.

Block, W. (1985). Arthropod interactions in an Antarctic terrestrial community. *Antarctic Nutrient Cycles and Food Webs* (Ed. by W. R. Siegfried, P. R. Condy & R. M. Laws), pp. 614–619. Berlin: Springer-Verlag.

Bornebusch, C. H. (1930). The fauna of forest soil. *Forstlige Forsøgsvæsen, Danmark*, **11**, 1–224.

Bradford, M. A., Jones, T. H., Bardgett, R. D., et al. (2002). Impacts of soil faunal community composition on model grassland ecosystems. *Science*, **298**, 615–618.

Breymeyer, A. I. & Van Dyne, G. M. (1980). *Grasslands Systems Analysis and Man*. Cambridge: Cambridge University Press.

Brussaard, L., Behan-Pelletier, V. M., Bignell, D. E., et al. (1997). Biodiversity and ecosystem functioning in soil. *Ambio*, **26**, 563–570.

Burtelow, A. E., Bohlen, P. J. & Goffman, P. M. (1998). Influence of exotic earthworm invasion on soil organic matter, microbial biomass and denitrification potential in forest soils of the northeastern United States. *Applied Soil Ecology*, **9**, 197–202.

Clements, F. E. (1936). Nature and structure of the climax. *The Journal of Ecology*, **24**, 252–284.

Cohan, F. M. (2002). Sexual isolation and speciation in bacteria. *Genetica*, **116**, 359–370.

Cole, L., Bardgett, R. D., Ineson, P. & Hobbs, P. J. (2002). Enchytraeid worm (Oligochaeta) influences on microbial community structure, nutrient dynamics and plant growth in blanket peat subjected to warming. *Soil Biology and Biochemistry*, **34**, 83–92.

Coleman, D. C. (1976). A review of root production processes and their influence on soil biota in terrestrial ecosystems. *The Role of Terrestrial and Aquatic Organisms in Decomposition Processes* (Ed. by J. M. Anderson & A. Macfadyen), pp. 417–434. Oxford: Blackwell Scientific.

Coleman, D. C. (1985). Through a ped darkly: an ecosystem assessment of root–soil–microbial–faunal interactions. *Ecological Interactions in Soil: Plants, Microbes and Animals* (Ed. by A. H. Fitter, D. Atlinson, D. J. Read & M. B Usher), pp. 1–21. Oxford: Blackwell Scientific.

Coleman, D. C. & Crossley, D. A. Jr. (1996). *Fundamentals of Soil Ecology*. San Diego, CA: Academic Press.

Coleman, D. C., Reid, C. P. P. & Cole, C. V. (1983). Biological strategies of nutrient cycling in soil systems. *Advances in Ecological Research*, **13**, 1–55.

Collins, H. P., Elliott, E. T., Paustian, K., *et al.* (2000). Soil carbon pools and fluxes in long-term corn belt agroecosystems. *Soil Biology and Biochemistry*, **32**, 157–168.

Cracraft, J. (2002). The seven great questions of systematic biology: an essential foundation for conservation and the sustainable use of biodiversity. *Annals of the Missouri Botanical Garden*, **89**, 127–144.

Davidson, M. M. & Broady, P. A. (1996). Analysis of gut contents of *Gomphiocephalus hodgsoni* Carpenter (Collembola: Hypogastruridae) at Cape Geology, Antarctica. *Polar Biology*, **7**, 463–467.

De Deyn, G. B., Raaijmakers, C. E., Zoomer, H. R., *et al.* (2003). Soil invertebrate fauna enhances grassland succession and diversity. *Nature*, **422**, 711–713.

Elton, C. (1927). *Animal Ecology*. New York: MacMillan.

EMBRAPA (2002). *International Technical Workshop on Biological Management of Soil Ecosystems for Sustainable Agriculture Programs, Abstracts and Related Documents*. Londina: Brazilian Agricultural Research Corporation. Ministry of Agriculture Livestock and Food Supply.

Fenchel, T. (2003). Biogeography for bacteria. *Science*, **301**, 925–926.

Finlay, B. J. (2002). Global dispersal of free-living microbial eukaryote species. *Science*, **296**, 1061–1063.

Fitter, A. H. (1985). Functional significance of root morphology and root system architecture. *Ecological Interactions in Soil: Plants, Microbes and Animals* (Ed. by A. H. Fitter, D. Atkinson, D. J. Read & M. B. Usher), pp. 87–106. Oxford: Blackwell Scientific.

Fitzsimmons, J. M. (1971). On the food habits of certain Antarctic arthropods from coastal Victoria Land and adjacent islands. *Pacific Insects Monograph*, **25**, 121–125.

Fox, C. A. & MacDonald, K. B. (2003). Challenges related to soil biodiversity research in agroecosystems: issues within the context of scale of observation. *Canadian Journal of Soil Science*, **83**, 231–244.

Freckman, D. W. (1994). *Life in the Soil. Soil Biodiversity: Its Importance to Ecosystem Processes*. Fort Collins, CO: Natural Resource Ecology Laboratory, Colorado State University.

Freckman, D. W. & Virginia, R. A. (1989). Plant-feeding nematodes in deep-rooting desert ecosystems. *Ecology*, **70**, 1665–1678.

Frey, S. D., Elliott, E. T. & Paustian, K. (1999). Bacterial and fungal abundance and biomass in conventional and no-tillage agroecosystems along two climatic gradients. *Soil Biology and Biochemistry*, **31**, 573–585.

Gause, G. F. (1934). *The Struggle for Existence*. New York: Hafner.

Giller, P. S. (1996). The diversity of soil communities, the 'poor man's tropical rainforest'. *Biodiversity and Conservation*, **5**, 135–168.

Gleason, H. A. (1926). The individualistic concept of the plant association. *Bulletin of the Torrey Botanical Club*, **53**, 7–26.

Grinnell, J. (1917). The niche-relationships of the California thrasher. *The Auk*, **34**, 427–433.

Groffman, P. M. & Bohlen, P. J. (1999). Soil and sediment biodiversity: cross-system comparisons and large-scale effects. *BioScience*, **49**, 139–148.

Groffman, P. M., House, G. J., Hendrix, P. F., Scott, D. E. & Crossley, D. A. Jr. (1986). Nitrogen cycling as affected by interactions of components in a Georgia piedmont agroecosystem. *Ecology*, **67**, 80–87.

Hawksworth, D. L. (2001). The magnitude of fungal diversity: the 1.5 million species estimate revisited. *Mycological Research*, **105**, 1422–1432.

Hawksworth, D. L. & Rossman, A. Y. (1997). Where are all the undescribed fungi? *Phytopathology*, **87**, 888–891.

Heal, O. W. & French, D. D. (1974). Decomposition of organic matter in tundra. *Soil Organisms and Decomposition in Tundra* (Ed. by A. J. Holding, O. W. Heal, S. F. MacLean & P. W. Flanagan), pp. 279–310. Stockholm: Tundra Biome Steering Committee.

Hendrix, P. F. & Bohlen, P. J. (2002). Exotic earthworm invasions in North America: ecological and policy implications. *BioScience*, **52**, 801–811.

Hendrix, P. F., Parmelee, R. W., Crossley, D. A., et al. (1986). Detritus food webs in conventional and no-tillage agroecosystems. *BioScience*, **36**, 374–380.

Hooper, D. U., Bignell, D. E., Brown, V. K., et al. (2000). Interactions between aboveground and belowground biodiversity in terrestrial ecosystems: patterns, mechanisms, and feedback. *BioScience*, **50**, 1049–1061.

Hopkins, D. W. & Gregorich, E. G. (2003). Detection and decay of the Bt endotoxin in soil from a field trial with genetically-modified maize. *European Journal of Soil Science*, **54**, 793–800.

Hunt, H. W. & Wall, D. H. (2002). Modeling the effects of loss of soil biodiversity on ecosystem function. *Global Change Biology*, **8**, 33–50.

Hunt, H. W., Coleman, D. C., Ingham, E. R., et al. (1987). The detrital food web in a shortgrass prairie. *Biology and Fertility of Soils*, **3**, 57–68.

Hutchinson, G. E. (1959). Homage to Santa Rosalia; or, why are there so many kinds of animals? *The American Naturalist*, **93**, 145–149.

Hyams, E. (1952). The soil community. *Soil and Civilization* (Ed. by E. Hyams), pp. 17–27. New York: Harper and Row.

Ingham, R. E., Trofymow, J. A., Ingham, E. R. & Coleman, D. C. (1985). Interactions of bacteria, fungi, and their nematode grazers: effects on nutrient cycling and plant-growth. *Ecological Monographs*, **55**, 119–140.

Jackson, R. B., Banner, J. L., Jobbagy, E. G., Pockman, W. T. & Wall, D. H. (2002). Ecosystem carbon loss with woody plant invasion of grasslands. *Nature*, **418**, 623–626.

Jacks, G. V. & Whyte, R. O. (1939). *The Rape of the Earth*. London: Faber and Faber.

Jamieson, B. G. (1988). On phylogeny and higher classification of Oligochaeta. *Cladistics*, **4**, 367–401.

Johnson, D., Leake, J. R., Ostle, N., Ineson, P. & Read, D. J. (2002). In situ $^{13}CO_2$ pulse-labelling of upland grassland demonstrates a rapid pathway of carbon flux from arbuscular mycorrhizal mycelia to the soil. *New Phytologist*, **153**, 327–334.

Kaufmann, R. (2001). Invertebrate succession on an alpine glacier foreland. *Ecology*, **82**, 2261–2278.

Killham, K. (1994). *Soil Ecology*. Cambridge: Cambridge University Press.

Knoll, A. H. (2003). *Life on a Young Planet: The First Three Billion Years of Evolution on Earth*. Princeton: Princeton University Press.

Knopf, F. L. & Rupert, J. R. (1999). The use of crop fields by breeding mountain plovers. *Studies in Avian Biology*, **19**, 81–86.

Lavelle, P. & Spain, A. V. (2001). *Soil Ecology*. Dordrecht: Kluwer Academic.

Mamiya, Y. (1983). Pathology of the pine wilt disease caused by *Bursaphelenchus xylophilus*. *Annual Review Phytopathology*, **21**, 201–220.

Manefield, M., Whiteley, A. S., Ostle, N., Ineson, P. & Bailey, M. J. (2002). Technical considerations for RNA-based stable isotope probing: an approach to associating microbial diversity with microbial community function. *Rapid Communications in Mass Spectrometry*, **16**, 2179–2183.

Marion, G. M., Henry, G. H. R., Freckman, D. W., *et al.* (1997). Open-top designs for manipulating field temperature in high-latitude ecosystems. *Global Change Biology*, 3, 20–32.

Masters, G. J., Brown, V. K. & Gange, A. C. (1993). Plant mediated interactions between above- and below-ground insect herbivores. *Oikos*, **66**, 148–151.

May, R. M. (1988). How many species are there on Earth? *Science*, **241**, 1441–1449.

May, R. M. (1997). Complex animal interactions introductory remarks. *Multitrophic Interactions in Terrestrial Ecosystems* (Ed. by A. C. Gange & V. K. Brown), pp. 305–306. Oxford: Blackwell Science.

Moore, J. C. & de Ruiter, P. C. (1997). Compartmentalization of resource utilization within soil ecosystems. *Multitrophic Interactions in Terrestrial Systems* (Ed. by A. C. Gange & V. K. Brown), pp. 375–393. Oxford: Blackwell Science.

Moore, J. C., Walter, D. E. & Hunt, H. W. (1988). Arthropod regulation of micro- and mesobiota in belowground detrital food webs. *Annual Review of Entomology*, **33**, 419–439.

Morris, C. E., Bardin, M., Berg, O., *et al.* (2002). Microbial diversity: approaches to experimental design and hypothesis testing

in primary scientific literature from 1975 to 1999. *Microbiology and Molecular Biology Reviews*, **66**, 592–616.

Mota, M. M., Braasch, H., Bravo, M. A., *et al.* (1999). First report of *Bursaphelenchus xylophilus* in Portugal and in Europe. *Nematology*, **1**, 727–734.

Nabonne, G. M. (2003). The crucial 80% of life's epic. *Science*, **301**, 919.

Naeem, S., Hahn, D. R. & Schuurman, G. (2000). Producer–decomposer co-dependency influences biodiversity effects. *Nature*, **403**, 762–764.

Naeem, S., Thompson, L. J., Lawler, S. P., Lawton, J. H. & Woodfin, R. M. (1994). Declining biodiversity can alter the performance of ecosystems. *Nature*, **368**, 734–737.

Odum, E. (1969). The strategy of ecosystem development. *Science*, **164**, 262–270.

Paul, E. A. & Clark, F. E. (1996). *Soil Microbiology and Biochemistry*. San Diego, CA: Academic Press.

Paul, E. A., Collins, H. P. & Leavitt, S. W. (2001). Dynamics of resistant soil carbon of midwestern agricultural soils measured by naturally occurring ^{14}C abundance. *Geoderma*, **104**, 239–256.

Petersen, H. & Luxton, M. (1982). A comparative analysis of soil fauna populations and their role in decomposition processes. *Oikos*, **39**, 287–388.

Polis, G. A., Anderson, W. B. & Holt, R. D. (1997). Toward an integration of landscape and food web ecology: the dynamics of spatially subsidized food webs. *Annual Review of Ecology and Systematics*, **28**, 289–316.

Radajewski, S., Webster, G., Reay, D. S., *et al.* (2002). Identification of active methylotroph populations in an acidic forest soil by stable isotope probing. *Microbiology*, **148**, 2331–2342.

Read, D. J. (1991). Mycorrhizas in ecosystems. *Experientia*, **47**, 376–391.

Rutherford, T. A., Mamiya, Y. & Webster, J. M. (1990). Nematode-induced pine wilt disease:

factors influencing its occurrence and distribution. *Forest Science*, **36**, 145–155.

Schenk, H. J. & Jackson, R. B. (2002). The global biogeography of roots. *Ecological Monographs*, **72**, 311–328.

Schimel, D. S. (1995). Terrestrial ecosystems and the carbon cycle. *Global Change Biology*, **1**, 77–91.

Schulze, E. D. & Mooney, H. A. (1994). *Biodiversity and Ecosystem Function*. Berlin: Springer-Verlag.

Sinclair, B. J. & Sjursen, H. (2001). Terrestrial invertebrate abundance across a habitat transect in Keble Valley, Ross Island, Antarctica. *Pedobiologia*, **45**, 134–145.

Sjursen, H. & Sinclair, B. J. (2002). On the cold hardiness of *Stereotydeus mollis* (Acari: Prostigmata) from Ross Island, Antarctica. *Pedobiologia*, **46**, 188–195.

Staddon, P. L., Ostle, N., Dawson, L. A. & Fitter, A. H. (2003a). The speed of soil carbon throughput in an upland grassland is increased by liming. *Journal of Experimental Botany*, **54**, 1461–1469.

Staddon, P. L., Ramsey, C. B., Ostle, N., Ineson, P. & Fitter, A. H. (2003b). Rapid turnover of hyphae of mycorrhizal fungi determined by AMS microanalysis of ^{14}C. *Science*, **300**, 1138–1140.

Stevens, M. I. & Hogg, I. D. (2002). Expanded distributional records of Collembola and Acari in southern Victoria Land, Antarctica. *Pedobiologia*, **46**, 485–495.

Stewart, W. D. P. (1976). *Nitrogen Fixation by Free Living Micro-Organisms*. Cambridge: Cambridge University Press.

Stotzky, G. & Bollag, J. M. (1992). *Soils, Plants and the Environment*. New York: Marcel Dekker.

Swift, M. J. & Anderson, J. M. (1994). Biodiversity and ecosystem function in agricultural systems. *Biodiversity and Ecosystem Function* (Ed. by E. D. Schulze & H. A. Mooney), pp. 15–41. Berlin: Springer-Verlag.

Swift, M. J., Heal, O. W. & Anderson, J. M. (1979). *Decomposition in Terrestrial Ecosystems*. Oxford: Blackwell.

Symstad, A. J., Chapin, F. S., Wall, D. H., *et al.* (2003). Long-term and large-scale perspectives on the relationship between biodiversity and ecosystem functioning. *BioScience*, **53**, 89–98.

Takacs, D. (1996). *The Idea of Biodiversity: Philosophies of Paradise*. Baltimore and London: The John Hopkins Press.

Tansley, A. G. (1935). The use and abuse of neglected concepts and terms. *Ecology*, **16**, 284–307.

Tiedje, J. M. & Stein, J. L. (1999). Microbial diversity: strategies for its recovery. *Manual of Industrial Microbiology and Biotechnology* (Ed. by A. L. Deamin & J. E. Davies), pp. 682–692. Washington, DC: American Society for Microbiology.

Tiedje, J. M., Cho, J. C., Murray, A., *et al.* (2001). Soil teeming with life: new frontiers for soil science. *Sustainable Management of Soil Organic Matter* (Ed. by R. M. Rees, B. C. Ball, C. D. Campbell & C. A. Watson), pp. 393–411. New York: CAB International.

Tilman, D., Knops, J., Wedin, D., *et al.* (1997a). The influence of functional diversity and composition on ecosystem processes. *Science*, **277**, 1300–1302.

Tilman, D., Naeem, S., Knops, J., *et al.* (1997b). Biodiversity and ecosystem properties. *Science*, **278**, 1866–1867.

Torsvik, V., Goksoyr, J. & Daae, F. (1990). High diversity in DNA of soil bacteria. *Applied and Environmental Microbiology*, **56**, 782–787.

US Fish and Wildlife Service (1999). Endangered and threatened wildlife and plants: proposed threatened status for the mountain plover. *Federal Register*, **64**, 7587.

Usher, M. B. (1985). Population and community dynamics in the soil ecosystem. *Ecological Interactions in Soil: Plants, Microbes and Animals* (Ed. by A. H. Fitter, D. Atkinson, D. J. Read & M. B. Usher), pp. 243–265. Oxford: Blackwell Scientific.

Usher, M. B., Davis, P. R., Harris, J. R. W. & Longstaff, B. C. (1979). A profusion of species? Approaches towards understanding the dynamics of the populations of the micro-arthropods in decomposer communities. *Population Dynamics* (Ed. by R. M. Anderson, B. D. Turner & L. R. Taylor), pp. 359–384. Oxford: Blackwell Scientific.

Van der Heijden, M. G. A., Klironomos, J. N., Ursic, M., *et al.* (1998). Mycorrhizal fungal diversity determines plant biodiversity, ecosystem variability and productivity. *Nature*, 396, 69–72.

Van Dyne, G. M. (1969). *The Ecosystem Concept in Natural Resource Management*. New York: Academic Press.

Virginia, R. A. & Wall, D. H. (1999). How soils structure communities in the Antarctic Dry Valleys. *BioScience*, 49, 973–983.

Virginia, R. A. & Wall, D. H. (2000). Ecosystem functioning. *Encyclopedia of Biodiversity* (Ed. by S. Levin), pp. 345–352. New York: Academic Press.

Wake, M. H. (1983). *Gymnopis multiplicata, Dermophis mexicanus,* and *Dermophis parviceps. Costa Rican Natural History* (Ed. by D. H. Janzen), pp. 400–401. Chicago, IL: University of Chicago Press.

Wake, M. (1993). The skull as a locomotor organ. *The Skull: Functional and Evolutionary Mechanisms* (Ed. by J. Hanken & B. K. Hall), pp. 240. Chicago, IL: University of Chicago Press.

Waksman, S. A. (1932). *Principles of Soil Microbiology*. Baltimore, MD: The Williams and Wilkins Company.

Wall, D. H. & Virginia, R. A. (1999). Controls on soil biodiversity: insights from extreme environments. *Applied Soil Ecology*, 13, 137–150.

Wall, D. H., Adams, G. & Parsons, A. N. (2001a). Soil biodiversity. *Global Biodiversity in a Changing Environment: Scenario for the 21st Century* (Ed. by F. S. Chapin & O. E. Sala), pp. 47–82. New York: Springer-Verlag.

Wall, D. H., Palmer, M. A. & Snelgrove, P. V. R. (2001b). Biodiversity in critical transition zones between terrestrial, freshwater, and marine soils and sediments: processes, linkages, and management implications. *Ecosystems*, 4, 418–420.

Wall, D. H., Snelgrove, V. R. & Covich, A. P. (2001c). Conservation priorities for soil and sediment invertebrates. *Conservation Biology: Research Priorities for the Next Decade* (Ed. by M. E. Soulé & G. H. Orians), pp. 99–123. Washington, DC: Island Press.

Wall Freckman, D. F. & Virginia, R. A. (1998). Soil biodiversity and community structure in the McMurdo Dry Valleys, Antarctica. *Ecosystem Dynamics in a Polar Desert: The McMurdo Dry Valleys, Antarctica* (Ed. by J. C. Priscu), pp. 323–336. Washington, DC: American Geophysical Union.

Wall Freckman, D. H., Blackburn, T. H., Hutchings, P., Palmer, M. A. & Snelgrove, P. V. R. (1997). Linking biodiversity and ecosystem functioning of soils and sediments. *Ambio*, 26, 556–662.

Wallwork, J. A. (1976). *The Distribution and Diversity of Soil Fauna*. London: Academic Press.

Wardle, D. A. (2002). *Communities and Ecosystems: Linking the Aboveground and Belowground Components*. Princeton, NJ: Princeton University Press.

Warnock, A. J., Fitter, A. H. & Usher, M. B. (1982). The influence of a springtail *Folsomia candida* (Insecta, Collembola) on the mycorrhizal association of leek *Allium porrum* and the vesicular–arbuscular mycorrhizal endophyte *Glomus fasciculatus. New Phytologist*, 90, 285–292.

Welker, J. M., Molau, U., Parsons, A., Robinson, C. H. & Wookey, P. A. (1997). Responses of *Dryas octopetala* to ITEX environmental manipulations: synthesis with circumpolar comparisons. *Global Change Biology*, 3, 61–73.

Whitaker, R. J., Grogan, D. W. & Taylor, J. W. (2003). Geographic barriers isolate endemic populations of hyperthermophilic archaea. *Science*, **301**, 976–978.

Williamson, J. & Harrison, S. (2002). Biotic and abiotic limits to the spread of exotic revegetation species. *Ecological Applications*, **12**, 40–51.

Wilson, E. O. (2002). *The Future of Life*. New York: Alfred A. Knopf.

Worster, D. (1994). *Nature's Economy: A History of Ecological Ideas*. Cambridge: Cambridge University Press.

PART II

The soil environment

CHAPTER TWO

The habitat of soil microbes

IAIN M. YOUNG
University of Abertay Dundee
KARL RITZ
Cranfield University

SUMMARY

1. This chapter deals with the impact of the soil's physical habitat on the operation of soil microbes.
2. The importance of the spatial and temporal nature of soil structural heterogeneity is emphasised.
3. The moisture characteristic is revealed as having a pre-eminent impact on soil biology.
4. The sensory ecology of nematodes is described in relation to the chemotaxis process.

Introduction

All terrestrial life lives and moves in the context of a more or less physically structured environment. The antelopes have their veldt, the mountain goats their crevasses, the rabbits their warrens and soil microbes their dark recesses of soil structure. How each individual, population and community operates is defined to a large extent by the physical landscape in which they live, which serves to partition substrate, mates, predators, water, gases and so forth. Geography, even the microgeography of the soil, sorts and drives the species in earth and on Earth.

Much has been published in relation to the so-called aggregate sizes and the presence of microorganisms. For example, Vargas and Hattori (1986) are convinced that there is order in soil, with bacteria living in the centre of aggregates more often than on the surface. Linn and Doran (1984) link the direct influence of soil structure to changes in microbial activity and community structure at the field scale. Work from The Netherlands has produced conclusive proof that the structure of soil, and the attendant moisture, are controlling factors in predator–prey interactions between bacteria and protozoa (e.g. Kuikman *et al.* 1989; Postma & van Veen 1990), an area that Young *et al.* (1994) and Young and

Biological Diversity and Function in Soils, eds. Richard D. Bardgett, Michael B. Usher and David W. Hopkins.
Published by Cambridge University Press. © British Ecological Society 2005.

Crawford (2001) revisited with respect to the impact of structure on substrate location and accessibility.

In this review we attempt to place soil microbes in the context of their physical environments, and to raise fundamental questions as to why the organisation of soil is so extremely spatially and temporally heterogeneous – the impact of larger animals on soil is well documented and we have discussed this elsewhere (Young et al. 1998). It turns out that soil would be better conceptualised as oases surrounded by desert, rather than biofilms dominated by rhizospheres. The driving philosophy in this paper is that the desert regions have just as much to do with functionality of the soil system as do the oases.

Many of the chapters in this publication highlight the undeniable fact that soil is teeming with life. The figures are now well rehearsed but remain remarkable: in one gram of grassland soil can reside 10^9 bacteria, 10^5 protozoa, 10^2 nematodes and 1 km of fungal hyphae. These impressive figures are only beaten by the large surface area found in the same soil. Making simple yet reasonable assumptions, a realistic estimate of surface area rises to 20 m^2. This of course depends upon the resolution of measurement, but the important point is that, *at the scale of the microbes*, they themselves cover less than 0.0001% of the total surface area theoretically available for colonisation. Coincidentally, this is the same percentage of the Earth's land surface that is currently estimated to be covered by humankind. And perhaps, not surprisingly, similar mechanisms and processes control the patchiness of microbes in soil and humans on Earth.

Soil is not a bed of aggregates

One of the most conceptually distorting and pervasive pieces of dogma in soil science is that soil is comprised of a collection of discrete aggregates arranged in a bed, and that the aggregate is a fundamental 'building block' of soil. That is not to say that aggregates do not exist. Rather, they only exist as discrete entities following mechanical disruption of the soil, whether it be by the plough, spade or sieve, or in limited self-mulching soils. Aggregates are essentially operationally defined units, and the size and number of them that can be derived from any soil is contingent on the energy used to obtain them. So perhaps the exception to our doggerel is a newly prepared arable seedbed – and incidentally, with *appropriate energetic inputs*, an optimal architecture for seedling establishment. But within days the discreteness of the aggregates disappears and the 'true' soil structure – the continuous labyrinthine pore network – ultimately always connected at some scale, prevails. That is certainly what subterranean life encounters. Aggregate size distributions tell us little about the structure of the soil habitat, and aggregates are best viewed as discrete masses of soil out of their natural context, occasionally convenient for some experimental studies and, alas, conceptually convenient. The 'surface' of an aggregate does not necessarily exist in the soil as a surface – *in situ*, it may have been an incipient line of weakness rather than lining a pore wall.

Soil is four dimensional

All soil systems are spatially heterogeneous across many orders-of-magnitude of scale, and exist in three dimensions of space and a fourth in time. This may seem a trite statement, but these facts are often ignored – witness the myriad of publications using sieved soil, or models 'working' in two dimensions. Some of this work was done for very good reasons: heterogeneity at some level had to be decreased so that a specific hypothesis could be tested; three-dimensional simulations were until recently beyond the computational power of most computers. Thus some good work has been published using such approaches (e.g. Arah & Smith 1989; Preston *et al.* 1999). However, there is a limit to the degree that the characteristics of the real soil system should be ignored.

Starting from two-dimensional datasets taken from thin-sections, we can now readily construct a picture of the importance of the physical diversity of the soil system to function. Figure 2.1 presents a range of soil thin-sections that highlight this variability. Recent work has attempted to use similar techniques, but with biological components intact (Fig. 2.2), and reveals a hitherto unseen world. How organisms sense this complex world is dealt with later on. However, the importance of spatial context in biological function is revealed by the recent work of Nunan *et al.* (2003) who quantified the spatial clustering of bacteria in agricultural soils.

The rationale for suggesting that the spatial position of bacteria is related to structure, as suggested by Nunan *et al.* (2003), is arrived at not simply by the fact that the physical framework of the soil impacts on how bacteria can move and be moved, but more from the way it interacts with the flow of oxygen and gases, and mediates the water-films surrounding bacterial communities. To function efficiently, aerobic bacteria require an appropriate oxygen diffusion rate towards themselves and a concomitantly appropriate carbon dioxide diffusion rate away from themselves. If either becomes suboptimal then microaerophilic and eventually anaerobic conditions prevail, even in relatively dry soils (Young & Ritz 1998). Thus, being close to air conduits, and obviously substrate, is important, as long as this is balanced by an avoidance of desiccation – as evidenced by the mucilage surrounding some bacterial communities. Nunan *et al.* (2002) found strong correlation between the spatial location of bacteria and pores in the subsoil of an arable soil. In further investigations focusing on mapping bacteria at multiple scales, they also found that spatial structure (i.e. spatial correlations between bacteria) were only present at the micrometre scale in the topsoil, while there were two distinct scales within the subsoil (micrometre scale and scales ranging over centimetres to metres). They suggest that these differences were linked to a, as yet unexplained, range of factors that impact on bacterial distribution.

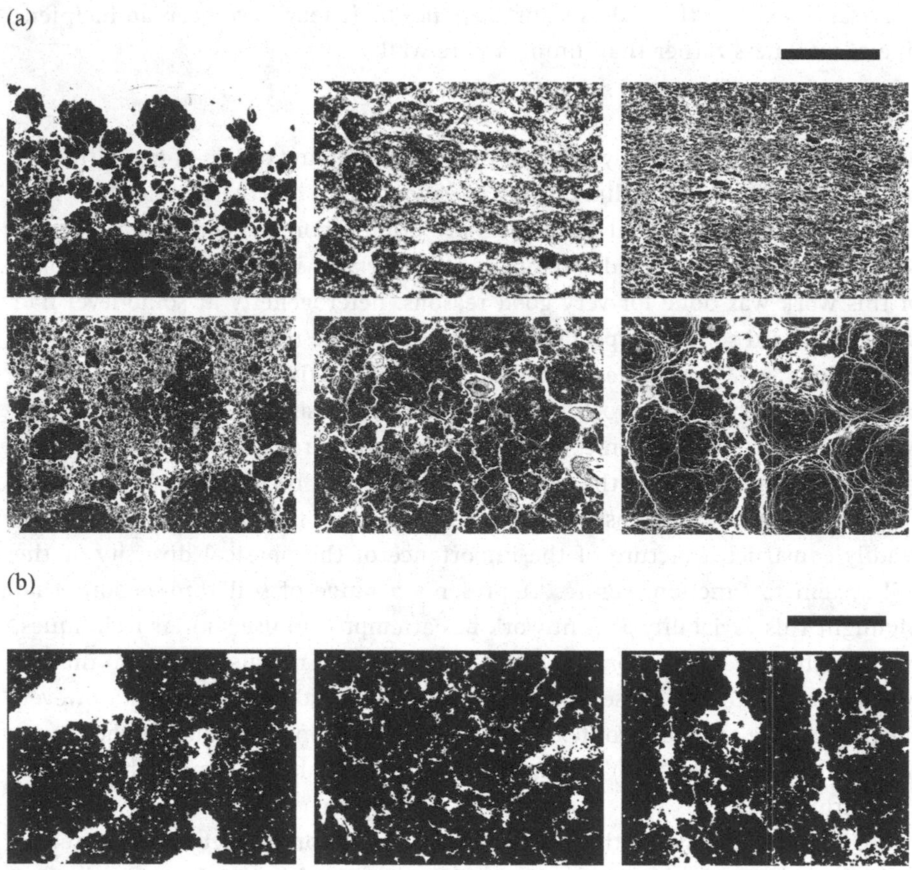

(a)

(b)

Figure 2.1. Soil structure as visualised by microscopy of soil thin-sections. (a) Variety of soil pore morphologies apparent in diverse soils. Scale bar, 10 mm. Montage derived from source images in Fitzpatrick (1993). (b) Three sections from regions within 1 m of each other in the topsoil of an arable field, demonstrating how pore morphology can show great spatial variation within an ostensibly uniform and frequently mixed soil, in this case by cultivation practices. Scale bar, 5 mm.

While measuring soil bacteria *in situ*, at least in two dimensions, has had limited development, the measurement of the physical architecture of the soil in three dimensions has developed rapidly. This development has been recently reviewed (Young *et al.* 2001) and will not be repeated here. Suffice to say a range of experimental and numerical techniques are now available to visualise and simulate the geometrical complexity of soil in three dimensions over a range of spatial scales. Now, for the first time, we have a real opportunity to examine the biological and physical diversity of soil systems in tandem. This is best illustrated by the recent application of computer-aided microtomography to visualise and quantify soil structure at scales relevant to microbes.

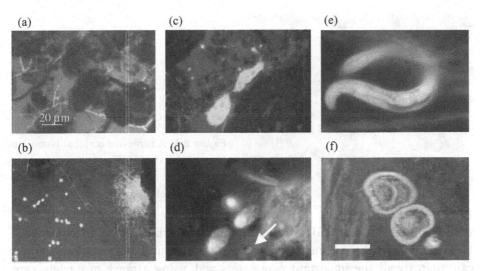

Figure 2.2. Soil structure and soil biota, as visualised in thin-sections of the same soil as in Fig. 2.1b, prepared appropriately to preserve biological tissue (cf. Nunan *et al.* 2001). (a) Mycelia of *Rhizoctonia solani* in sterilised soil. Note preferential growth in larger pores, and apparent lack of penetration of denser masses of soil. (b) Unidentified mycelium in field soil, with abundant sporangia in main region of pore. (c) Naked amoeba passing through narrow pore in arable soil. Note fidelity of preservation of this delicate organism. (d) Testate amoebae accumulating near pore wall of arable soil. Bacterial colony arrowed. (e) Nematode in arable soil, obviously distorted by the pore it is inhabiting. (f) Enchytraeid worm adjacent to decomposing root (to left). Image shows tangential section through body of the worm, which is looping through plane of section. Note soil material in lumen of gut in upper portion, and how the body of this animal is distorted by the confines of the pore. Scale bars, 20 μm for (a)–(e); 100 μm for (f).

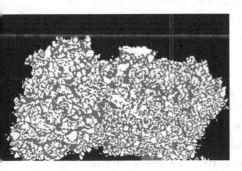

Figure 2.3. Cross-section from high-resolution X-ray tomography scan. See text for details. Grey represents internal porosity, white represents solid particles. The length of the aggregate is 1.1 mm.

Figure 2.3 shows a cross-section of a natural aggregate taken from the UK's Soil Biodiversity site in Sourhope, Scotland. The image was captured using the microtomography facility at the Advanced Photon Source, University of Chicago. The smallest feature observed is 4.4 μm. In Fig. 2.4, we see a visualisation of the same aggregate, but partially excised to expose the inner architecture together

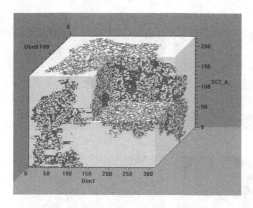

Figure 2.4. A three-dimensional reconstructed image from the same aggregate used in Fig. 2.3. Black represents internal porosity, light shading represents solid particles.

with the outer geometry. Here we glimpse the importance of connectivity and heterogeneity in three dimensions. Using three-dimensional image analysis we can quantify all the important parameters and, using a range of models, can start to predict energy flows and biological functions. Once these methodologies have become commonly available and are presented in an accessible way to the biologists, many fundamental questions concerning the link between biological function and habitat will be able to be addressed. However, it is rare that structure, in terms of biological function, acts in isolation; there is a crucial further component in the three-dimensional soil jigsaw, i.e. water.

Water water everywhere?

The performance and survival of individual organisms and communities is driven by their ability to sense, locate and access substrate and mates, and to avoid predators. Within soil where habitats are characterised by spatio-temporal heterogeneities at the scale of the bacteria and above, there is a set of complex processes that determine the efficacy of any individual or community in performing their functions. All soil microbes are aquatic in the sense that they require the presence of moisture to function. As such, the volume and distribution of moisture within soil is a key determinant of the function of all soil microbes. Soil water acts as a canal for motile organisms and as a valve shutter in relation to gas transport. Coleman *et al.* (2003) make an intriguing point that the inherent ability of soils to physically partition water and air may have been a launch pad for the evolution of terrestrial animals, as it provided an important adaptive bridge for aquatic life forms, a sort of evolutionary test-bed for leaving aquatic domains.

A key determinant of the volume and distribution of water in soil is the moisture release curve, which is determined by the physical structure of the soil. This has been shown many times, but emphasis has often mistakenly been placed on the estimations of 'equivalent spherical pore sizes', rather than on the fact that the moisture release characteristic is an energy curve. This curve describes all

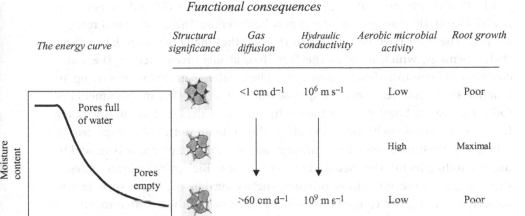

Figure 2.5. A summary of the impact of the moisture release characteristic on a range of soil functions.

the hydraulic and gaseous pathways in soil, and thus has an impact on the operation of the biota. Figure 2.5 illustrates the qualitative and quantitative impact the moisture release curve has on many soil processes. A crucial distinction is that it is the matric potential of the soil rather than the volume/mass of water that is the important parameter. The matric potential determines the thickness of the water-films around organic and inorganic particles, thus determining the connectivity of water in the soil system and, if the size of the water-film is large enough to block or reduce pore pathways, the gaseous diffusion of particles. At large scales, this is why measures of water-filled pore space (degree of saturation) have been significantly correlated with nitrogen transformations (Linn & Doran 1984). The variability often associated with measures of gas diffusion is linked to the variations in water-films at the small scale, due to the heterogeneities of the small-scale structure.

All these characteristics are neatly summarised by the general expression:

$$\Theta \text{ vs. } \Psi_m^b, \tag{2.1}$$

where Θ, Ψ_m, and b represent volumetric moisture content, matric potential and a power-law constant b, respectively. There are several important variants of this equation, but for simplicity we will use this form. Many biologists have used this curve to determine an array of parameters that hopefully correlate to biological function. These range from attempts to predict pore volume available to specific microbes (Wright *et al.* 1993) through to predator–prey interactions (Heijnen & van Veen 1991). But generic relations have not been forthcoming. The peculiarities of the moisture release curve are that it provides a characteristic energy curve for a soil under a particular state, and that there is a twist in the tail in that it exhibits hysteresis. Simply put, if you wet a dry soil, rather than dry a wet soil, the energetic relations may be different. There are several reasons for this. The most cited one relates to the fact that many pores are larger than their openings, which leads to the fact that, at any given suction, the water content of a pore full of water drying, will be greater than that on rewetting. In practice such 'hystereses' may simply be due to the fact that measurements of soil moisture are taken prior to the equilibration, or the fact that on drying air fails to enter pores via bottlenecks, which then remain wetter than expected. In short, care must be taken when utilising the soil moisture characteristic, and in understanding its full significance in soil processes.[1] Biology complicates physics still further where organisms produce surface-tension modifiers such as the powerful hydrophobic compounds of some microbes or the assorted mucilages that gum up pores.

So, rather than a simple construct through which moisture is drained or added to the soil system, the moisture characteristic in three-dimensional structure is revealed as a key determinant of all energy flows through soil. Figure 2.5 shows published data relating gas diffusion and hydraulic conductivity. In practical terms, at saturation, gas will move at around $1\,cm\,d^{-1}$, compared to $1\,cm\,h^{-1}$ in a desaturated soil. The implications for O_2 diffusion, greenhouse gas productions, etc., are clear.

The moisture characteristic also determines the accessibility of reactive sites to various agents, be they biological or chemical. The hydraulic flow paths determine the connectivity of the soil system and thus, in combination with the accessibility of reactive sites, the probability of pollutant/pathogen dispersal to waterways. It is also worthy to note that if the total number of bacteria at any one time cover less than $10^{-4}\%$ of the total surface area, at low suctions (i.e. under dry conditions), the total number of active bacteria will cover far less of the surface area. This emphasises the importance of spatial context in the measurement of all soil biophysical processes and the special role that the moisture characteristic plays in soil biological processes. To emphasise this point, we present a biophysical case study.

The importance of soil structure: a case study from sensory ecology

Sensory ecology relates to how an organism acquires and responds to information about its environment. Consider the questions: what makes a nematode move in soil, and what is it responding to? All organisms respond to external stimuli, both positively and negatively. Within the soil fabric, light becomes absent at depth and senses other than sight take precedence. The blind nematode has therefore developed senses more attuned to the opaque, feast-and-famine cycles that prevail in soil systems. The utility of investigating

---

[1] An excellent teaching tool for biologists to understand the basic principles of the moisture release curve, and indeed other relevant areas, can be found at http://aquarien.com.sptutor.

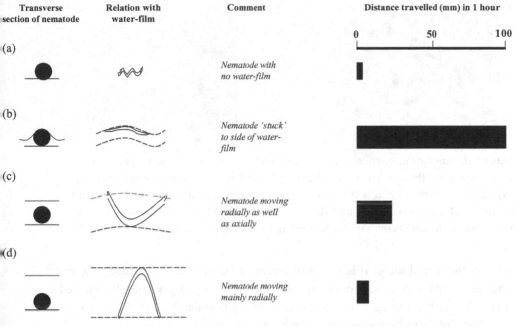

Transverse section of nematode	Relation with water-film	Comment	Distance travelled (mm) in 1 hour

Figure 2.6. The impact of water-films on nematode movement. The thickness of the water-film (as prescribed by the structure of soil and the matric potential) defines the degrees of freedom for nematode movement. (a) In the absence of water, no movement occurs. (b) Maximum forward movement occurs when a thin film surrounds the nematode. (c, d) Too thick a film allows the nematode to move in all directions, thus significantly decreasing forward movement. Adapted from Wallace (1958).

such a soil-borne organism is highlighted by the multitude of papers dealing with the bacterial-feeding nematode *Caenorhabditis elegans* in the scientific literature (see Blaxter 1998). A crucial model multi-cellular eukaryote, and the first to have its whole genome sequenced, this species is in fact a soil organism. Ironically, the majority of the huge number of studies on this animal seem to ignore its natural habitat and study a vast array of attributes in the sensory-deprivation chambers that are agar plates (Hodgkin & Donaich 1997).

As roots deposit carbon substrates into the rhizosphere, bacteria will start to digest such material and emit a wide array of chemicals into the surrounding soil, ranging from CO_2 from respiration to volatile organic carbon compounds that are secondary by-products of metabolism. Some distance from the root, nematodes will be randomly foraging, the path of movement determined by the combination of structure and water within the local soil volume. This scenario forms the basis of the chemotaxis model of Anderson *et al.* (1998b). Figure 2.6 illustrates how the variability of water-films impacts on the movement of nematodes. As explained previously, this relates to the moisture characteristic.

After a finite period, the chemicals emitted from the bacteria (along with those 'leaked' from roots, etc.) will diffuse through the soil and establish gradients

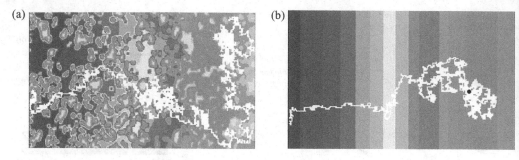

Figure 2.7. Simulation of chemotaxis process within (a) a simple non-structured environment and (b) a structured environment. The nematode (•) moves (white trails) in response to a chemical gradient (darkest shades represent highest concentrations, and lightest shades lowest concentrations). Heterogeneity in (b) is introduced by imposing a two-dimensional structural barrier. Adapted from Anderson *et al.* (1998b).

towards the nematode(s). It is worth observing here that such diffusion is again modulated by the moisture characteristic, and that CO_2, a gas often implicated in the directional movement of soil organisms, is heavier than air. The result is that horizontal gradients are more likely than vertical gradients.

The chemical gradient emanating from the rhizosphere will, above a certain concentration and perhaps only with a certain cocktail of compounds, be sensed by the nematode. In the case of *C. elegans*, chemosensory neurons make up roughly 10% of its total nervous system; with the genome encoding around 1000 orphan receptors that Bargmann (1998) suggests may be chemoreceptors. Once sensed, the nematode will move towards the strongest gradient, switching from a random to a biased random foraging strategy and thus, all things being equal, reach the bacteria. Several workers have, with various small modifications, observed this scenario using experimental and theoretical methods (Anderson *et al.* 1998a, b). Young *et al.* (1996, 1998) show how nematode movement is affected by particle size and moisture, and available substrate, respectively. Figure 2.7 illustrates the impact of structure on both nematode movement and the diffusion of chemicals. The interesting observation from these simulated results is that the one structure impacts directly on the chemotaxis process operating over molecular scales and up to scales of tens of centimetres; the structure modulates the rate of diffusion of gas away from the bacteria (molecular scale) and retards the rate of movement of the nematode (mm–cm), thus impacting on the nematode's ability to sense gradients (molecular) and reach and eat bacteria (μm).

Change is constant

A characteristic of the physical habitats of soil is that they are constantly changing. Dependent on the physico-chemical make up of soil, simple additions of water can significantly alter the pore sizes in soil, through swelling mechanisms.

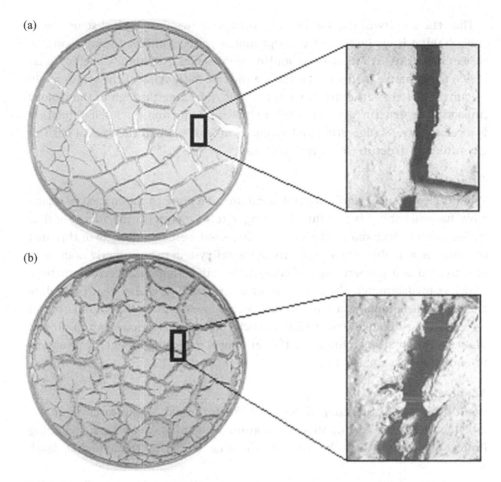

Figure 2.8. The impact of adding fungi to soil structural generation from a soil slurry:
(a) control, (b) control + fungi. A fungus creates a more varied structure: greater
variability in topography and in crack formation. The latter is illustrated in the enlarged
boxed images from each system.

This is well known. Less well accepted is the important and functionally significant impact of biology on the habitat. Figure 2.8 shows the great impact that
fungi can have on both large- and small-scale structures in soil. What seems
to happen is that fungi introduce a multiple-scale physical diversity into the
system, possibly due to their filamentous growth form, which has unique spatial properties in that the organism can transcend scales from micrometres to
kilometres. For example, fungi can alter cracking pathways, making the whole
soil system more heterogeneous. Fungi also affect soil hydrophobicity. Recent
work has suggested that the glycoprotein glomalin, exuded exclusively from
arbuscular mycorrhizal (AM) fungi, retards water ingress into soil, thus reducing
structural disintegration through slaking. Concentrations of glomalin have been
observed to be highly correlated with aggregate stability (Rillig et al. 2001).

That the activity of the soil biota in turn regulates the physical structure of their habitat is an important concept and is perhaps one of the most under-researched areas in soil science, and hence deserves attention. Crawford *et al.* (1995) using simple model systems make the case for looking at the soil as a self-organising system: structure impacts on biology, feeding through to the biology impacting on structure, and onto future biological function. Such self-regulation is a relatively new suggestion and promises at least a sound theoretical basis to examine soil structure–biota interactions.

Conclusions

What is clear is that examining biological function outwith the context of organisms' natural habitat is fraught with danger. It is timely with the advent of new molecular, physical and modelling techniques to invest significantly in this area of research, as it allows heterogeneity of the soil system to be properly examined, and treated as a property of soil ecosystems rather than as an error function, which is predominately the case in most soil research. It is also clear that while the rhizosphere remains an important area for investigation, the volumes of soil outwith the rhizosphere also play a crucial role in the functioning of all soil biota, as the physical architecture that envelopes it controls all energy functions to and from rhizosphere soil.

Acknowledgements

Work described in this chapter has been funded by the UK Research Councils, and has been carried out in collaboration with many co-workers, including Professor John Crawford, Dr Katherine Helming, Dr Narise Nunan and Dr Mark Rivers.

References

Anderson, A. R. A., Sleeman, B. D., Young, I. M. & Griffiths, B. S. (1998b). Nematode movement along a chemical gradient in a structurally heterogeneous environment: 2. Theory. *Fundamental and Applied Nematology*, **20**, 165–172.

Anderson, A. R. A., Young, I. M., Sleeman, B. D., Griffiths, B. S. & Roberston, W. M. (1998a). Nematode movement along a chemical gradient in a structurally heterogeneous environment: 1. Experiment. *Fundamental and Applied Nematology*, **20**, 157–163.

Arah, J. R. M. & Smith, K. A. (1989). Steady-state denitrification in aggregated soils: mathematical model. *Journal of Soil Science*, **40**, 139–149.

Bargmann, C. I. (1998). Neurobiology of the *Caenorhabditis elegans* genome. *Science*, **282**, 2028–2033.

Blaxter, M. (1998). *Caenorhabditis elegans* is a nematode. *Science*, **282**, 2041–2046.

Coleman, D. C., Crossley, D. A. & Hendrix, P. F. (2003). *Fundamentals of Soil Ecology*. London: Academic Press.

Crawford, J. W., Verrall, S. & Young, I. M. (1995). The origin and loss of fractal scaling in soil aggregates. *European Journal of Soil Science*, **48**, 643–650.

Fitzpatrick, E. A. (1993). *Soil Microscopy and Micromorphology*. Chichester: Wiley.

Heijnen, C. E. & van Veen, J. A. (1991). A determination of protective microhabitat for bacteria introduced into soil. *FEMS Microbiology Ecology*, **85**, 73–80.

Hodgkin, J. & Donaich, T. (1997). Natural variation and copulatory plug formation in *Caenorhabditis elegans*. *Genetics*, **146**, 149–164.

Kuikman, P. J., van Vuuren, M. M. I. & van Veen, J. A. (1989). Effect of soil moisture on predation by protozoa of bacteria biomass and the release of bacterial nitrogen. *Agriculture, Ecosystems and Environment*, **27**, 271–279.

Linn, D. M. & Doran, J. W. (1984). Aerobic and anaerobic microbial populations in no-till and plowed soils. *Soil Science Society of America Journal*, **48**, 794–799.

Nunan, N., Ritz, K., Crabb, D., *et al*. (2001). Quantification of the *in situ* distribution of soil bacteria by large-scale imaging of thin sections of undisturbed soil, *FEMS Microbiology Ecology*, **37**, 67–77.

Nunan, N., Wu, K., Young, I. M., Crawford, J. W. & Ritz, K. (2002). *In situ* spatial patterns of soil bacterial populations, mapped at multiple scales, in an arable soil. *Microbial Ecology*, **44**, 296–305.

Nunan, N., Wu, K., Young, I. M., Crawford, J. W. & Ritz, K. (2003). Spatial distribution of bacterial communities and their relationships with the micro-architecture of soil. *FEMS Microbiology Ecology*, **44**, 203–215.

Postma, J. & van Veen, J. A. (1990). Habitable pore space and survival of *Rhizobium leguminosarum* biovar *trifolii* introduced into soil. *Microbial Ecology*, **19**, 149–161.

Preston, S., Griffiths, B. S. & Young, I. M. (1999). Links between substrate additions, native microbes and the structural complexity and stability of soils. *Soil Biology and Biochemistry*, **31**, 1541–1547.

Rillig, M. C., Wright, S. F., Kimball, B. A., *et al*. (2001). Elevated carbon dioxide and irrigation effects on water stable aggregates in a sorghum field: a possible role for arbuscular mycorrhizal fungi. *Global Change Biology*, **7**, 333–337.

Vargas, R. & Hattori, T. (1986). Protozoan predation of bacterial cells in soil aggregates. *FEMS Microbiology Ecology*, **38**, 233–242.

Wallace, H. R. (1958). Movement of eelworms: I. The influence of pore size and moisture content of the soil on the migration of larvae of the beet eelworm, *Heterodera schachtii* Schmidt. *Annals of Applied Biology*, **46**, 74–85.

Wright, D. A., Killham, K., Glover, L. A. & Prosser, J. I. (1993). The effect of the location in soil on protozoan grazing of a genetically modified inoculum. *Geoderma*, **56**, 633–640.

Young, I. M., Blanchart, E., Chenu, C., *et al*. (1998). The interaction of soil biota and soil structure under global change. *Global Change Biology*, **4**, 703–712.

Young, I. M. & Crawford, J. W. (2001). Protozoan life in a fractal world. *Protist*, **152**, 123–126.

Young, I. M., Crawford, J. W. & Rappoldt, C. (2001). New methods and models for characterising structural heterogeneity of soil. *Soil and Tillage Research*, **61**, 33–45.

Young, I. M., Griffiths, B. G. & Robertson, W. M. (1996). Continuous foraging by bacterial-feeding nematodes. *Nematologica*, **43**, 378–382.

Young, I. M., Griffiths, B. S., Robertson, W. M. & McNicol, J. W. (1998). Nematode (*Caenorhabditis elegans*) movement in sand as affected by particle size, moisture and the presence of bacteria (*Escherichia coli*). *European Journal of Soil Science*, **49**, 237–241.

Young, I. M. & Ritz, K. (1998). Tillage, habitat space and microbial function. *Soil and Tillage Research*, **53**, 201–213.

Young, I. M., Roberts, A., Griffiths, B. S. & Caul, S. (1994). Growth of a ciliate protozoan in model ballotini systems of different particle sizes. *Soil Biology and Biochemistry*, **26**, 1173–1178.

CHAPTER THREE

Twenty years of molecular analysis of bacterial communities in soils and what have we learned about function?

A. G. O'DONNELL
University of Newcastle upon Tyne
S. R. COLVAN
University of Newcastle upon Tyne
E. MALOSSO
University of Newcastle upon Tyne
S. SUPAPHOL
Kasetsart University

SUMMARY

1. Soils support a taxonomically and physiologically diverse biota widely regarded as more extensive than that of any other group of organisms. However, the limits of this diversity and its importance in delivering soil function remain unclear.
2. The last 20 years have seen renewed interest in soil microbiology and in the application of molecular methods to explore what is there and how it changes over time or in response to environmental stimuli. For the first time microbiologists have been able to open the microbial 'black box' in soils.
3. The justification often given for opening this 'black box' is that the diversity of the contents therein is vitally important to the maintenance and sustainability of the biosphere. However, despite almost 20 years of detailed sifting through the box contents, there is little evidence to support this, suggesting that many of the thousands of microorganisms in soils are functionally redundant and that many of the major functions of the microbial biomass are unaffected by its exact species composition.
4. This chapter looks at some of the approaches used to open the microbial 'black box' and discusses whether we are any closer to understanding the relationship between diversity and function in soils.

Introduction

Soils are an important natural resource and have a key role in the biosphere with most of the annual carbon and nutrient fluxes occurring in the top 5–10 cm of

Biological Diversity and Function in Soils, eds. Richard D. Bardgett, Michael B. Usher and David W. Hopkins.
Published by Cambridge University Press. © British Ecological Society 2005.

the soil profile. This relatively thin covering of soil is essentially a non-renewable resource that provides and regulates many of the functions and processes that support life. At the heart of this life support system is the soil biota without which there would be no biogeochemical cycling of carbon and nitrogen, no organic matter transformation or carbon sequestration, no gas exchange and no pollutant mitigation. As the engine that drives soil function, the use of land management practices to manipulate the soil biota offers the possibility of regulating soil biological activity to mitigate the consequences of global change. Understanding the relationships between biodiversity and function has become one of the major challenges facing soil biology and in particular soil microbiology. This challenge is made all the more difficult by the fact that the microbial biomass is both directly and indirectly influenced by a highly heterogeneous and complex, physico-chemical environment (O'Donnell & Goerres 1999). This soil environment supports a taxonomically and physiologically diverse biota widely regarded as more extensive than that of any other group of organisms (Kennedy & Gewin 1997). However, the limits of this diversity (Curtis et al. 2002) and its importance remain the subject of much discussion (Finlay & Clarke 1999).

One of the key considerations in relating microbial diversity to function is the ability to determine accurately and reproducibly the size, activity and diversity of the soil microbial biomass. However, prior to the introduction of molecular techniques, reliably measuring these parameters and relating them to individual organisms or taxa was virtually impossible with a clear mismatch between our ability to make process measurements, such as respiration (Hopkins & Ferguson 1994) and soil enzyme activity (Nannipieri et al. 1978; Colvan et al. 2001), and our ability to identify which organisms or groups of organisms were involved in these processes.

Microbial ecology is a relatively young science that until the introduction of molecular techniques relied on the recovery and analysis of the pure culture, a clonal population originating from a single cell by asexual reproduction. However, it is well established that the ability of microorganisms from natural environments to grow on artificial culture media is low and that many prokaryotic microorganisms are extremely difficult to grow axenically in the laboratory (Connon & Giovannoni 2002; Rappe et al. 2002). This is probably due to a plethora of reasons from the activity of lysogenic phage, quorum sensing, syntrophic metabolism and the availability of growth factors (Rodriguez-Valera 2002) to our lack of understanding of the physical characteristics of the microniche (Young et al. this volume) and their importance in regulating growth in natural environments. The limitation of culture approaches and the inability to measure confidently even the dominant components of the community was clearly a problem for microbial ecology.

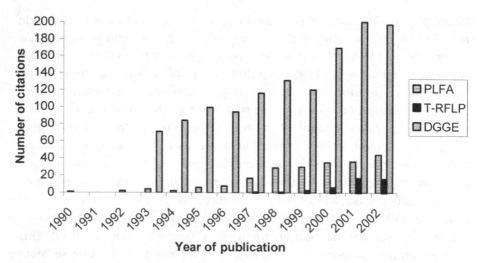

Figure 3.1. Number of citations (web of science) for selected microbial community fingerprinting methods over the period 1990 to 2003. PLFA, phospholipid fatty acid; T-RFLP, terminal restriction fragment length polymorphism; DGGE, denaturing gradient gel electrophoresis

Microbial diversity and the impact of small subunit 16S rRNA

Molecular ecology is ecology at its most reductionist where molecules and changes in the relative concentrations of these molecules are used to infer the importance of individual taxa at the system level. Although other signature molecules have been used as surrogates for microbial groups (e.g. phospholipid fatty acids, PLFA, Steenwerth *et al.* 2002; ergosterol, Malosso *et al.* 2004; teichoic acid and lipopolysaccharide; Mummey *et al.* 2002) 'molecular microbial ecology' is often seen as synonymous with the extraction, amplification and analysis of small subunit rRNA, particularly bacterial 16S rRNA. This is evidenced by the vast literature on the use of these techniques and their now routine application to studies of diversity and community analysis (Fig. 3.1) in natural environments.

Although one might argue over the usefulness of 16S rRNA gene methods in understanding function, there can be no doubt that over the last 20 years they have contributed significantly to our understanding of the extent of microbial diversity in soils. These culture-independent, molecular ecological analyses have been partly responsible for the increase in the number of major bacterial lineages from the 12 originally described by Woese (1987) using cultured isolates to over 36 based on culture and culture-independent approaches (Tiedje *et al.* 1999). The availability of these data have made possible the detailed description of microbial community structure, including the number of species present (α-diversity) and the number of individuals of each species (evenness), and in doing so have provided microbial ecologists with new tools with which to explore microbial communities. These tools include physiological (Whiteley *et al.* 2001), genetic (Whiteley *et al.* 1996; Griffiths *et al.* 2000) and stable isotope probing

(Manefield *et al.* 2002) techniques that may ultimately provide insight into the specific activities of individual cells *in situ*.

While 16S rRNA gene analyses have undoubtedly contributed to our understanding of the extent of bacterial diversity in soils, they have yet to impact significantly on our understanding of important processes, their regulation or how they might be manipulated. It is therefore apposite to ask whether there is sufficient justification for continuing such studies. Is the goal to inventory all soil bacteria and if so what do we expect to achieve? Even when furnished with a complete catalogue of 16S rDNA, is the microbial ecologist likely to be any closer to understanding function as defined by expressed phenotypes, behaviour and activity? The current situation is nicely illustrated by Rodriguez-Valera (2002) who states that the 'black-box is now dimly lit but we have yet to see the characters or the play'. Indeed, the need for molecular microbial ecology to progress beyond cataloguing diversity is clearly illustrated in the work by Hagström *et al.* (2002) who re-analysed the bacterioplankton-derived 16S rDNA sequences deposited at GenBank and showed that over the period between 1990 and 2001 the rate of new additions to the data-base peaked in 1999 and has since levelled off. When the data were de-replicated using 97% similarity as a cut-off, 1117 unique ribotypes were found with 508 derived from cultured organisms and 609 originating from as yet uncultured lineages. These authors concluded that in marine environments the apparent species richness is relatively low and that, in the absence of cloning biases or diminishing interest in marine diversity, the current databases may already provide good coverage of the 16S rDNA sequences of marine environments.

Given that functional redundancy is expected to be the norm among soil bacteria with all but a limited number of functions (e.g. nitrification, methanogenesis) carried out by physiologically and taxonomically diverse organisms and the existence of 'keystone' microbial species highly unlikely, it is pertinent to ask whether the 16S rRNA gene is a relevant molecular marker for functional analyses. The 16S rRNA gene was originally identified as a prime candidate for understanding microbial evolution because the rRNA molecule is a mosaic of highly conserved sequence domains interspersed with more variable regions (Woese 1987). The strength of the approach lies in the fact that through comparative sequence analysis different levels of taxonomic rank can be compared facilitating culture-independent characterisation of signature sequences from diverse environments. The development of microbial community analysis is firmly rooted in molecular systematics where the increased availability of primers for the 16S rDNA gene has resulted in molecular microbial ecology being dominated by studies of 16S rRNA gene diversity. However, the 16S rDNA gene is, on average, only 0.05% of the bacterial genome and in all but a few cases (e.g. in nitrifying bacteria) is of limited value in predicting the physiology of the organism or its interaction with its environment (Kowalchuk *et al.* 1997; Schramm *et al.* 1998). As such, one must ask whether it is reasonable to expect that soil communities catalogued using 16S rRNA genes are likely to reveal any relationship whatsoever to function.

Microbial community structure and function

Despite the advantages of amplification, cloning and sequencing approaches, routinely used by taxonomists to characterise cultured isolates, they are too laborious for use on all but the simplest of microbial communities and as a result many community-fingerprinting methods have been developed. These include denaturing gradient gel electrophoresis (DGGE; Muyzer 1999), thermal gradient gel electrophoresis (TGGE; Muyzer 1999), single strand conformational polymorphism (SSCP; Liu & Sommer 1995), restriction fragment length polymorphism (RFLP; Barns *et al.* 1999), amplified ribosomal DNA restriction analysis (ARDRA; Wenderoth & Reber 1999) and terminal restriction fragment length polymorphism (T-RFLP; Marsh 1999). Like the cloning and sequencing approaches, these community-fingerprinting techniques start with the extraction and PCR amplification of small subunit ribosomal DNA, and the different separation methods are used as an alternative to cloning for the resolution of complex PCR amplicons. Since all of the techniques employ gel electrophoresis to affect this resolution, the success of the technique and the complexity of the banding patterns can depend on the separation method used. A detailed review of the relative advantages and disadvantages of the different methods is beyond the scope of this article but have been widely reported previously (Muyzer & Smalla 1998; Tiedje *et al.* 1999).

The major advantage of fingerprinting methods such as DGGE, TGGE and T-RFLP is that they afford a means of monitoring community changes in response to external stimuli such as changes in land management or pesticide application. Furthermore, extraction of the ribosomal rRNA, followed by reverse transcription then amplification of the 16S cDNA, can be used to quantify changes in specific components of the bacterial community in response to land management (Felske *et al.* 2000). We have used this approach to study changes in the structure of the microbial community in a Thai soil contaminated with petroleum hydrocarbons following the addition of inorganic nitrogen and phosphorus and used the data to develop an ecologically relevant bioaugmentation strategy for remediation. Changes in bacterial community structure were first determined using denaturing gradient gel electrophoresis (DGGE) of both the DNA and the rRNA following reverse transcription PCR using primers specific to the V3 region of the 16S rRNA gene. The resultant 193 bp amplicons were then resolved using DGGE (Fig. 3.2) and the banding patterns analysed using stepwise discriminant function analysis (SDA). Stepwise discriminant analysis of the rRNA showed that the DGGE patterns changed systematically over the incubation period and further statistical analysis demonstrated that the number of bands needed to recover the difference between groups over time could be reduced to five. The

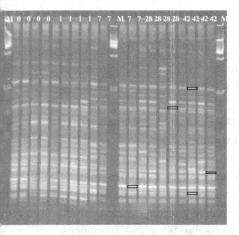

Figure 3.2. 16S rRNA (following RT-PCR) of Thai soil contaminated with petroleum hydrocarbons. Changes in bacterial communities were assessed 0, 1, 7, 28 days after addition of nitrogen and phosphorus. Lane numbers refer to days after amendment and M indicates the internal standards. Marked bands show those selected following stepwise analysis (SDA) of the banding patterns as major contributors to the discrimination between communities over time.

latter sequences were recovered in clades containing known isolates of *Bacillus*, *Microbacterium* and *Pseudomonas*. Subsequent studies have shown that all of these cloned sequences can be related to isolates from these soils with all of the organisms able to degrade petroleum hydrocarbons. Thus, as used here, the molecular analyses provided a mechanism for investigating the responses of taxonomically diverse organisms that was then used to guide the selective isolation. Doing such studies without the a-priori information from the molecular analysis would be difficult if not impossible because of a lack of information on the range of selective media needed to encompass the diversity in taxonomic response.

Microbial diversity and the impact of small subunit 18S rRNA analysis

As indicated above, the application of molecular biology to the analysis of soil bacterial community diversity is already well established. However, molecular techniques are also being used increasingly to study soil fungal communities (Vainio & Hantula 2000; Pennanen *et al.* 2001) and Kowalchuk (1999) has highlighted the improvements in soil fungal community studies following the introduction of molecular techniques.

The next few years are likely to see an extension of the amplification, cloning and sequencing approaches to include other soil taxa including protozoa (Ekelund 2002; Mo *et al.* 2002) and nematodes (Blaxter *et al.* 1998; Waite *et al.* 2003). Nematodes represent an important part of the soil microfauna that affect directly and indirectly the size, activity and diversity of the soil microflora (Ekschmitt *et al.* 1999). Morphological analyses (de Goede & Bongers 1998) have shown nematode diversity to be extensive but unlike bacteria or fungi (O'Donnell *et al.* 1994) the phylogenetic robustness of such classifications is largely untested and some studies have suggested that the circumscription of nematode taxa using morphological characters is not consistent with their genetic (McKeand 1998) and phylogenetic affiliation (Blaxter *et al.* 1998). While work has been done using large subunit (LSU) 28S (Livaitis *et al.* 2000) and internal transcribed spacer (ITS) regions of the ribosomal RNA gene (Zijlstra *et al.* 1997) to determine differences between species and differences at subspecies levels, there are few reports of studies on the use of PCR, cloning and sequencing techniques for the direct analysis of nematode diversity and community structure in soils (Foucher & Wilson 2002).

Waite *et al.* (2003) used consensus nematode 18S ribosomal DNA primers to amplify nematode 18S rDNA from whole soil community DNA extracted from a range of European grassland types. Cloning of the PCR amplicons followed by restriction digest analysis resulted in the recovery of 34 unique nematode sequences from the four grasslands studied. Comparison of these data with the comparatively small number of 18S rDNA nematode sequences currently held in on-line databases (4244 compared to 227 687 bacterial 16S rDNA sequences) revealed that all of the sequences could be assigned to known nematode taxa, albeit tentatively in some cases. To assess whether primers used to amplify 18S rDNA might be used to fingerprint genetic diversity in nematode communities in soil, the environmental sequence data were used to design a second set of primers (469 bp) for DGGE that carried a GC-clamp (a 30–40 base pair sequence of DNA rich in guanine and cytosine, Waite *et al.* 2003). To confirm that this second set of primers was amplifying nematode sequences, selected bands on the DGGE gels were extracted, PCR amplified and sequenced. The final alignment was 337 bases and the analyses revealed the presence of sequence signatures from the genera *Paratrichodorus, Plectus, Steinernema, Globodera, Cephalobus* and *Pratylenchoides*.

In addition to the successful recovery of nematode sequence signatures directly from soil, these studies suggested that unlike bacterial DGGE (Felske & Akkermans 1998) the extent of the diversity (taxon richness) recovered is closely linked to the size of the sample, making the analysis of larger samples more critical in establishing nematode taxon diversity. Thus, the effective application of molecular fingerprinting approaches to the study of nematode diversity will need to address questions of spatial heterogeneity and sample size either by processing larger soil samples (e.g. by pooling DNA from many subsamples) or possibly by indirect methods where 'soil community DNA' is recovered following extraction of the nematodes from soils.

Microbial diversity and the impact of phospholipid fatty acid profiling techniques

While much of the work on assessing microbial diversity in soils has been on the analysis of small subunit rRNA molecules, other molecules have also been used as surrogates to study microbial community structure in soils. These have generally been treated suspiciously by many microbial ecologists because of possible changes in the chemical composition of microorganisms during growth in soils and other natural environments (Tunlid & White 1992). However, despite these concerns, chemical markers have been used in many environments to study changes in the size, evenness and structure of microbial communities (Frostegård & Bååth 1996; Malosso *et al.* 2004). Phospholipid fatty acids (PLFA) are the most commonly used chemical markers and give a fingerprint of the entire microbial community structure. PLFA analysis, combined with multi-variate statistical analysis (Malosso *et al.* 2004), affords a detailed assessment of changes in the microbial biomass at the community level (Bossio & Scow 1998) in response to environmental perturbations. In soils, PLFA has been used to study shifts in microbial community structure in response to changes in soil pH (O'Donnell *et al.* 2001), the impact of residue management and ash inputs on the growth of ectomycorrhizae in forest soils (Hagerberg & Wallander 2002) and on the effects of management practices in long-term field trials (Lawlor *et al.* 2000; O'Donnell *et al.* 2001).

When used on soils, PLFA analysis provides a fingerprint of the total microbial community and includes fatty acid signature lipids from bacteria and from fungi. This has led to its widespread use in investigating the effects of management and fertility status on bacterial:fungal ratios in soils (Frostegård & Bååth 1996). These studies are based on the use of octadecadienoic acid (18:2ω6) to indicate changes in fungal biomass and the use of *iso-* and *anteiso*-branched chain acids (e.g. *i*-15:0, *ai*-15:0) to indicate bacterial biomass. Such studies have suggested that changes from high fertility, high input systems to low fertility, low input systems are characterised by an increase in soil fungi and a decrease in the bacterial:fungal ratio (Bardgett *et al.* 1999). However, recent studies (Malosso *et al.* 2004) in which the transfer of carbon from ^{13}C-labelled *Deschampsia antarctica* into 18:2ω6 and ergosterol (a fungal sterol that has also been used as a biomarker for fungal biomass) have questioned the widespread acceptability of this approach. In this work, changes in microbial community structure were assessed using PLFA and an analysis of the fatty acids associated with the neutral lipid fraction (NLFA). These studies showed that there were no significant changes in PLFA or NLFA profiles during residue decomposition suggesting no change in microbial community structure. There was a marked increase, however, in ergosterol levels in these soils indicative of growth of the fungal biomass. Analysis of this ergosterol using gas chromatography-mass spectrometry (GC-MS) confirmed the transformation of the plant residue by showing the incorporation of ^{13}C plant carbon into the ergosterol. This incorporation of ^{13}C into ergosterol increased over the incubation period and importantly was not evident in the analysis of either the PLFA or NLFA fractions nor in 18:2ω6. This suggests that the correlation between ergosterol and 18:2ω6 does not hold for all fungal communities (Zeller *et al.* 2001).

Microbial diversity in soils: what next?

A justification often presented for studying microbial diversity in soils and for opening the microbial 'black box' (Andren & Balandreau 1999; Tiedje *et al.* 1999) is that the diversity of the contents therein is vitally important to the maintenance and sustainability of the biosphere. However, despite almost 20 years of detailed sifting through the box contents, there is little evidence to support this, suggesting that many of the thousands of microorganisms in soils are functionally redundant (Yin *et al.* 2000) and that many of the major functions of the microbial biomass are unaffected by its exact species composition (Andren & Balandreau 1999). Furthermore, the advent of full genomic sequencing has seen the availability of a number of completed prokaryotic genomes. This has occurred concurrently with advances in computing science and data analysis such that comparative analyses of genomes from taxonomically diverse bacteria, although labour intensive and far from routine, is now possible. These studies have provided evidence of significant horizontal gene transfer between prokaryotes and are questioning yet again the definition and existence of bacterial species (O'Donnell *et al.* 1994; Rosselló-Mora & Amann 2001). The extent of horizontal gene transfer in both prokaryotes and eukaryotes is strongly correlated with gene function with marked differences in observed transfer frequencies evident between 'informational genes' and 'operational genes' (*sensu* Jain *et al.* 1999). 'Informational genes', which include 16S rDNA, are involved in transcription, translation and related processes (e.g. tRNA synthetases, GTPases, ATPases), and are far less likely to be horizontally transferred and maintained than 'operational genes' involved in the synthesis of amino acids, cell envelope proteins, co-factors, or fatty acids and phospholipids (Jain *et al.* 1999). This has been described as the 'complexity hypothesis' (Jain *et al.* 1999) and has obvious implications for the use of sequence data from 'informational genes' to study diversity:function relationships in soils where function (the expression of phenotype) is largely coded for by operational genes. As we learn more about bacterial genomes and their organisation, it is becoming apparent that they are often ordered into clusters of linked genes that provide 'adaptive properties' to the cell (Rodriguez-Valera 2002). With the availability of additional genomic sequences from closely related organisms, the extent and importance of these 'islands' (Ochman *et al.* 2000; Karlin 2001, Finan 2002) and their role in adaptation to growth in spatially and temporally separated environments will become increasingly apparent.

The addition of full genomic sequences to the prokaryotic database will allow us to evaluate the significance and distribution of these 'adaptive islands' and their role in bacterial speciation and phenotypic expression in natural environments. This will see significant growth in environmental genomics and the very real possibility of finally understanding the relationship between diversity and function in natural systems by enabling us to co-locate phenotype and, using mRNA analysis coupled to microarrays, an expressed phenotype in soil microorganisms. The technologies for doing this are already here and will soon become less costly and more widely applied. Microarray technologies will enable us to assess routinely community diversity and function in soils by direct hybridisation of oligonucleotides fixed on slides which, when used with functional gene arrays (Wu *et al.* 2001), should enable community structure and function to be directly related with the potential of providing custom-designed chips for different applications and different soil environments. Similarly, the development of methods for cloning and sequencing large segments of DNA from soils (e.g. the soil metagenome, Rondon *et al.* 2000) will provide vast amounts of information on cell physiology, metabolism and genotype (Rondon *et al.* 2000), and importantly how these are organised on the chromosome and how they might be controlled *in situ*. If the clustering of key functional genes in operons and 'adaptive islands' proves to be widespread, then the task of assigning function and of assessing functional redundancy will become easier.

The last 20 years have seen the rebirth of microbial ecology as a discipline that for the most part has been driven by developments in molecular biology. Many of these techniques are based on developments in understanding taxonomic complexity in microorganisms and in using these to explore microbial communities without the need for culture. Although easy to dismiss the contribution made by these community analysis techniques to understanding function, it is important to understand that the field is still very much in a discovery phase as researchers have adapted and applied molecular technologies to soils. Our understanding of diversity and our abilities to determine diversity at species, subspecies and genetic levels mean that microbial ecology is no longer limited by an inability to determine what is present and how it changes. Hopefully, this will see the development of a theoretical framework for microbial ecology that will provide a better understanding of microbial systems and an ability to model and predict the impact of environmental perturbations on community structure and function in soils.

References

Andren, O. & Balandreau, J. (1999). Biodiversity and soil functioning: from black box to can of worms? *Applied Soil Ecology*, **13**, 105–108.

Bardgett, R. D., Lovell, R. D., Hobbs, P. J. & Jarvis, S. C. (1999). Seasonal changes in microbial communities along a fertility gradient of temperate grasslands. *Soil Biology and Biochemistry*, **31**, 1021–1030.

Barns, S. M., Takala, S. L. & Kuske, C. R. (1999). Wide distribution and diversity of members of the bacterial kingdom Acidobacterium in the environment. *Applied and Environmental Microbiology*, **65**, 1731–1737.

Blaxter, M., De Ley, P., Garey, J. R., *et al.* (1998). A molecular evolutionary framework for the phylum Nematoda. *Nature*, **392**, 71–75.

Bossio, D. A. & Scow, K. M. (1998). Impacts of carbon and flooding on soil microbial communities: phospholipid fatty acid profiles and substrate utilisation patterns. *Microbial Ecology*, **35**, 265–278.

Colvan, S. R., Syers, J. K. & O'Donnell, A. G. (2001). Effect of long-term fertiliser use on acid and alkaline phosphomonoesterase and phosphodiesterase activities in managed grassland. *Biology and Fertility of Soils*, **34**, 258–263.

Connon, S. A. & Giovannoni, S. J. (2002). High-throughput methods for culturing microorganisms in very-low-nutrient media yield diverse new marine isolates. *Applied and Environmental Microbiology*, **68**, 3878–3885.

Curtis, T. P., Sloan, W. T. & Scannell, J. W. (2002). Estimating prokaryotic diversity and its limits. *Proceedings of the National Academy of Sciences, USA*, **99**, 10 494–10 499.

De Goede, R. G. M. & Bongers, T. (1998). *Nematode Communities of Northern Temperate Grassland Ecosystems*. Giessen: Focus.

Ekelund, F. (2002). Estimation of protozoan diversity in soil. *European Journal of Protistology*, **37**, 361–362.

Ekschmitt, K., Bakonyi, G., Bongers, M., *et al.* (1999). Effects of the nematofauna on microbial energy and matter transformation rates in European grassland soils. *Plant and Soil*, **212**, 45–61.

Felske, A. & Akkermans, A. D. L. (1998). Spatial homogeneity of abundant bacterial 16S rRNA molecules in grassland soils. *Microbial Ecology*, **36**, 31–36.

Felske, A., Wolterink, A., van Lis, R., de Vos, W. M. & Akkermans, A. D. L. (2000). Response of soil bacterial community to grassland succession as monitored by 16S rRNA levels of the predominant ribotypes. *Applied and Environmental Microbiology*, **66**, 3998–4003.

Finan, T. M. (2002). Evolving insights: symbiosis islands and horizontal gene transfer. *Journal of Bacteriology*, **184**, 2855–2856.

Finlay, B. J. & Clarke, K. J. (1999). Ubiquitous dispersal of microbial species. *Nature*, **400**, 828.

Foucher, A. & Wilson, A. (2002). Development of a polymerase chain reaction-based denaturing gradient gel electrophoresis technique to study nematode species biodiversity using the 18S rDNA gene. *Molecular Ecology Notes*, **2**, 45–48.

Frostegård, A. & Bååth, E. (1996). The use of phospholipid fatty acid analysis to estimate bacterial and fungal biomass. *Biology and Fertility of Soils*, **22**, 59–65.

Griffiths, R. I., Whiteley, A. S., O'Donnell, A. G. & Bailey, M. J. (2000). Rapid method for co-extraction of DNA and RNA from natural environments for analysis of ribosomal DNA- and rRNA-based microbial community composition. *Applied and Environmental Microbiology*, **66**, 5488–5491.

Hagerberg, D. & Wallander, H. (2002). The impact of forest residue removal and wood ash amendment on the growth of the ectomycorrhizal external mycelium. *FEMS Microbiology Ecology*, **39**, 139–146.

Hagström, A., Pommier, T., Rohwer, F., *et al.* (2002). Use of 16S ribosomal DNA for delineation of marine bacterioplankton species. *Applied and Environmental Microbiology*, **68**, 3628–3633.

Hopkins, D. W. & Ferguson, K. E. (1994). Substrate induced respiration in soil amended with different amino acid isomers. *Applied Soil Ecology*, **1**, 75–81.

Jain, R., Rivera, M. C. & Lake, J. A. (1999). Horizontal gene transfer among genomes: the complexity hypothesis. *Proceedings of the National Academy of Sciences, USA*, **96**, 3801–3806.

Karlin, S. (2001). Detecting anomalous gene clusters and pathogenicity islands in diverse bacterial genomes. *Trends in Microbiology*, **9**, 335–343.

Kennedy, A. C. & Gewin, V. L. (1997). Soil microbial diversity: present and future considerations. *Soil Science*, **162**, 607–617.

Kowalchuk, G. A. (1999). New perspectives towards analysing fungal communities in terrestrial environments. *Current Opinion in Biotechnology*, **10**, 247–251.

Kowalchuk, G. A., Stephen, J. R., DeBoer, W., *et al.* (1997). Analysis of ammonia-oxidizing bacteria of the beta subdivision of the class Proteobacteria in coastal sand dunes by denaturing gradient gel electrophoresis and sequencing of PCR-amplified 16S ribosomal DNA fragments. *Applied and Environmental Microbiology*, **63**, 1489–1497.

Lawlor, K., Knight, B. P., Barbosa-Jefferson, V. L., et al. (2000). Comparison of methods to investigate microbial populations in soils under different agricultural management. *FEMS Microbiology Ecology*, **33**, 129–137.

Liu, Q. & Sommer, S. S. (1995). Restriction-endonuclease fingerprinting (REF): a sensitive method for screening mutations in long, contiguous segments of DNA. *BioTechniques*, **18**, 470–477.

Livaitis, M. K., Bates, J. W., Hope, W. D. & Moens, T. (2000). Inferring a classification of the Adenophorea (Nematoda) from nucleotide sequences of the D3 expansion segment (26/28S rDNA). *Canadian Journal of Zoology*, **78**, 911–922.

Malosso, E., English, L., Hopkins, D. W. & O'Donnell, A. G. (2004). Use of ^{13}C-labelled plant materials and ergosterol and FAMEs to analyse organic matter decomposition in Antarctic soil. *Soil Biology and Biochemistry*, **36**, 165–175.

Manefield, M., Whiteley, A. S., Griffiths, R. I. & Bailey, M. J. (2002). RNA stable isotope probing, a novel means of linking microbial community function and phylogeny. *Applied and Environmental Microbiology*, **62**, 5367–5373.

Marsh, T. L. (1999). Terminal restriction fragment length polymorphism (T-RFLP): an emerging method for characterizing diversity among homologous populations of amplification products. *Current Opinion in Microbiology*, **2**, 323–327.

McKeand, J. B. (1998). Molecular diagnosis of parasitic nematodes. *Parasitology*, **117**, S87–96.

Mo, C. Q., Douek, J. & Rinkevich, B. (2002). Development of a PCR strategy for thraustochytrid identification based on 18S rDNA sequence. *Marine Biology*, **140**, 883–889.

Mummey, D. L., Stahl, P. D. & Buyer, J. S. (2002). Microbial biomarkers as an indicator of

ecosystem recovery following surface mine reclamation. *Applied Soil Ecology*, **21**, 251–259.

Muyzer, G. (1999). DGGE/TGGE a method for identifying genes from natural ecosystems. *Current Opinion in Microbiology*, **2**, 317–322.

Muyzer, G. & Smalla, K. (1998). Application of denaturing gradient gel electrophoresis (DGGE) and temperature gradient gel electrophoresis (TGGE) in microbial ecology. *Antonie van Leeuwenhoek*, **73**, 127–141.

Nannipieri, P., Johnson, R. L. & Paul, A. E. (1978). Criteria for measurement of microbial growth and activity in soil. *Soil Biology and Biochemistry*, **10**, 223–229.

Ochman, H., Lawrence, J. G. & Groisman, E. A. (2000). Lateral gene transfer and the nature of bacterial innovation. *Nature*, **405**, 299–304.

O'Donnell, A. G. & Goerres, H. E. (1999). 16S rDNA methods in microbiology. *Current Opinion in Biotechnology*, **10**, 225–229.

O'Donnell, A. G., Goodfellow, M. & Hawksworth, D. L. (1994). Theoretical and practical aspects of the quantification of biodiversity among microorganisms. *Philosophical Transactions of the Royal Society of London, Series B, Biological Sciences*, **345**, 65–73.

O'Donnell, A. G., Seasman, M., Macrae, A., Waite, I. & Davies, J. T. (2001). Plants and fertilisers as drivers of change in microbial community structure and function in soils. *Plant and Soil*, **232**, 135–145.

Pennanen, T., Paavolainen, L. & Hantula, J. (2001). Rapid PCR based method for the direct analysis of fungal communities in complex environmental samples. *Soil Biology and Biochemistry*, **33**, 697–699.

Rappe, M. S., Connon, S. A., Vergin, K. L. & Giovannoni, S. J. (2002). Cultivation of the ubiquitous SAR11 marine bacterioplankton clade. *Nature*, **418**, 630–633.

Rodriguez-Valera, F. (2002). Approaches to prokaryotic biodiversity: a population genetics perspective. *Environmental Microbiology*, **4**, 628–633.

Rondon, M. R., August, P. R., Bettermann, A. D., *et al.* (2000). Cloning the soil metagenome: a strategy for accessing the genetic and functional diversity of uncultured microorganisms. *Applied and Environmental Microbiology*, 66, 2541–2547.

Rosselló-Mora, R. & Amann, R. (2001). The species concept for prokaryotes. *FEMS Microbiological Reviews*, 25, 39–67.

Schramm, A., de Beer, D., Wagner, M. & Amann, R. (1998). Identification and activities *in situ* of *Nitrosospira* and *Nitrospira* spp. as dominant populations in a nitrifying fluidized bed reactor. *Applied and Environmental Microbiology*, 64, 3480–3485.

Steenwerth, K. L., Jackson, L. E., Calderon, F. J., Stromberg, M. R. & Scow, K. M. (2002). Soil microbial community composition and land use history in cultivated and grassland ecosystems of coastal California. *Soil Biology and Biochemistry*, 34, 1599–1611.

Tiedje, J. M., Asuming-Brempong, S., Nusslein, K., Marsh, T. L. & Flynn, S. J. (1999). Opening the black box of soil microbial diversity. *Applied Soil Ecology*, 13, 109–122.

Tunlid, A. & White, D. C. (1992). Biochemical analysis of biomass, community structure, nutritional status and metabolic activity of microbial communities in soil. *Soil Biochemistry*, Vol. 7 (Ed. by G. Stotzky & J. M. Bollag), pp. 229–262. New York: Marcel Dekker.

Vainio, E. & Hantula, J. (2000). Direct analysis of wood-inhabiting fungi using denaturing gradient gel electrophoresis of amplified ribosomal DNA. *Mycological Research*, 104, 927–936.

Waite, I. S., O'Donnell, A. G., Harrison, A., *et al.* (2003). Design and evaluation of nematode 18S rDNA primers for PCR and denaturing gradient gel electrophoresis (DGGE) of soil community DNA. *Soil Biology and Biochemistry*, 35, 1165–1173.

Wenderoth, D. F. & Reber, H. H. (1999). Correlation between structural diversity and catabolic versatility of metal-affected prototrophic bacteria in soil. *Soil Biology and Biochemistry*, 31, 345–352.

Whiteley, A. S., O'Donnell, A. G., Macnaughton, S. J. & Barer, M. R. (1996). Cytochemical co-localization and quantitation of phenotypic and genotypic characteristics in individual bacterial cells. *Applied and Environmental Microbiology*, 62, 1873–1879.

Whiteley, A. S., Wiles, S., Lilley, A. K., Philip, J. & Bailey, M. (2001). Ecological and physiological analyses of pseudomonad species within a phenol remediation system. *Journal of Microbiological Methods*, 44, 79–88.

Woese, C. R. (1987). Bacterial evolution. *Microbiological Reviews*, 51(2), 221–227.

Wu, L., Thompson, D. K., Li, G., *et al.* (2001). Development and evaluation of functional gene arrays for detection of selected genes in the environment. *Applied and Environmental Microbiology*, 67, 5780–5790.

Yin, B., Crowley, D., Sparovek, G., DeMelo W. J., & Borneman, J. (2000). Bacterial functional redundancy along a soil reclamation gradient. *Applied and Environmental Microbiology*, 66, 4361–4365.

Zeller, V., Bardgett, R. D. & Tappeiner, U. (2001). Site and management effects on soil microbial properties of subalpine meadows: a study of land abandonment along a north–south gradient in the European Alps. *Soil Biology and Biochemistry*, 33, 639–649.

Zijlstra, C., Uenk, B. J. & Van Silfhout, C. H. (1997). A reliable, precise method to differentiate species of root-knot nematodes in mixtures on the basis of ITS-RFLPs. *Fundamental and Applied Nematology*, 20, 59–63.

CHAPTER FOUR

Carbon as a substrate for soil organisms

D. W. HOPKINS
University of Stirling
E. G. GREGORICH
Agriculture and Agri-Food Canada

SUMMARY

1. In many ecological studies, soil carbon is regarded as a barely differentiated whole with little attention paid to its underlying characteristics.
2. Although it is widely appreciated that decomposer organisms are nearly infallible as degraders of organic molecules, there are marked differences in the utilisation of different components of organic matter by organisms depending on chemical and physical characteristics, location and availability in time in soil.
3. We discuss the characteristics of soil carbon as a substrate and emphasise a 'soil metabolomic' approach for characterising the range of molecules in complex, composite substrates, and the potential that stable isotope probing offers for linking organisms to their substrates via enrichment of their biomolecules as they exploit isotopically enriched substrates.
4. Using selected examples, we examine the influence of the chemical characteristics/quality, quantity, location and timing of supply of organic matter on the amount, activity and, where possible, the diversity of soil organisms.
5. We are some way from unifying relationships between the quality, quantity, location and timing of delivery or availability of soil carbon on the size, activity and diversity of soil organisms. However, we point ways forward in which the information on the physics, chemistry and management are linked to the biology of soils.

Introduction
Currency of soil carbon

Humans view soil carbon in various physical (e.g. aggregates, density fractions), chemical (e.g. carbohydrates, aromatic compounds), biological (e.g. microbial biomass) and even economic (e.g. dollars per tonne or carbon credits) ways which

Biological Diversity and Function in Soils, eds. Richard D. Bardgett, Michael B. Usher and David W. Hopkins.
Published by Cambridge University Press. © British Ecological Society 2005.

are not usually ecological. Indeed, even among environmental scientists, the application of increasingly sophisticated analytical approaches for characterising soil carbon has probably contributed to the ecological relevance of soil carbon becoming obscured. The soil organisms do not, however, have these perspectives, but regard organic compounds simply as substrates supplying resources and energy.

The vast majority of soil organisms are heterotrophs that rely almost entirely on organic carbon fixed above-ground and transported into the soil by roots or deposited at or close to the soil surface as detritus. Globally, the estimates of the total amount of organic carbon in soils is between 1500 and 2000 Pg, the vast majority of which comes from plants. Assuming steady state, the annual input of organic carbon to soils of about 60 Pg a^{-1} (IPCC 2000, 2001) gives a mean residence time in decades. This is not to diminish the important role of below-ground autotrophs, which have key roles in soil biological processes such as nitrification. Given the central role of soils in the global carbon cycle, and the importance of plants as sources of soil carbon, we wish to focus on the role of plant-derived carbon as a substrate for heterotrophic soil organisms. The specific roles of plants and their symbionts, and of other autotrophs, in the ecological functioning of soils are covered elsewhere (Leake *et al.* this volume; Schimel *et al.* this volume; Standing *et al.* this volume; Wardle this volume).

Fractions of soil carbon: babies and bathwater

Understanding the scientific literature on organic carbon in soils is handicapped by a range of overlapping and somewhat arbitrary terms. The organic carbon in soils is usually denoted by the terms *soil organic matter* and *humus*, which are often used synonymously. The other relevant, and probably most ecologically appropriate term, *detritus* is unfortunately rarely used in the soil science literature. Most of the research on the distribution and characteristics of organic carbon in soils has been done with pedology or agricultural chemistry objectives. This has often led to important components of soil organic carbon being excluded from analysis and the development of systems of classifying soil organic matter that cannot be linked to ecological processes.

Soil organic matter is the 'total' organic fraction of the soil including plant and animal residues at various stages of decomposition, cells and tissues of soil organisms, and their metabolites (Gregorich *et al.* 2001). The inclusion of living organisms with the soil organic matter is an operational necessity, as complete separation of organisms from the other organic carbon is impossible (Hopkins *et al.* 1991). This means that organic carbon analysis of soil bulks together the organisms with their source of energy and nutrients. Neither the baby nor the bathwater is thrown out. However, it is significant that the organic matter fraction typically analysed, while including carbon in root exudates and sloughed-off cells, excludes the fraction of the soil that is larger than 2 mm, the larger particulate matter (macroorganic matter), the 'light' fraction (i.e. buoyant in aqueous solutions) and surface litter (Gregorich *et al.* 2001). From the standpoint of functional ecology, this means that fresh plant residues, which are among the most energetically and nutritionally rich resources and, therefore, the site of the most intense biological activity and largest biomass, are discarded. To wit, the baby is thrown out with the bathwater.

Ecological functions of soil organic matter

The reservoir of organic carbon in soils globally (1500 to 2000 Pg) is a larger carbon reservoir than the biological compartment (500 Pg) and the atmosphere (760 Pg) combined (Schimel 1995; IPCC 2000, 2001). Most of the biological compartment is composed of the residues from plants, animals and microorganisms at different stages of decay and modification by both biological and abiotic means. In the global context, a reservoir of this magnitude is a tremendous biogeochemical buffer between the land and the atmosphere. At this scale, and from a human perspective, the storage of carbon in soil is important because of concern about the potential to reduce carbon flux to the atmosphere. At smaller scales, the interactions of soil organisms with organic matter have profound roles in the regulation of soil fertility, the maintenance of soil structure and the disposal of wastes. It is the utilisation of carbon in soils as a substrate of energy for soil organisms that underpins all these ecological functions. Assessments of total energy flow through terrestrial ecosystems show that between 60 and 90% of the net primary production is dissipated by respiration of decomposer organisms (Swift *et al.* 1979; Brady & Weil 1999).

Characteristics of organic carbon in soil

Chemical and functional characteristics: glimpses through keyholes

Chemical characterisation of soil organic matter has been an objective of soil scientists for many centuries and has led to a bewildering array of fractions (Fig. 4.1). This means, for example, that the carbon in recently arrived plant litter may simultaneously appear in the particulate matter, any of the non-humic biomolecules, the dissolved organic matter and in any of the various kinetically defined fractions. In Fig. 4.1, we summarise many of the different fractions of soil organic matter to illustrate the lack of a common framework and the difficulties of working with highly complex mixtures. Our summary is far from exhaustive and fuller definitions of the different fractions can be found in other sources (e.g. Baldock & Nelson 2000; Gregorich *et al.* 2001).

There is a massive body of literature on the chemical characteristics of the mysterious humic substances because the extraction procedure involving strong alkalis and acids provides one of the few methods of preparing a sample of soil organic matter that is amenable to chemical characterisation. The ecological

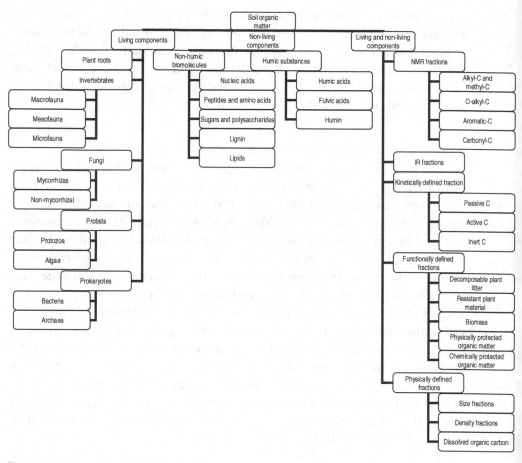

Figure 4.1. Summary of different soil organic matter fractions. The different compartments arise from a variety of different schemes for fractionation and approaches to analysis. The different compartments are not mutually exclusive and the divisions between different branches cannot therefore be regarded as rigid; we have attempted to present the most rational classification. See Baldock and Nelson (2000) for further information on many of the compartments.

relevance of these substances has been questioned by soil science 'dissidents' (Tate 2001). It is now well documented that such preparations contain artefacts produced during sample preparation. According to Baldock *et al.* (1991), the 'use of alkaline reagents to extract soil organic carbon is inefficient and may create artefacts. In addition, extraction schemes are entirely non-selective with respect to biological entities in soils'. Nevertheless, deeply cherished terms such as humic acid have, like some of the molecules it purports to represent, proved remarkably resistant to decay. Burdon (2001) proposes a more ecologically helpful approach for understanding humic substances: they are simply 'mixtures of plant and microbial constituents plus the same constituents at various stages of degradation'. This theme has developed further, and the interpretation of analyses of soil organic matter fractions are being performed with due recognition of the plant and microbial biochemicals from which they were derived (Kögel-Knabner 2002).

Understanding the chemical composition of the different fractions of soil carbon varies enormously with context. In relation to soil biodiversity, two components of the non-humic biomolecules, lipids and nucleic acids, are receiving particular attention because of the taxonomic information they carry, with the result that detailed molecular structures have been resolved. This provides little direct information about the determinants of diversity and how the organisms interact with the organic resources of the soil, but does provide signals of that diversity. Stable isotope probing, in which the biomolecules of organisms metabolise an isotopically enriched substrate, provides a powerful technique to link specific organisms to particular substrates against the background of many potential substrates and/or many capable but inactive organisms. This approach owes much to one of the most elegant pieces of experimental work in biochemistry: Meselson and Stahl's incorporation of ^{15}N into newly synthesised DNA and separation of the denatured DNA into two distinct bands by density gradient centrifugation to establish the semi-conservative replication mechanism, nearly 50 years ago (Holmes 2002). In a modern manifestation of this labelling approach, taxonomic biomarker molecules containing ^{13}C incorporated from an enriched substrate are isolated and analysed. Heavy RNA from methylotrophic bacteria and phenol-degrading bacteria has been isolated by density gradient centrifugation after supplying $^{13}CH_4$ or ^{13}C phenol, respectively, which was then used to determine which organisms were active in the metabolism of $^{13}CH_4$ or ^{13}C phenol, respectively (Bull et al. 2000; Manefield et al. 2002). The approach is not restricted to stable isotope probing of nucleic acids: 'heavy' phospholipid fatty acids were analysed by gas chromatography-mass spectrometry (GC-MS) after supplying ^{13}C ethanoate (acetate; Arao 1999), and ^{13}C from labelled grass residues was traced into the fungal biomarker, ergosterol, and fatty acid methyl esters by Malosso et al. (2004). In addition to labelling the biomolecules of the first consumers of an organic substrate, the approach offers the potential to track the route, or routes, ^{13}C travels through food webs by exploiting the sequence in which the pulse appears in different organisms (Bruneau et al. 2002; Johnson et al. 2002). Undoubtedly, stable isotope probing offers tremendous potential to track a substrate carbon into organisms for many of the more labile organic compounds in soils. Understanding the biological utilisation of the more stable carbon will require a better understanding of the chemical identity of molecules so that suitable substrates can be identified as labels. Chemical details of the stabilised organic carbon in soil remain elusive and the most likely way forward is probably the use of analogues (Haider & Martin 1975).

Decomposition and turn-over

Decomposition is the progressive breakdown of organic materials, ultimately into inorganic constituents, and is mediated mainly by soil microorganisms, which derive energy and nutrients from the diverse range of molecules in the soil organic matter. Microorganisms are the major *direct* contributors to the flux of carbon from soil to the atmosphere (Hassink *et al.* 1994). Soil animals play a major role in facilitating the decomposition process by physically fragmenting and mixing organic residues into the soil, thereby increasing the surface area of substrates and their exposure to microbial activity. They also influence decomposition by inoculating substrates with microbes, altering microbial activity (e.g. by deposition of mucus or wastes), altering the composition of the decomposer community and maintaining the soil structure. However, the *direct* faunal contribution to soil to atmosphere carbon flux is typically between 5 and 15% of the total flux (Hassink *et al.* 1994; Alphei *et al.* 1996). The net effect is the release of carbon and nutrients back into circulation at both local and global scales.

The substrates for decomposition include a wide range of materials, forming a continuum from recently added plant litter and carbon transferred into the soil as root exudates and via mycorrhizas to very stable, highly altered organic matter. Many components of fresh plant litter decompose very quickly because they are rich in energy, readily accessible to organisms and, particularly in the case of sugars and peptides, rapidly assimilated. Consequently, though it represents only a small fraction of carbon in soil, about half of the CO_2 output from soil on a global basis comes from decomposition of the annual litter inputs to the soil (Coûteaux *et al.* 1995) – a fraction of organic matter that is not typically included in soil organic matter. Stable organic matter, at the other extreme, decomposes very slowly over centuries or millennia (Campbell *et al.* 1967); the size of this pool is very large.

The closest we have come to a unified concept for summarising the interaction of carbon and organisms in soils is embodied in models of soil carbon turn-over. The conceptual model summarised in Fig. 4.2 owes much to the Rothamsted carbon model (Jenkinson & Rayner 1977). Unlike some other models, which perform equally well in their ways, this model is based on compartments with functional relevance. In later versions, an inert carbon compartment has been added to improve the performance of the model. This is not a unique feature of soil carbon models: for example the Century model has a 'passive organic matter' compartment (Parton *et al.* 1987), which also functions to slow down the model. Clearly, there cannot be a completely inert soil carbon fraction in soil (how could it have arisen?), and modelling simply points to the presence of a very stable organic matter fraction. Making the link between the presence of this 'inert' soil carbon and its biological role remains elusive, but even soils that have been deprived of fresh organic matter inputs for prolonged periods

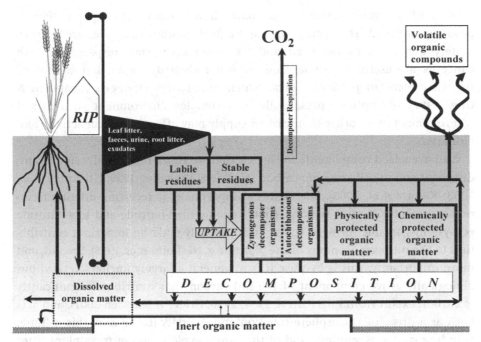

Figure 4.2. Decomposition and carbon turn-over in soil. A conceptual diagram summarising the main elements of the initial Rothamsted carbon model (Jenkinson & Rayner 1977). To this we have added other small, but potentially functionally important, compartments: the volatile organic carbon and the dissolved organic carbon derived during both decomposition of litter and exudation from plants. An inert organic matter pool is added as this appears in latter versions of the Rothamsted model. © 2003 D. W. Hopkins and E. G. Gregorich.

(decades) are still biologically active, and activity must be being sustained by more-degraded components of the soil organic matter (Lawson *et al.* 2000).

Interestingly, there is increasing evidence of a nearly inert carbon pool similar to and probably derived from charcoal, referred to as 'black carbon' in soils (Kuhlbusch 1998; Schmidt *et al.* 2001, 2002). This may be part of the 'missing link' in stable soil carbon, and given the absorptive properties of charcoal, this form of C may have wider functional roles in regulating nutrient flow (Lehmann *et al.* 2003) and absorbing toxic contaminants (Wilson *et al.* 2001). Furthermore, the Rothamsted model (Fig. 4.2) distinguishes between organisms that notionally respond to addition of fresh substrate (the zymogenous biomass; *sensu* Winogradsky (1924), and those that notionally eek out an existence on the older, more stable organic matter (the autochthonous biomass; *sensu* Winogradsky (1924)). The soil community is probably not as sharply divided as Winogradsky's definitions would imply. Similarly, the distinction between *r*-selected and *K*-selected organisms (i.e. showing rapid proliferation following a pulse of substrate availability, or maintaining near constant population) is probably not rigid. The zymogenous and autochthonous categories are approximately analogous to *r*-selected and *K*-selected organisms, respectively. Both concepts are useful for understanding soil carbon dynamics, but cannot be related directly to particular taxa, which may switch strategies (Chapman & Gray 1981), or applied unreservedly in a complex environment such as soil where many factors other than carbon supply may affect biological activity and biomass.

We have added components to Fig. 4.2 that we think functionally significant: dissolved and volatile organic carbon (Mackie & Wheatley 1999; Gregorich *et al.* 2003; Kalbitz *et al.* 2003). The former compartment is receiving attention currently, probably for two reasons. First, in some high-latitude and high-altitude ecosystems, organic nitrogen compounds probably make an important contribution to the nitrogen economy of the plants (e.g. Näsholm *et al.* 1998). Second, and more importantly, this is because it is in general a largely uncharacterised but biologically active component of the soil carbon that contributes significantly to carbon export from soils (Grieve 1984, 1990; Grieve & Marsden 2001), and acts both as a driver of rhizosphere function and diversity (Leake *et al.* this volume; Standing *et al.* this volume), and of the initial exploitation of fresh plant litter (Marstorp 1996a, b; Webster *et al.* 2000).

For modelling purposes, it has invariably been deemed adequate to treat the decomposer organisms as an undifferentiated entity in soil carbon turn-over. This is clearly an over simplification and, of course, a major theme of this whole volume.

Substrate quality

The substrate quality of soil organic matter can be regarded as the combined properties that influence the supply of carbon and energy to the heterotrophic soil organisms. Although this is a simple concept, the ability to assess substrate quality is not easy. Early studies recognised that different components of plant litter were lost at different rates, a reflection of their resource value to the decomposer organisms (Tenney & Waksman 1929; Minderman 1968), and, more recently, theoretical approaches to consider detritus as a continuum of molecules from the recalcitrant to the highly labile have been developed (Bossatta & Ågren 1985; Ågren & Bossatta 1996). However, for empirical purposes, what is required are analytical approaches free from extraction artefacts that can be applied across a wide range of soils to assess the proportions of different groups of organics in the soil organic matter (i.e. a 'soil metabolomic' approach to include the full complement of metabolites in the soil), or indicators of these compounds that reflect their relative stability. Substantial progress has been made with infrared spectroscopy, pyrolysis-GC-MS or pyrolysis-MS, nuclear magnetic resonance (NMR) and Ramon spectroscopy techniques. However, we are still a long way from being able to characterise all potential biological substrates in soil organic matter, and for key groups of organic compounds, most notably those containing nitrogen, the biochemical characteristics remain remarkably sparse even with the application of modern spectroscopic techniques (Clinton et al. 1995; Knicker & Lüdemann 1995) and the refinement of classical wet chemical techniques (Greenfield 2001; Kahn et al. 2001).

Fourier-transform infrared (FT-IR) spectroscopy of whole soils, or fractions separated by relatively gentle physical procedures such as sieving, provides information about the bonding of carbon in organic compounds (e.g. Hopkins & Shiel 1991; Niemeyer et al. 1992; Coûteaux et al. 2003) and for simple compounds identification of specific structure is possible. However, complexity and overlapping signals from the mineral component of the soil limits the application of FT-IR, but this is less of a problem if the organic-rich fractions of the soil are isolated from plant materials.

Pyrolysis to fragment complex molecules at points of molecular weakness by heating (either programmed or 'flash' combustion by Curie point combustion) yields products which can be analysed by MS or GC-MS on the basis of the mass-to-charge (m/z) ratio. These fragments can be resolved and assigned to known biochemicals in soils (Schnitzer 1990) and plant residues (Nierop et al. 2001). However, pyrolysis fragments with different structure and origin but, coincidentally, the same m/z cannot be resolved. Problems due to the loss of carbon by volatilisation during pyrolysis can be reduced by pyrolysis methylation using tetramethylammonium chloride (TMAH; Schulten & Leinweber 1996) and, for example, allows identification of specific aliphatic components of plant cuticles by GC-MS (del Rio & Hatcher 1998).

Carbon-13 NMR has been widely used to characterise soil organic matter and offers the potential advantage of being able to provide information specific to carbon without the need for any extraction (Wilson 1987; Preston 1996; Baldock & Nelson 2000). However, soils are a hostile environment for NMR analysis because of the low total carbon concentration, the low natural abundance of ^{13}C and the presence in most soils of large concentrations of paramagnetic iron which reduces the resolving power of the techniques. Some of these problems are surmountable: many researchers (e.g. Oades et al. 1987; Hopkins et al. 1993) have used relatively gentle fractionation to concentrate the carbon relative to both the bulk soil and the iron; ^{13}C enriched substrates either as pure compounds (Baldock et al. 1991; Webster et al. 1997) or enriched plant materials (Hopkins et al. 1997) can be added to the soil to increase resolution in general or provide pulses that can be followed; the paramagnetic iron may be selectively removed from the soil (Arshad et al. 1988) or the inorganic component of the soil may be cautiously dissolved with hydrogen fluoride (Skjemstad et al. 1994); and pulse sequences or other operating parameters can be manipulated to emphasise signals from ^{13}C in different chemical or physical environments

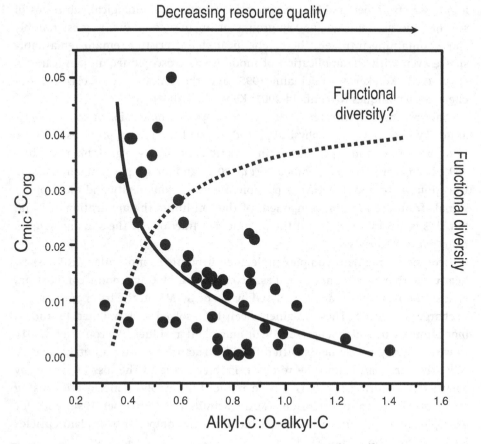

Figure 4.3. Relationship of substrate quality as assessed by solid-state ^{13}C NMR and soil microbial biomass. Adapted from Webster *et al.* (2001).

at the molecular level (Tate *et al.* 1990; Golchin *et al.* 1997; Webster *et al.* 1997). However, absolute quantification of solid-state ^{13}C NMR spectra remains illusive (Kinchesh *et al.* 1995).

Webster *et al.* (2001) proposed that solid-state ^{13}C NMR spectroscopy may provide an approach for assessing the substrate quality of soil carbon: the ratio of O-alkyl-C (–OCH$_2$) signal, indicative of polysaccharides, representing the labile fraction, to the alkyl-C (–CH$_2$), indicative of predominantly aliphatic compounds including lipids of both plant and microbial origin (Wilson 1987; Baldock *et al.* 1990; Kögel-Knabner 2002), representing the more stable components (Hopkins *et al.* 1997; Webster *et al.* 1997). The relationship in Fig. 4.3 indicates that the microbial biomass content of the soil declines with declining substrate quality, as expressed by increasing ratio of O-alkyl-C to alkyl-C. It is notable also that those soils not conforming to the relationship undershoot the biomass prediction, suggesting that a factor other than the quantity and quality of soil organic matter limits the size of the microbiological community. This analytical approach may offer an objective way to assess the functional, biological health of soils across a wide range of soil carbon contents and qualities of organic matter as a biological substrate. Furthermore, as the diversity of compounds increases with increasing decay, it may also be expected that the functional diversity of the soil community will be greater at lower substrate quality.

The dissolved organic carbon (DOC) in soils is a transient pool, considered an important substrate for soil organisms (Marschner & Bredow 2002) because microbial uptake mechanisms require an aqueous medium (Marschner & Kalbitz 2003). Intermediates in the decomposition of plant materials are likely major contributors to DOC. Thus, soil organisms contribute simultaneously both to the production and consumption of DOC. This may explain why Gregorich et al. (2003) observed that the amount of labile DOC as a proportion of total soil carbon did not differ between soils under different management having large differences in the quantities and composition of organic carbon inputs (Table 4.1). Subsequent analyses have indicated that the carbohydrate content of the DOC was a good predictor of the turn-over of DOC (Table 4.1). This observation is consistent with those of Kalbitz et al. (2003), who reported that threshold carbohydrate concentrations (as well as a range of other instrumental properties) might exist for decomposition of DOC from a range of different soils.

More subtle effects of the composition of organic residues on decomposition have been detected in plants modified genetically to alter the properties of the lignin. Plant residue decomposition is affected by relatively small changes in the lignin composition induced by down-regulation of a single plant gene (Hopkins et al. 2001; Pilate et al. 2002). Although lignin modification led to an increase in initial decomposition rate (Hopkins et al. 2001), the total mass lost during 77 days was relatively small and the main differences were due to rapid loss of water-soluble plant material (Fig. 4.4), which was dominated by carbohydrates (Webster et al. submitted). The proposed explanation is that the modified lignin affords less physical protection from microbial and enzymatic attack to the carbohydrates (Hopkins et al. 2001; Webster et al. 2004). This is supported by an investigation of the effects of earthworms on the decomposition of lignin-modified plants which showed that earthworms physically condition the plant material by communition and masceration, thereby reducing the physical protection afforded to labile components to a greater extent in the modified than the unmodified material (Hopkins et al. 2004).

Substrate quantity

The ratio of soil carbon in organisms to total soil carbon is usually fairly constant (Powlson et al. 1987; Wardle 1992). However, since the organisms are alive, they are more responsive to environmental changes than is the bulk soil carbon. Hence, subtle differences in the ratio of carbon in organisms to the total soil

Table 4.1. *Proportion of labile dissolved organic carbon (DOC), biodegradation characteristics in a bioassay, and quantities of carbohydrate carbon, protein carbon and amino acid nitrogen in cold- and hot-water extracts of soils from beneath different crop rotations and amendments. MRT, mean residence time; k, first-order decay constant.*

Crop rotation – amendment	Labile DOC[a] (% of DOC)	Labile decay constant, k[a] (days⁻¹)	Labile MRT[a] (days)	Carbohydrate C[b] (mg C kg⁻¹ soil)	Protein C[b] (mg C kg⁻¹ soil)	Amino acid N[b] (mg N kg⁻¹ soil)
Cold-water extracts						
Maize monoculture						
manured	28	2.2	0.5	120	39	2.6
unamended	32	2.0	0.5	61	8	0.9
Maize-Soy						
manured	26	1.7	0.6	84	26	2.0
fertilised	28	1.5	0.7	60	10	0.3
Hot-water extracts						
Maize monoculture						
manured	35	0.86	1.2	170	120	27.0
unamended	38	0.98	1.0	130	48	4.0
Maize-Soy						
manured	35	0.87	1.1	160	81	14.0
fertilised	38	1.3	0.8	120	50	7.0

[a] Data in columns 2, 3 and 4 from Gregorich et al. (2003).
[b] Data in columns 5, 6 and 7 from Gregorich & Hopkins (unpublished).

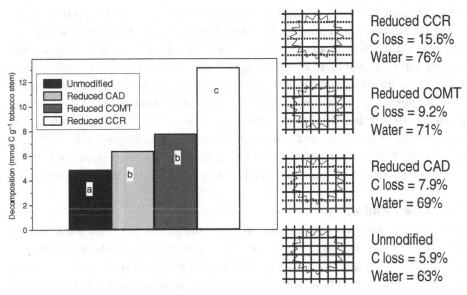

Figure 4.4. Left: Comparison of the decomposition in soil of sections of tobacco stem with different genetic modifications to lignin biosynthesis leading to reduced activity of selected enzymes in the biosynthesis pathway. Adapted from Hopkins *et al.* (2001). Right: Model diagram of the effect of the genetic modifications on the amount and conformation of lignin and associated polysaccharide (centre) and a summary of the properties of the decomposing plant residues. Columns with different letters are significantly different. CAD, cinnamyl alcohol dehydrogenase; COMT, caffeic acid O-methyl transferase; CCR, reduced cinnamoyl CoA reductase. In the model diagrams, the vertical lines represent the amount of lignin, the horizontal lines represent the cross-linking in the lignin and the star shape represents the labile polysaccharide components of ligno-cellulose.

carbon may provide an early signal of important changes in carbon cycling processes (Powlson *et al.* 1987) or a response to stress from, for example, long-term exposure to toxic metals (Chander & Brookes 1991). Accompanying changes in diversity are less easily interpreted, but, for example, Bentham *et al.* (1992) produced a scheme to interpret changes in diversity, represented predominantly by an increase in fungal biomass relative to bacteria, alongside increased biomass during secondary succession. Similarly, Frostegård *et al.* (1996) report shifts in community composition as revealed by lipid analysis under metal-stressed conditions. Such observations should be linked to the relationship between size of microbial community and the substrate quality of soil carbon (Fig. 4.3).

Where available substrate concentrations are very small, subtle interactions between biological activity and substrate addition have been reported. De Nobili *et al.* (2001) reported disproportionately large respiratory responses to the addition of small, non-saturating amounts of organic substrates, which led to larger losses of carbon as CO_2 carbon than had been added. This 'priming effect' represents non-sustainable resource utilisation, which may be explained as a response by some of the microorganisms to a small amount of substrate heralding a larger supply later. De Nobili *et al.* (2001) hypothesised that this is an evolutionary response to life in a resource-limited environment, where r-selected decomposer microorganisms are able to respond to the low concentration of soluble substrate at the advancing edge of a diffusion front arising from the presence of, for example, a newly arrived morsel of substrate or an approaching root. It is notable that the response from a chemical cocktail collected from roots led to a larger 'trigger effect' than that produced by pure glucose or amino acids.

Substrate location

Soils are not homogeneous in space. As other authors in this volume discuss, the spatial variability of soil likely has profound effects on diversity (Bardgett *et al.* this volume; Young *et al.* this volume). The spatial variability of substrate exists as nested units (i.e. colloid, aggregate, horizon, profile, landscape scales) separated by transition zones rather than sharply demarcated zones. At the microsite level, variation leads to non-uniform distributions of microorganisms (Jones & Griffiths 1964) with heterotrophic microorganisms being clustered around pieces of organic matter (Gray 1990). 'A single cellulose fiber provides a specialised environment with its own characteristic microflora, yet may occupy a volume not more than a cubic millimetre' (Stanier 1953). This has important functional significance as demonstrated by Parkin (1987) who related the main share of denitrifying activity to organic-rich 'hot spots'. The location of organic carbon at different positions in the soil on these size scales also influences the distribution and activity of macroorganisms (Guild 1952; Briones *et al.* 1992; Mueller *et al.* 1999).

It is worth noting that most of our understanding of carbon as a substrate for soil organisms relates to studies at or near the soil surface and it is often assumed that subsoil biology and carbon cycling are fundamentally similar to that in surface soils. But surface and subsurface soils have distinct biological responses to nutrient additions (Fierer *et al.* 2003a, b). Subsurface soils have low concentrations of organic carbon and oxygen, and have relatively few microorganisms (10^4 to 10^8 g^{-1}; Ghiorse & Wilson 1988; Harvey & Widdowson 1992), with the number limited by the supply of carbon from the surface (Smith & Duff 1988), and obligately oligotrophic bacteria being well represented (Stetzenbach *et al.* 1986). The numbers of protozoa and fungi are similarly small at depth (Sinclair & Ghiorse 1989; Kinner *et al.* 1990).

At the profile scale, the functioning of no-till and conventional-till agricultural ecosystems have been the subject of recent detailed investigation because of the potential for the former to increase terrestrial carbon storage and thereby offset increases in atmospheric CO_2. No-till agriculture leads to increased vertical stratification of organic matter. Hendrix *et al.* (1986) argued that this increased stratification increased the role of fungi in the food web at or near the soil surface of no-till systems, with more bacteria deeper in the profile. This is the consequence of the fungi being better able to exploit the patchy distribution of organic resources (Beare *et al.* 1992) and serving as 'bridges' for the translocation of nutrients between soil and surface residues (Hendrix *et al.* 1986). The contrast in carbon storage under no-till agriculture across Canada provides an intriguing example of·how differences in soil biodiversity link to global-scale biogeochemistry. VandenBygaart *et al.* (2003) report that in western Canada the rate of soil carbon storage under no-till was 32 ± 15 g C m^{-2} ya^{-1}, whereas in eastern Canada, it was -7 ± 27 g C m^{-2} ya^{-1}. They suggested that the population sizes and diversity of the earthworm communities were one of the factors responsible for the lower storage of carbon in eastern Canada. The biomass and diversity of indigenous earthworms in northern North America are limited, but non-native species of earthworms have arrived from Europe in mud-packed rootballs of transported plants and ships' ballast. *Lumbricus* spp. have flourished in eastern Canada (Tomlin & Fox 2003), but are rare in western Canada (Clapperton *et al.* 1997; Tomlin & Fox 2003). In the eastern Canadian soils, the decay of crop residues left on the surface is facilitated particularly by *Lumbricus terrestris* L., whereas in the west where there are relatively few surface-feeding earthworms, more residue remains on the soil surface and decomposition is limited by dryness in the summer and coldness in the winter (Clapperton *et al.* 1997; VandenBygaart *et al.* 2003).

At the aggregate scale, not all the organic carbon in soil is equally accessible as a substrate, and microorganisms are not uniformly distributed throughout aggregates, leading to interactions between the placement of carbon and the properties of the microorganisms at the particular site. Killham *et al.* (1993) showed that glucose carbon placed in small (<6 µm diameter) pores was incorporated into microbial biomass and turned over more slowly than carbon placed in pores with a larger (<30 µm) average diameter. Limited accessibility of organic carbon to soil organisms or their enzymes leads to physical protection by partial or complete encapsulation of the organic resource by mineral soil components. According to Skene *et al.* (1997), this type of physical protection is the predominant mechanism limiting microbial untilisation of readily decomposable (i.e. energy-rich, high resource quality) organic carbon which, if exposed, would be rapidly mineralised. Carbon in more degraded forms is, by contrast, relatively stable because of poor resource quality and is thus regarded as chemically protected (Skene *et al.* 1997). Whether a soil aggregate contains predominantly fresh or decomposed organic matter is a major determinant of aggregate stability (Golchin *et al.* 1994). Fresh organic matter in aggregates fuels more biological activity, which contributes to greater aggregate stability than occurs in aggregates that contain more decomposed organic matter (Golchin *et al.* 1994).

At the colloid scale, the availability and subsequent utilisation of soluble organic carbon are regulated by adsorption/desorption mechanisms. Stabilisation of dissolved organic C by sorption is not unlimited and depends on the type and quantity of the minerals (Baldock & Skjemstad 2000), as well as the chemical composition of the dissolved organic carbon (Kalbitz *et al.* 2003). While sorption to colloid surfaces may protect organic molecules from microbial utilisation to some extent, thus regulating substrate supply, the same surfaces may also sorb microorganisms and act as refugia from desiccation and predation. These sorbed microorganisms prime a location to intercept mobile substrate molecules moving through the soil pore network (Bundt *et al.* 2001). Reactions at colloid surfaces may potentially influence carbon transformations, i.e. function in contrasting ways, but how these feed into, or are affected by, the biological diversity of soil is largely unknown.

Substrate timing

The time at which a substrate arrives in soil, relative to the time of growth or life stage of the soil organisms, has important implications for carbon cycling. Soil organisms may, however, subsist in the absence of conspicuous primary producers by running down ancient deposits, as has been proposed for the specialised soil communities of the Antarctic dry valleys (Burkins *et al.* 2000), with supplements of autochthonous detritus (Moorhead *et al.* 2003). Indeed, Hodkinson *et al.* (2002) proposed that organic inputs including detritus represent an unrecognised heterotrophic phase in early succession that is sufficient to sustain initial functioning communities comprised of scavenging detritivores and predators, which are instrumental in facilitating the establishment of plants; in continental Antarctica, climatic conditions, of course, prevent this development.

Leaving aside the initial triggers, over long time periods, carbon accumulates in soils because the rate of productivity exceeds that of decomposition. Chronosequences have been investigated to estimate rates of carbon accumulation in soils (Haslam *et al.* 1998; Wardle *et al.* 2003). The accumulation of soil carbon through succession is associated with an increase in soil microbial biomass and increase in the respiratory quotient (Anderson 1994), but a decline of the biomass carbon to total carbon ratio, which has been interpreted in terms of Odum's (1969) theory of ecosystem development (Anderson 1994). Whether these differences always signal ecosystem development is doubtful because similar effects may also be related to stress (Chander & Brookes 1991; Wardle & Ghani 1995).

Soil contains a large number of inactive organisms, particularly among the microorganisms (Jenkinson & Ladd 1981). This arises, in part, because of separation in time or space from utilisable, energetic substrate. However, larger members of the soil community modify the environmental conditions and redistribute resources by, for example, burrowing activity or exudation of organic carbon. This can promote activity of *r*-selected microorganisms in what Lavelle *et al.* (1994) refer to as the macroorganisms playing the role of Prince Charming to the 'sleeping (microbial) beauties'. Bringing the energetic substrate into proximity to dormant microorganisms may be an important element in the macroorganisms' 'kiss'.

Conclusions

Is there a cause-and-effect relationship between carbon and biodiversity? Does carbon have direct or indirect effects on soil biodiversity? Does soil biodiversity significantly affect carbon cycling? Are carbon and biodiversity independent? Few clear-cut effects are shown in the scientific literature that can answer these questions. There are many examples of changes in gross processes, and in some cases differences in diversity are inferred, but usually it is implied that carbon supply and environmental conditions are the predominant influence on soil organisms. But there are relatively few clear-cut cases where cause and effect can be unequivocally established when the interactions of soil organisms and carbon as a substrate are considered. By combining the techniques of the physical and organic chemists, that have been most employed in search of chemical structures of soil organic matter, with those of ecologists, microbial biochemists and physiologists, the potential exists to understand better how the organic substrate properties and the diversity of the soil community interact. It is, however, essential that this is done with a clear view of how soils exist in the field.

Acknowledgements

We are grateful to the Royal Society of Edinburgh and the UK Biotechnology and Biological Sciences Research Council (Underwood Fund for International Collaboration) for support to facilitate the preparation of this chapter.

References

Ågren, G. I. & Bossatta, E. (1996). Quality: a bridge between theory and experiment in soil organic matter experiments. *Oikos*, **76**, 522–528.

Alphei, J., Bonkowski, M. & Scheu, S. (1996). Protozoa, Nematoda and Lumbricidae in the rhizosphere of *Hordelymus europeaus* (Poaceae): faunal interactions, response of microorganisms and effects on plant growth. *Oecologia*, **106**, 111–126.

Anderson, T. H. (1994). Physiological analysis of microbial communities in soil: applications and limitations. *Beyond the Biomass: Compositional and Functional Analysis of Soil Microbial Communities* (Ed. by K. Ritz, J.

Dighton & K. E. Giller), pp. 67–76. Chichester: Wiley.

Arao, T. (1999). *In situ* detection of changes in soil bacterial and fungal activities by measuring ^{13}C incorporation into soil phospholipid fatty acids from ^{13}C acetate. *Soil Biology and Biochemistry*, **31**, 1015–1020.

Arshad, M. A., Ripmeester, J. A. & Schnitzer, M. (1988). Attempts to improve solid-state ^{13}C NMR spectra of whole mineral soils. *Canadian Journal of Soil Science*, **68**, 593–602.

Baldock, J. A., Currie, G. J. & Oades, J. M. (1991). Organic matter as seen by solid state ^{13}C NMR and pyrolysis tandem mass spectrometry. *Advances in Soil Organic Matter Research: The*

Impact on Agriculture and the Environment (Ed. by W. S. Wilson), pp. 45–60. Cambridge: Royal Society of Chemistry.

Baldock, J. A. & Nelson, P. N. (2000). Soil organic matter. *Handbook of Soil Science* (Ed. by M. E. Sumner), pp. B25–B84. Boca Raton, FL: CRC Press.

Baldock, J. A., Oades, J. M., Vassallo, A. M. & Wilson, M. A. (1990). Solid-state CP/MAS ^{13}C NMR analysis of bacterial and fungal cultures isolated from a soil incubated with glucose. *Australian Journal of Soil Research*, **28**, 213–225.

Baldock, J. A. & Skjemstad, J. O. (2000). Role of the soil matrix and minerals in protecting natural organic materials against biological attack. *Organic Geochemistry*, **31**, 697–710.

Bentham, H., Harris, J. A., Birch, P. & Short, K. C. (1992). Habitat classification and soil restoration assessment using analysis of soils microbiological and physico-chemical characteristics. *Journal of Applied Ecology*, **29**, 711–718.

Beare, M. H., Parmlee, R. W., Hendrix, P. F., et al. (1992). Microbial and faunal interactions and effects on litter nitrogen and decomposition in agroecosystems. *Ecological Monographs*, **62**, 569–591.

Bossatta, E. & Ågren, G. I. (1985). Theoretical analysis of decomposition of heterogeneous substrates. *Soil Biology and Biochemistry*, **17**, 601–610.

Brady, N. C. & Weil, R. R. (1999). *Nature and Properties of Soils*. London: Prentice-Hall.

Briones, M. J. I., Mascato, R. & Mato, S. (1992). Relationships of earthworms with environmental factors studied by detrended canonical correspondence analysis. *Acta Oecologia*, **13**, 617–626.

Bruneau, P. M. C., Ostle, N., Davidson, D. A., Grieve, I. C. & Fallick, T. (2002). Determination of rhizosphere ^{13}C pulse signals in soil thin sections by laser ablation isotope ratio mass spectrometry. *Rapid Communications in Mass Spectrometry*, **16**, 2190–2194.

Bull, I. D., Parekh, N. R., Hall, G. H., Ineson, P. & Evershed, R. P. (2000). Detection and classification of atmospheric methane oxidizing bacteria in soil. *Nature*, **405**, 175–178.

Bundt, M., Widmer, F., Pesaro, M., Zeyer, J. & Blaser, P. (2001). Preferential flow paths: biological 'hot spots' in soils. *Soil Biology and Biochemistry*, **33**, 729–738.

Burdon, J. (2001). Are the traditional concepts of the structures of humic substances realistic? *Soil Science*, **166**, 752–769.

Burkins, M. B., Virginia, R. A., Chamberlain, C. P. & Wall, D. H. (2000). Origin and distribution of soil organic matter in Taylor Valley, Antarctica. *Ecology*, **81**, 2377–2391.

Campbell, C. A., Paul, E. A., Rennie, D. A. & McCallum, K. J. (1967). Applicability of the carbon-dating method of analysis to soil humus studies. *Soil Science*, **104**, 217–224.

Chander, K. & Brookes, P. C. (1991). The effects of heavy metals from past applications of sewage-sludge on microbial biomass and organic matter accumulation in a sandy loam and a silty loam UK soil. *Soil Biology and Biochemistry*, **23**, 927–932.

Chapman, S. J. & Gray, T. R. G. (1981). Endogenous metabolism and macromolecular composition of *Arthrobacter globisformis*. *Soil Biology and Biochemistry*, **13**, 11–18.

Clapperton, M. J., Miller, J. J., Larney, F. J. & Lindwall, C. W. (1997). Earthworm populations as affected by long-term tillage practices in southern Alberta, Canada. *Soil Biology and Biochemistry*, **29**, 631–633.

Clinton, P. W., Newman, R. H. & Allen, R. B. (1995). Immobilization of ^{15}N in forest litter studied by ^{15}N CP MAS NMR spectroscopy. *European Journal of Soil Science*, **46**, 551–556.

Coûteaux, M.-M., Berg, B. & Rovira, P. (2005). Near infrared reflectance spectroscopy for determination of organic matter fractions including biomass in coniferous forest soils. *Soil Biology and Biochemistry*, **35**, 1587–1600.

Coûteaux, M.-M., Bottner, P. & B. Berg. (1995). Litter decomposition, climate and litter quality. *Trends in Ecology and Evolution*, **10**, 63–66.

De Nobili, M., Contin, M., Mondini, C. & Brookes, P. C. (2001). Soil microbial biomass is triggered into activity by trace amounts of substrate. *Soil Biology and Biochemistry*, **33**, 1163–1170.

Del Rio, J. C. & Hatcher, P. G. (1998). Analysis of aliphatic biopolymers using thermo-chemolysis with tetramethylammonium hydroxide (TMAH) and gas chromatography mass spectrometry. *Organic Geochemistry*, **29**, 1441–1451.

Fierer, N., Allen, A. S., Schimel, J. P. & Hoden, P. A. (2003a). Controls on microbial CO_2 production: a comparison of surface and subsurface soil horizons. *Global Change Biology*, **9**, 1–11.

Fierer, N., Schimel, J. P. & Hoden, P. A. (2003b). Variations in microbial community composition through two soil depth profiles. *Soil Biology and Biochemistry*, **35**, 167–176.

Frostegård, A., Tunlid, A. & Bååth, E. (1996). Changes in microbial community structure during long-term incubation in two soils experimentally contaminated with metals. *Soil Biology and Biochemistry*, **28**, 55–63.

Ghiorse, W. C. & Wilson, J. T. (1988). Microbial ecology of the terrestrial subsurface. *Applied and Environmental Microbiology*, **33**, 107–173.

Golchin, A., Baldock, M. A., Clarke, P., Higashi, T. & Oades, J. M. (1997). The effects of vegetation and burning on the chemical composition of soil organic matter in a volcanic ash soil. II. Density fractions. *Geoderma*, **76**, 175–192.

Golchin, A., Oades, J. M., Skjemstad, J. O. & Clarke, P. J. (1994). Study of free and occluded particulate organic matter in soils by solid state ^{13}C CP/MAS NMR spectroscopy and scanning electron microscopy. *Australian Journal of Soil Research*, **32**, 285–309.

Gray, T. R. G. (1990). Soil bacteria. *Soil Biology Guide* (Ed. by D. L. Dindal), pp. 15–31. New York: Wiley.

Greenfield, L. G. (2001). The origin and nature of organic nitrogen in soil as assessed by acidic and alkaline hydrolysis. *European Journal of Soil Science*, **52**, 575–584.

Gregorich, E. G., Beare, M. H., Stoklas, U. & St-Georges, P. (2003). Biodegradability of soluble organic matter in maize-cropped soils. *Geoderma*, **113**, 237–252.

Gregorich, E. G., Turchenek, L. W., Carter, M. R. & Angers, D. A. (eds) (2001). *Soil and Environmental Science Dictionary*. Boca Raton, FL: CRC Press.

Grieve, I. C. (1984). Concentrations and annual loadings of dissolved organic matter in a small moorland stream. *Freshwater Biology*, **14**, 533–537.

Grieve, I. C. (1990). Seasonal, hydrological and land management factors controlling dissolved organic carbon concentrations in the Loch Fleet catchment, SW Scotland. *Hydrological Processes*, **4**, 231–239.

Grieve, I. C. & Marsden, R. L. (2001). Effects of forest cover and topographic factors on TOC and associated metals at various scales in western Scotland. *Science of the Total Environment*, **265**, 143–151.

Guild, W. J. McL. (1952). Variation in earthworm numbers within field populations. *Journal of Animal Ecology*, **21**, 169–183.

Haider, K. & Martin, J. P. (1975). Decomposition of specifically ^{14}C labelled benzoic and cinnamic acid derivatives in soil. *Soil Science Society of America Proceedings*, **39**, 657–662.

Harvey, R. W. & Widdowson, M. A. (1992). Microbial distributions, activities and movement in the terrestrial subsurface: experimental and theoretical studies. *Interacting Processes in Soil Science* (Ed. by P. Baveye & B. A. Stewart), pp. 185–225. Boca Raton, FL: Lewis.

Haslam, S. F. I., Chudek, J. A., Goldspink, C. R. & Hopkins, D. W. (1998). Organic carbon accumulation in a moorland soil chronosequence. *Global Change Biology*, **4**, 305–313.

Hassink, J., Neutel, A. M. & de Ruiter, P. (1994). C and N mineralization in sandy and loamy grassland soils: the role of microbes and microfauna. *Soil Biology and Biochemistry*, **26**, 1565–1571.

Hendrix, P. F., Parmlee, R. W., Crossley Jr., D. A., *et al.* (1986). Detritus food webs in conventional and no-tillage agroecosystems. *BioScience*, **36**, 374–380.

Hodkinson, I. D., Webb, N. R. & Coulson, S. J. (2002). Primary community assembly on land: missing stages: Why are the heterotrophic organisms always there first? *Journal of Ecology*, **90**, 569–577.

Holmes, F. L. (2002). *Meselson, Stahl, and the Replication of DNA: A History of "The Most Beautiful Experiment in Biology"*. New Haven, CT: Yale University Press.

Hopkins, D. W., Chudek, J. A. & Shiel, R. S. (1993). Chemical characterization and decomposition of organic matter from two contrasting grassland soil profiles. *Journal of Soil Science*, **44**, 147–157.

Hopkins, D. W., Chudek, J. A., Webster, E. A. & Barraclough, D. (1997). Following the decomposition of ryegrass labelled with ^{13}C and ^{15}N in soil by solid state nuclear magnetic spectroscopy. *European Journal of Soil Science*, **48**, 623–631.

Hopkins, D. W., Macnaughton, S. J. & O'Donnell, A. G. (1991). A dispersion and differential centrifugation technique for representatively sampling microorganisms from soil. *Soil Biology and Biochemistry*, **23**, 217–225.

Hopkins, D. W. & Shiel, R. S. (1991). Spectroscopic characterization of organic matter from soil with mull and mor humus forms. *Advances in Soil Organic Matter Research: The Impact on Agriculture and the Environment* (Ed. by W. S. Wilson), pp. 71–90. Cambridge: Royal Society of Chemistry.

Hopkins, D. W., Tilston, E. L., Webster, E. A., *et al.* (2004). Decay in soil of residues from plants with genetic modifications to lignin biosynthesis. *Genetically Modified Crops: Ecological Dimensions* (Ed. by H. F. van Emben & A. Gray). Cambridge: Cambridge University Press.

Hopkins, D. W., Webster, E. A., Chudek, J. A. & Halpin, C. (2001). Decomposition in soil of tobacco plants with genetic modifications to lignin biosynthesis. *Soil Biology and Biochemistry*, **33**, 1455–1462.

Intergovernmental Panel on Climate Change (IPCC) (2000). *Land Use, Land Use Change, and Forestry*. Cambridge: Cambridge University Press.

Intergovernmental Panel on Climate Change (IPCC) (2001). *Climate Change 2001: The Scientific Basis*. Cambridge: Cambridge University Press.

Jenkinson, D. S. & Ladd, J. N. (1981). Microbial biomass in soil: measurement and turnover. *Soil Biochemistry, Vol. 5* (Ed. by E. A. Paul & J. N. Ladd), pp. 415–471. New York: Marcel Dekker.

Jenkinson, D. S. & Rayner, J. H. (1977). The turnover of soil organic matter in some of the Rothamsted classical experiments. *Soil Science*, **123**, 298–305.

Johnson, D., Leake, J. R., Ostle, N., Ineson, P. & Read, D. J. (2002). *In situ* ^{13}CO$_2$ pulse-labelling of upland grassland demonstrates a rapid transfer pathway of carbon flux from arbuscular mycorrhizal mycelia to the soil. *New Phytologist*, **153**, 327–334.

Jones, D. & Griffiths, E. (1964). The use of soil thin sections for the study of soil microorganisms. *Plant and Soil*, **20**, 232–240.

Kalbitz, K., Schmetwitz, J., Schwesig, D. & Matzer, E. (2003). Biodegradation of soil-derived dissolved organic matter as related to its properties. *Geoderma*, **113**, 273–291.

Killham, K., Amato, M. & Ladd, J. N. (1993). Effect of substrate location in soil and soil pore-water regime on carbon turnover. *Soil Biology and Biochemistry*, **25**, 125–138.

Kinchesh, P., Powlson, D. S. & Randall, E. W. (1995). ^{13}C NMR studies of soil organic matter in whole soils: I. Quantitation possibilities. *European Journal of Soil Science*, **46**, 123–138.

Kinner, N. E., Bunn, A. L., Meeher, L. D. & Harvey, R. W. (1990). Enumeration and variability in the distribution of protozoa in an organically contaminated subsurface environment. *Transactions of the American Geophysical Union*, **71**, 1319–1320.

Knicker, H. & Lüdemann, H.-D. (1995). N-15 and C-13 CPMAS and solution NMR studies of N-15 enriched plant material during 600 days of microbial degradation. *Organic Geochemistry*, **23**, 119–126.

Kögel-Knabner, I. (2002). The macromolecular organic composition of plant and microbial residues as inputs to soil organic matter. *Soil Biology and Biochemistry*, **34**, 139–162.

Kuhlbusch, T. A. J. (1998). Black carbon and the carbon cycle. *Science*, **280**, 1903–1904.

Lavelle, P., Lattaud, C. & Trigo, D. (1994). Mutualism and biodiversity in soils. *Plant and Soil*, **170**, 23–33.

Lawson, T., Hopkins, D. W., Chudek, J. A., Janaway, R. C. & Bell, M. G. (2000). Interactions of the soil organisms with materials buried for up to 33 years in the Wareham archaeological experimental earthwork. *Journal of Archaeological Science*, **27**, 273–285.

Lehmann, J., Pereira da Silva, J., Steiner, C., et al. (2003). Nutrient availability and leaching in an archaeological Anthrosol and a Ferralsol of the Central Amazon basin: fertilizer, manure and charcoal amendments. *Plant and Soil*, **249**, 343–357.

Mackie, A. E. & Wheatley, R. E. (1999). Effects and incidence of volatile organic compound interactions between fungal and bacterial isolates. *Soil Biology and Biochemistry*, **31**, 375–385.

Malosso, E., English, L. C., Hopkins, D. W. & O'Donnell, A. G. (2004). Use of ^{13}C-labelled plant materials and ergosterol and FAMEs to analyse organic matter decomposition in Antarctic soil. *Soil Biology and Biochemistry*, **36**, 165–175.

Manefield, M., Whiteley, A. S., Griffiths, R. I. & Bailey, M. R. (2002). RNA stable isotope probing, a novel means of linking microbial community function to phylogeny. *Applied and Environmental Microbiology*, **68**, 5367–5373.

Marschner, B. & Bredow, A. (2002). Temperature effects on release and ecologically relevant properties of dissolved organic carbon in sterilized and biologically active soil samples. *Soil Biology and Biochemistry*, **34**, 459–466.

Marschner, B. & Kalbitz, K. (2003). Controls of bioavailability and biodegradability of dissolved organic matter in soils. *Geoderma*, **113**, 211–235.

Marstorp, H. (1996a). Influence of soluble carbohydrates, free amino acids and protein content on the decomposition of *Lolium multiflorum* shoots. *Biology and Fertility of Soils*, **21**, 257–263.

Marstorp, H. (1996b). Interactions in microbial uses of soluble plant components. *Biology and Fertility of Soils*, **22**, 45–52.

Minderman, G. (1968). Addition, decomposition, and accumulation of organic matter in forests. *Journal of Ecology*, **56**, 355–362.

Moorhead, D. L., Barrett, J. A., Virginia, R. A., Wall, D. H. & Porazinska, D. (2003). Organic matter and soil biota of upland wetlands in Taylor Valley, Antarctica. *Polar Biology*, **26**, 567–576.

Mueller, G., Broll, G. & Tarnocai, C. (1999). Biological activity as influenced by microtopography in a cryosolic soil, Baffin Island, Canada. *Permafrost and Periglacial Processes*, **10**, 279–288.

Näsholm, T., Ekblad, A., Nordin, A., et al. (1998). Boreal forest plants take up organic nitrogen. *Nature*, **392**, 914–916.

Niemeyer, J., Chen, Y. & Bollag, J.-M. (1992). Characterization of humic acids, composts, and peat by diffuse reflectance Fourier transform infrared spectroscopy. *Soil Science Society of America Journal*, **56**, 135–140.

Nierop, K. G. P., van Lagen, B. & Buurman, P. (2001). Composition of plant tissue and soil organic matter in the first stages of a vegetation succession. *Geoderma*, **100**, 1–24.

Oades, J. M., Vassallo, A. M., Waters, A. G. & Wilson, M. A. (1987). Characterisation of organic matter in particle size and density fractions from a red-brown earth by solid-state ^{13}C NMR. *Australian Journal of Soil Research*, **26**, 287–299.

Odum, E. P. (1969). The strategy of ecosystem development. *Science*, **164**, 262–270.

Parkin, T. B. (1987). Soil microsites as a source of denitrification variability. *Soil Science of America Journal*, **51**, 1194–1199.

Parton, W. J., Schimel, D. S., Cole, C. V. & Ojima, D. S. (1987). Analysis of factors controlling soil organic matter in Great Plains grasslands. *Soil Science Society of America Proceedings*, **53**, 1173–1179.

Pilate, G., Guiney, E., Holt, K., et al. (2002). Field and pulping performances of transgenic trees with altered lignification. *Nature Biotechnology*, **20**, 607–612.

Powlson, D. S., Brookes, P. C. & Christensen, B. T. (1987). Measurement of soil microbial biomass provides an early indication in changes in total organic matter due to straw incorporation. *Soil Biology and Biochemistry*, **19**, 159–164.

Preston, C. M. (1996). Applications of NMR to soil organic matter analysis: history and prospects. *Soil Science*, **161**, 144–166.

Schimel, D. S. (1995). Terrestrial ecosystems and the carbon cycle. *Global Change Biology*, **1**, 77–91.

Schmidt, M. W. I., Skjemstad, J. O., Czimczik, C., et al. (2001). Comparative analysis of black carbon in soils. *Global Biogeochemical Cycles*, **15**, 163–167.

Schmidt, M. W. I., Skjemstad, J. O. & Jäger, C. (2002). Carbon isotope geochemistry and nanomorphology of soil black carbon: black chernozemic soils in central Europe originate from ancient biomass. *Global Biochemical Cycles*, **16**, 1123–1130.

Schnitzer, M. (1990). Selected methods for the characterization of soil humic substances. *Humic Substances in Crop and Soil Sciences: Selected Readings* (Ed. by P. MacCarthy, R. L. Malcolm, C. E. Clapp & P. R. Bloom), pp. 65–89. Madison, WI: Soil Science Society of America.

Schulten, H.-R. & Leinweber, P. (1996). Characterization of humic and soil particles by analytical pyrolysis and computer modelling. *Journal of Analytical and Applied Pyrolysis*, **38**, 1–53.

Sinclair, J. T. & Ghiorse, W. C. (1989). Distribution of aerobic bacteria, protozoa, algae, and fungi in deep subsurface environments. *Geomicrobiology Journal*, **7**, 15–31.

Skene, T. M., Skjemstad, J. O., Oades, J. M. & Clarke, P. J. (1997). The influence of inorganic matrices on the decomposition of Eucalyptus litter. *Australian Journal of Soil Research*, **35**, 73–87.

Skjemstad, J. O., Clarke, P., Taylor, J. M., Oades, J. M. & Newman, R. H. (1994). The removal of magnetic materials from surface soils: a solid-state 13C CP/MAS study. *Australian Journal of Soil Research*, **32**, 1215–1229.

Smith, R. L. & Duff, J. H. (1988). Denitrification in a sand and gravel aquifer. *Applied and Environmental Microbiology*, **54**, 1071–1078.

Stanier, R. Y. (1953). Adaptation, evolutionary and physiological: or Darwinism among the microorganisms. *Adaptation in Microorganisms* (Ed. by R. Davies & E. F. Gale), pp. 1–14. Cambridge: Society for General Microbiology 3rd Symposium/Cambridge University Press.

Stetzenbach, L. D., Kelley, L. M. & Sinclair, N. A. (1986). Isolation, identification and growth of well-water bacteria. *Ground Water*, **24**, 6–10.

Swift, M. J., Heal, O. W. & Anderson, J. M. (1979). *Decomposition in Terrestrial Ecosystems*. Oxford: Blackwell Scientific.

Tate, K. R., Yamanoto, K., Churchman, G. J., Meinhold, R. & Newman, R. H. (1990). Relationships between the type and carbon chemistry of humic acids from New Zealand and Japanese soils. *Soil Science and Plant Nutrition*, **36**, 611–621.

Tate III, R. L. (2001). Soil organic matter: evolving concepts. *Soil Science*, **166**, 721–722.

Tenney, F. & Waksman, S. A. (1929). Composition of natural organic compounds and their decomposition in soil: IV. The nature and rapidity of decomposition of various organic complexes in different plants under aerobic conditions. *Soil Science*, **28**, 55–84.

Tomlin, A. D. & Fox, C. A. (2003). Earthworms and agricultural systems: status of knowledge and research in Canada. *Canadian Journal of Soil Science*, **83**, 265–278.

VandenBygaart, A. J., Gregorich, E. G. & Angers, D. A. (2003). Influence of agricultural management on soil organic carbon: a compendium and assessment of Canadian studies. *Canadian Journal of Soil Science*, **83**, 363–380.

Wardle, D. A. (1992). A comparative assessment of the factors which influence microbial biomass carbon and nitrogen in soil. *Biological Reviews*, **67**, 321–358.

Wardle, D. A. & Ghani, A. (1995). A critique of the microbial metabolic quotient (qCO2) as a bioindicator of disturbance and ecosystem development. *Soil Biology and Biochemistry*, **27**, 1601–1610.

Wardle, D. A., Hörnberg, G., Zackrisson, O., Kalela-Brundin, M. & Coomes, D. A. (2003). Long-term effects of wildfire on ecosystem properties across an island area gradient. *Science*, **300**, 972–975.

Webster, E. A., Chudek, J. A. & Hopkins, D. W. (1997). Fates of ^{13}C from enriched glucose and glycine in an organic soil determined by NMR. *Biology and Fertility of Soils*, **25**, 389–395.

Webster, E. A., Chudek, J. A. & Hopkins, D. W. (2000). Carbon transformations during decomposition of different components of plant leaves in soil. *Soil Biology and Biochemistry*, **32**, 301–314.

Webster, E. A., Halpin, C., Chudek, J. A. & Hopkins, D. W. (2004). Decomposition in soil of soluble, insoluble and lignin-rich fractions of plant material from tobacco with genetic modifications to lignin biosynthesis. *Soil Biology and Biochemistry*, **37**, 751–760.

Webster, E. A., Hopkins, D. W., Chudek, J. A., et al. (2001). The relationship between the size of the soil microbial community and the resource quality of soil organic matter. *Journal of Environmental Quality*, **30**, 147–150.

Wilson, J. A., Demis, J., Pulford, I. D. & Thomas, S. (2001). Sorption of Cr(III) and Cr(VI) by natural (bone) charcoal. *Environmental Geochemistry and Health*, **23**, 291–295.

Wilson, M. A. (1987). *NMR Techniques and Applications in Geochemistry and Soil Chemistry*. Oxford: Pergamon Press.

Winogradsky, S. (1924). Sur la microflora autochthone de la terre arable. *Compte Rendu Academie Science, Paris*, **178**, 1236–1239.

PART III

Patterns and drivers of soil biodiversity

CHAPTER FIVE

The use of model *Pseudomonas fluorescens* populations to study the causes and consequences of microbial diversity

PAUL B. RAINEY
University of Auckland and University of Oxford
MICHAEL BROCKHURST
University of Oxford
ANGUS BUCKLING
University of Bath
DAVID J. HODGSON
University of Exeter
REES KASSEN
University of Oxford

SUMMARY

1. The microbial world is tremendously diverse. This fact was established in the early days of microbiology and is supported by ever increasing lists of 16S rDNA sequences and more recently by whole genome comparisons.
2. It is now time to divert attention from lists of organisms – even though these lists are undoubtedly incomplete – to questions such as the evolutionary and ecological causes of diversity; the ecological factors maintaining diversity and the significance of diversity in terms of ecosystem function.
3. Recognising the inherent difficulties of addressing these questions within the soil environment we have chosen to use experimental populations of bacteria maintained in simple laboratory environments. These populations have allowed us to reduce complexity to the point where insights into mechanistic processes become possible and have permitted rigorous empirical tests of fundamental ecological and evolutionary concepts.
4. Particularly significant has been clear demonstrations of the importance of ecological opportunity and competition in driving diversification of microbial populations. In addition, it has been possible to show how productivity, disturbance and predation can shape patterns of diversity by affecting the outcome of competition and how the observed patterns of diversity depend upon environmental complexity.

Biological Diversity and Function in Soils, eds. Richard D. Bardgett, Michael B. Usher and David W. Hopkins.
Published by Cambridge University Press. © British Ecological Society 2005.

5. Most recently we have begun to explore the consequences of microbial diversity in terms of ecosystem properties and have been able to show, at a mechanistic level, how diversity, productivity and invasibility are connected.

Introduction

Recent technological advances have confirmed a long-held suspicion that soils are biologically diverse. On the basis of both empirical (Torsvik *et al.* 1998) and theoretical studies (Curtis *et al.* 2002), soils are estimated to contain in the order of 7000 different taxa at an abundance of approximately 10^9 cells per cubic centimetre. While further work is likely to see a refinement of these estimates (Torsvik *et al.* 2002), it is timely to focus attention on the significance of this diversity: to understand its evolutionary origins, its ecological maintenance and its significance in terms of soil and ecosystem function.

Undoubtedly these are challenging questions, and especially so given the complexity and technical difficulties associated with working on soils and associated biota (both above and below-ground). Never-the-less, progress on various fronts can be and has been made. Our own approach involves the use of simple experimental bacterial populations propagated in highly controlled environments. The use of such populations allows complexity to be reduced to the point where mechanistic insights into fundamental processes governing the origin, maintenance and functional significance of diversity become possible. Here we present an overview of these studies, which while unashamedly devoid of a soil-specific context, we hope will stimulate the design and execution of experiments to test, in more complex systems, operation of the ecological and evolutionary processes outlined here. For the most part, our emphasis is on ecologically significant variation, i.e. variation shaped by history and natural selection.

Experimental bacterial populations

The use of experimental bacterial populations in ecology and evolution has been much reviewed (Dykhuizen 1990; Lenski & Travisano 1994; Lenski *et al.* 1998; Rainey *et al.* 2000; Jessup *et al.* 2004). Desirable features of these populations include rapid generation time, large population size, clonal reproduction, ability to store ancestors in a state of suspended animation, and physiological and genetic tractability. These features combine in such a way as to render bacterial populations ideally suited to the design of experiments that test mechanistic hypotheses – many of them of central importance in ecology and evolution.

Evolutionary emergence of diversity

The majority of phenotypic and ecological diversity on the planet has arisen during successive adaptive radiations, i.e. periods in which a single lineage rapidly diverges to generate multiple niche-specialist types. Microbiologists tend not to think of bacteria as undergoing adaptive radiation, but there is no reason to exclude them from this general statement – in fact rapid generation times and large population sizes suggest that bacteria, especially soil-inhabiting types, may be particularly prone to bouts of rapid ecological diversification. This being so, then insight into the evolutionary emergence of diversity requires an understanding of the causes of adaptive radiation.

The causes of adaptive radiation are many and complex, but at a fundamental level there are just two: one genetic and the other ecological. Put simply, heritable variation arises primarily by mutation, while selection working via various ecological processes shapes this variation into the patterns of diversity evident in the world around us – including the much-neglected forms in soil. Discussion of the genetic causes of diversity is beyond the scope of this article, but interested readers are referred to Maynard Smith (1989) for introduction to the ecological and evolutionary implications of mutation, recombination and migration, and Raff (1996) for elaboration of ideas arising from the interplay between developmental mechanisms and evolutionary patterns. Detail on genetic aspects of the model *Pseudomonas fluorescens* radiation can be found in Spiers *et al.* (2002).

The ecological causes of adaptive radiation are embodied in theory that stems largely from Darwin's insights into the workings of evolutionary change (Darwin 1859), but owes much to developments in the 1940s and 1950s attributable to Lack (1947), Dobzhansky (1951) and Simpson (1953). Recent work has seen a reformulation of the primary concepts (Schluter 2000a, b).

At its most basic, the ecological theory of adaptive radiation postulates that ecological opportunity (vacant niche space) and competition are necessary conditions for the emergence and maintenance of diversity. Imagine a single lineage, of any given species, in a pristine environment replete with ecological opportunity such that the environment provides alternate 'fitness peaks'. The population grows geometrically until the primary resource becomes limiting at which point competition – the so-called 'engine' of adaptive radiation – becomes a significant factor. Variant types (arising by mutation) are driven by competition to exploit new resource types where they are subject to different selective conditions. Under this scenario, the genotypes most favoured by selection are those that occupy niches different to those inhabited by the dominant type (species or genotype). Continual exposure to divergent natural selection promotes further divergence in phenotype leading, if unrestrained, to ecological speciation.

Ecological opportunity

Support for the ecological theory comes from studies on a wide range of organisms (see for example, Schluter 1994, 2000a, b; Wilson *et al.* 2000; Grant & Grant 2002), but arguably the most direct evidence comes from experimental bacterial populations (Rainey *et al.* 2000; Travisano & Rainey 2000). While the earliest studies were not direct tests of the theory, they none-the-less provide insights into the importance of ecological opportunity for lineage divergence and, more profoundly, into how the activities of one life form can create ecological opportunity for the evolutionary emergence of new types.

One of the most elegant examples of this kind stems from the work of Julian Adams and colleagues who performed long-term selection experiments with chemostat-propagated populations of *Escherichia coli* grown on a single limiting carbon source (Helling *et al.* 1987). Regular sampling from the chemostat populations revealed phenotypically distinct colony types on agar plates, indicating that the once genetically uniform populations had become polymorphic. Analysis of competitive interactions among genotypes showed that the polymorphism was stable and maintained by density dependent processes. This surprising finding appeared at first to contradict the niche exclusion principle (Hardin 1960), which states that the number of species cannot exceed the number of distinct resources. Further physiological and genetic studies were therefore undertaken which found evidence of resource partitioning of the limiting glucose resource by the numerically dominant genotype (Rosenzweig *et al.* 1994). Partitioning of glucose into acetate and glycerol, metabolic by-products arising from metabolism of glucose, provided ecological opportunity for the evolutionary emergence of mutants with enhanced capacities to metabolise these by-products. Indeed, the variant types that arose (and arose repeatably – see Treves *et al.* 1998) had, in one instance, evolved an enhanced capacity to metabolise acetate and, in the other, an improved capacity to recover glycerol. Rather than contradict the niche exclusion principle, these results demonstrate the fine scale at which the 'one organism, one niche' principle operates and shows how niche complementarity can promote the stable co-existence of competing types.

Our own work has tested more directly the roles of ecological opportunity and competition in evolutionary diversification by examining the patterns of diversity emerging after genetically identical founding-populations of *P. fluorescens* were propagated in two environments that differed solely in the amount of ecological opportunity afforded to the evolving populations (Rainey & Travisano 1998). Heterogeneous (spatially structured) environments were achieved by incubating broth-containing microcosms without shaking, whereas homogeneous (spatially unstructured) environments, devoid of ecological opportunity, were generated by incubating identical microcosms under a continuously shaken regime. Diversity, in the form of niche-specialist genotypes, emerged rapidly in spatially heterogeneous microcosms, but not in spatially homogeneous microcosms (Figs. 5.1 and 5.2).

It is reasonably easy to envisage ecological opportunity in the form of different chemical resources, spatially distinct niches, or arising from temporal variation in the availability of resources (see Levin 1972; Stewart & Levin 1973), but certain kinds of interactions can promote the emergence of diversity even in the absence of physically or chemically distinct niches. Interactions most likely to do this are those that occur over small spatial scales and where strict competitive

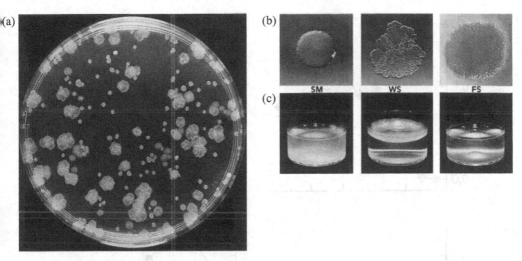

Figure 5.1. Phenotypic diversity and niche specificity among *P. fluorescens* SBW25 colonies evolved in a spatially heterogeneous environment. Microcosms were incubated without shaking to produce a spatially heterogeneous environment. (a) After seven days, populations show substantial phenotypic diversity which is seen after plating. (b) Most phenotypic variants can be assigned to one of three principal morph classes: smooth morphs (SM), wrinkly spreader (WS) and fuzzy spreader (FS). (c) Evolved morphs showed marked niche preferences. Reproduced from Rainey and Travisano (1998).

hierarchies are absent, such as in non-transitive communities (Czaran *et al.* 2002). One example with real relevance to soil microbial communities comes from experimental studies of bacteriocin (colicin)-producing (colicinogenic), colicin-resistant and colicin-sensitive derivatives of *E. coli* (Kerr *et al.* 2002). In an environment that allows for localised interactions, such as the surface of an agar plate, where the relationship between neighbouring cells is determined by spatial positioning, all three genotypes are maintained. In an environment where opportunity for localised interactions is negligible, such as in a shaken (spatially homogeneous) broth culture, diversity is lost.

Competition

The operation of competition, as described above, is readily perceived, but evidence of its moment-by-moment effects is difficult to obtain (Connell 1961; Hopf *et al.* 1993; Schluter 1994). Studies seeking to understand the role of competition in driving divergence in wild populations have first to provide convincing evidence that the species under consideration are indeed sympatric (Schluter 2000a, b). In the *P. fluorescens*-based microcosm experiments, there can be no doubt that competitive interactions between the different niche-specialist genotypes occurred during the course of their evolution; after all, these adaptive genotypes evolved in a common environment from a common ancestral genotype. Never-the-less, evidence that competition plays a role in divergence was

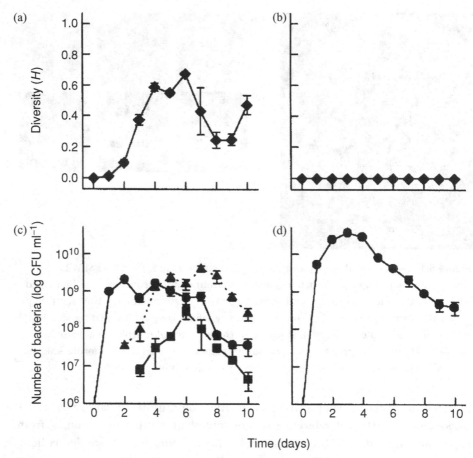

Figure 5.2. Effect of ecological opportunity on the evolution of genetic diversity. Two sets of identical populations were founded from the ancestral smooth morph and incubated either with or without shaking. Every 24 h, three replicate microcosms were destructively collected and diversity was determined by scoring the frequencies of morphs. (a) and (b) Diversity increased rapidly in the heterogeneous environment (a), whereas no diversity was detected in environments lacking spatial structure (b). (c) and (d) Evolutionary dynamics of the principal morph classes in the heterogeneous (c) and homogeneous (d) environments. Values are the means ± s.e.m. ($n = 3$). Circles represent smooth morphs (SM), triangles represent wrinkly spreader (WS) morphs and squares represent fuzzy spreader (FS) morphs. Reproduced from Rainey and Travisano (1998).

obtained by testing for the existence of tradeoffs and was achieved by experiments in which the ability of each niche-specialist genotype to invade from rare, a population dominated by another type, was determined. Evidence that different niche-specialists could invade from rare reveals competitive tradeoffs that are fully consistent with predictions from theory concerning the role of competition as the engine of divergence (Levene 1953; Abrams 1987).

Ecological maintenance of diversity

There is no guarantee that diversity, once present, will be maintained; how-ever, in a heterogeneous environment replete with ecological opportunity, stable maintenance of diversity is possible (Levene 1953; Tilman 2000). Theory pre-dicts that in a heterogeneous environment selection will favour the evolution of ecological specialists, with different types being favoured in different patches (niches). Ecological specialists by definition have a narrow niche breadth (com-pared to generalists) and trade enhanced competitive advantage in one niche against reduced competitive ability in others. The existence of tradeoffs pre-vents competitive exclusion by one type and renders co-existence of organisms possible.

An extensive discussion of mechanisms maintaining diversity is not feas-ible here, suffice to say that tradeoffs alone are a necessary, but not sufficient, condition for the stable maintenance of diversity in a heterogeneous environ-ment (for more detailed discussion see Chesson 2000; Rainey et al. 2000; Kassen 2002). Never-the-less, given the existence of tradeoffs, then diversity will be sta-bly maintained by negative frequency-dependent selection, provided dispersal among niches is limited (relative to the strength of selection), population size is regulated in a density-dependent manner at the level of individual niches and the total number of individuals contributed by each niche to the population is not too unbalanced (Levene 1953).

Evidence of tradeoffs and the concomitant operation of negative frequency-dependent selection can be difficult to determine in the wild, but has been detected and measured in laboratory populations (Lenski 1988; Rainey et al. 2000; Bohannan et al. 2002; Kassen 2002). In our own studies, niche specialisation is readily observable by eye and visual inspection of static microcosms also reveals clear evidence of fitness tradeoffs (see Fig. 5.1c): genotypes that colonise the air–broth interface clearly trade off their ability to inhabit the nether regions of the microcosm and vice versa (Rainey & Travisano 1998). The existence of such tradeoffs (arising from the effects of competition – see above) led to experi-ments which confirmed that the diversity is stable and maintained by negative frequency-dependent selection; the fitness of each genotype being inversely pro-portional to its frequency, with genotypes being favoured when rare (because resources are most abundant), but not when common (resources are rare and competition intense). The diversity is therefore protected and co-existence of genotypes assured.

These findings suggested further experiments to determine whether envir-onmental heterogeneity itself is an explanation for the diversity that appears during the course of selection in spatially heterogeneous microcosms. The fact that diversity failed to emerge in spatially unstructured microcosms strongly suggested that this was the case, but just how important this heterogeneity is to the continued persistence of diversity was unclear. We therefore took P. fluorescens populations that had already evolved substantial ecological diversity and split these in two: one set was propagated under the spatially heterogeneous regime and the other under the spatially homogeneous regime. Diversity was rapidly lost once heterogeneity was reduced; supporting the idea that heterogeneity itself is a primary cause of diversity and necessary for the continued maintenance of diversity (Rainey & Travisano 1998).

Processes shaping patterns of diversity

Various ecological processes affect patterns of diversity by changing relative demographic properties of populations. Three processes of particular importance are (1) the rate of production of organic matter (productivity), (2) the frequency or intensity at which organic matter is removed (disturbance) and (3) the intensity and magnitude of exploiters (parasites and predators).

In the case of productivity and disturbance, survey data from a range of taxa suggest that at the scale of a collection of communities across a landscape, diversity peaks at intermediate rates of both factors (Connell 1978; Abramsky & Rosenzweig 1984). Support for the generality of these patterns comes from both field and laboratory experiments, including studies with *P. fluorescens* populations (Figs. 5.3 and 5.4). Significantly though, in studies with *P. fluorescens*, unimodal (hump-shaped) relationships with diversity are detected solely in spatially heterogeneous microcosms and not in homogeneous microcosms (Buckling *et al.* 2000; Kassen *et al.* 2000). This suggests that environmental heterogeneity, while being crucial for the emergence and maintenance of diversity within populations, can also explain patterns of diversity at the scale of collections of populations or communities across a landscape (Kassen 2002).

Of further significance is the fact that the results observed in our experimental studies (Figs. 5.3 and 5.4) are explained by an extension to a simple model (Levene 1953) that outlines the conditions for the maintenance of diversity in a spatially heterogeneous environment (Kassen *et al.* 2000). Essential features of the model are antagonistic selection in a spatially heterogeneous environment composed of many niches and the proportion of individuals contributed by each niche to the total population. So long as there is genotype-by-environment interaction for fitness, and the number of individuals contributed by each niche to the total population is not too dissimilar, diversity will be stably maintained by negative frequency-dependent selection. The primary effect of both productivity and disturbance is to modulate the fraction of individuals contributed by each niche to the total population, such that at low and high levels of productivity, or disturbance, the number of individuals supported by each niche becomes unequal. This being so, then diversity falls because one niche contributes many more individuals to the community than the other, and the type that is better adapted to that more productive niche dominates the community.

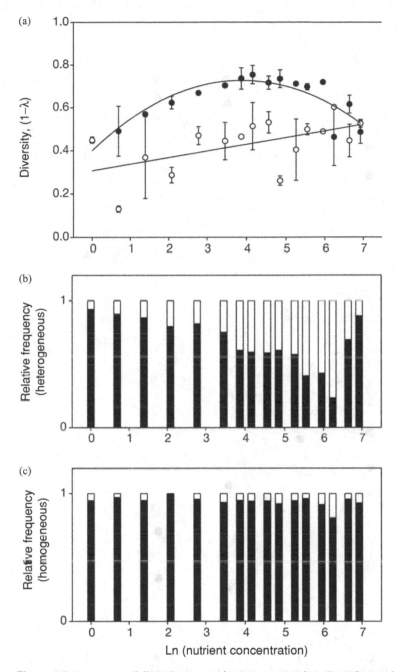

Figure 5.3. Response of diversity to nutrient concentration. Experimental culture media were mixtures of KB nutrients (glycerol and proteose peptone) and M9 mineral salts, made up by serial dilution in M9 salts to concentrations of 8x, . . . , 1/128x standard King's Medium B nutrient concentration. (a) Diversity expressed as $1 - \lambda$. Open circles are populations propagated in homogeneous, and solid circles populations propagated in heterogeneous, microcosms, with bars marking ± 1 s.e. of two replicates. (b) Relative frequency of different colony morphotypes in the heterogeneous environment; smooths (SM) (solid bars) versus wrinkly spreaders (WS) and fuzzy spreaders (FS) (open bars). (c) As in (b), for the homogeneous environment. Reproduced from Kassen *et al.* (2000).

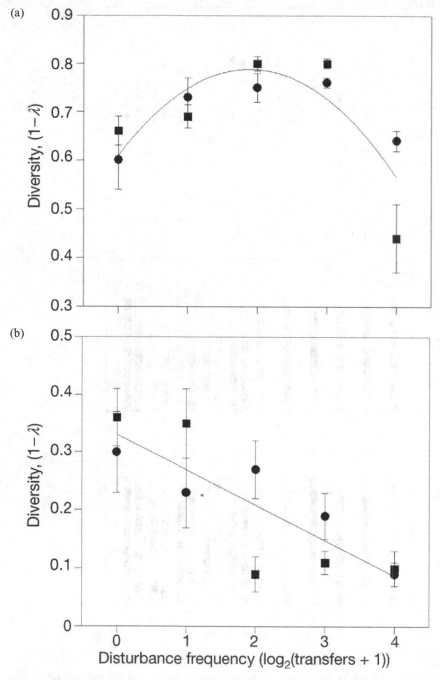

Figure 5.4. Mean diversity $(1 - \lambda)$ at different frequencies of disturbance in heterogeneous and homogeneous environments, after 16 days. Disturbances were wrought by vortex mixing immediately prior to transfer. Values of the mean are given as ± 1 s.e.m., $n = 8$. (a) Heterogeneous environment. (b) Homogeneous environment. Cultures were initiated with either isogenic (circles) or diverse (squares) base populations. Reproduced from Buckling *et al.* (2000).

Recently, we have examined the combined effects of productivity and disturbance on patterns of diversity and found that the two processes combine in a simple manner to generate a peak of diversity at intermediate levels of both factors. As is the case when each factor is treated separately, their combined effects on patterns of diversity seem to be understandable in terms of selection among niche specialists in a heterogeneous environment (Kassen et al. 2004).

Exploiters can also shape patterns of diversity, playing significant roles in diversification and ultimately speciation of their hosts and prey. Explicit tests of the effects of bacteriophage on diversity in heterogeneous environments show that when co-cultivated, strong selection for host resistance leads to genetic bottlenecks that greatly decrease sympatric diversity, but increase diversity among spatially distinct (allopatric) populations (Buckling & Rainey 2002a; Fig. 5.5). Diversification further escalates when interactions between host and parasite lead to an arms race of infectivity and resistance, sending populations along divergent evolutionary trajectories (Buckling & Rainey 2002a, b). Once again, the effects of bacteriophage on diversity are dependent on the heterogeneity of the environment. Whereas in a spatially heterogeneous environment bacteriophage decrease sympatric diversity while increasing allopatric diversity, in a spatially homogeneous environment bacteriophage increase both sympatric and allopatric diversity (Brockhurst et al. 2004).

Functional consequences of diversity

The functional significance of diversity has long been debated (Darwin 1859, Elton 1958): the central issue being whether diversity and ecosystem function are causally related (Huston 1997; Tilman et al. 1997; Loreau et al. 2001). Support for a causal connection stems from studies on a wide range of species that show that diverse communities are usually more productive and more resistant to invasion by foreign species. The causes of these effects, and whether they are linked by a common mechanism, are not fully understood, but three explanatory mechanisms are usually invoked. First, more diverse communities are likely to contain species that contribute disproportionately to community invasibility or productivity (dominance or selection effect). Second, more diverse communities will occupy more niches and monopolise more resources. Such niche complementarity will exclude most invading species and optimise the use of a varied resource base. Third, particular species may facilitate each other's growth or invasion resistance (positive interactions).

The model P. fluorescens populations have once again proved useful for obtaining insights into the consequences of diversity in terms of relating diversity to ecosystem function (Hodgson et al. 2002). Microcosms with varying genotypic and functional group diversities (based on colony morphology and resource use) were established by reconstructing communities to generate a range of combinations, from those with limited genotypic and functional diversity (two

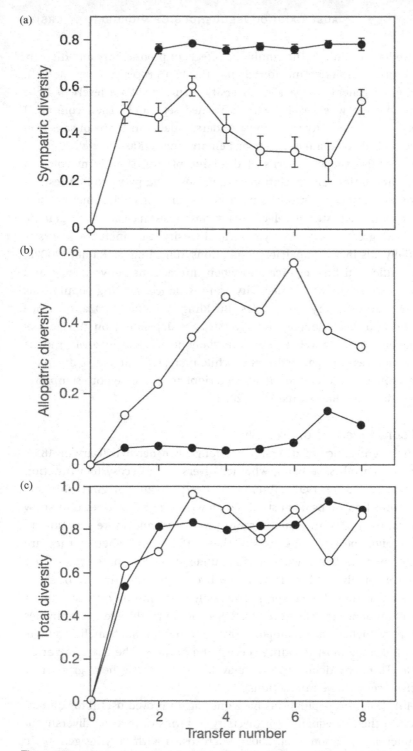

Figure 5.5. Effect of parasites (bacteriophages) on the partitioning of diversity. Mean (±1 s.e.m., $n = 12$) sympatric (within population) bacterial diversity (a), allopatric (between population) diversity (b), and total diversity (c) through time for populations evolving in isolation (filled symbols) and evolving with phages (open symbols). Reproduced from Buckling and Rainey (2002a).

independent genotypes occupying the same functional niche), through to those with high genotypic and functional diversity (six independent genotypes occupying three distinct functional niches). The productivity of each community was then examined in conjunction with its ability to withstand invasion by a foreign type. Consistent with expectations, more diverse communities were more productive and less susceptible to invasion. However, once the dominance effect was taken into consideration (by removing the effect due to the most numerically dominant type from each community) the effect of diversity on production and invasibility was no longer significant. Further investigations showed that the interactions within these supposedly simple communities are highly complex with examples of niche complementarity and positive and negative interactions among genotypes. The existence of such a range of complex interactions within the supposedly simple *Pseudomonas* populations is surprising and while there is scope for the furtherance of these studies they already point to the difficulties in trying to predict the response of an ecosystem to changes in diversity given the unpredictable nature of interactions among genotypes.

Conclusions

It is reasonable to ask what these studies contribute to our understanding of the causes and consequences of microbial diversity in soil. The answer is little in terms of the specific details, but in terms of general ecological and evolutionary processes that determine patterns of diversity and the relationship between this diversity and the functioning of ecosystems, the insights are likely to have some relevance. However, as with insights from any laboratory system, caution is necessary when extrapolating to higher orders of scale and complexity.

Our ability to manipulate the spatial structure of the environment has demonstrated the central role of environmental heterogeneity for the emergence and maintenance of diversity at both local and regional levels of scale. Soils are among the most ecologically complex environments on the planet: the variety of resources available for exploitation is so vast as to be almost unknowable and the uneven distribution of these resources across space and time is likely to generate substantial ecological opportunity. The bacteria inhabiting such an environment are likely to be subject to intense and continuous bouts of diversifying selection – made all the more intense through the action of parasites and predators. A likely outcome is the evolutionary emergence of ecologically and genetically diverse populations of precisely the kinds that are found in soils.

While such a scenario for the emergence of diversity is readily imaginable, there is no doubt that this is an over simplification. For a start, interactions between genotypes, which are key aspects of adaptive radiation, do not have systematic effects and while competition is often the engine of divergence, it can, when competing types are generalists (competitive over a range of niches), inhibit adaptive radiations (Travisano & Rainey 2000). Similarly, mutualistic and predatory/parasitic interactions can have contrasting effects on diversification.

To a large extent, the effect of species interactions on divergence is dependent upon the demographic properties of populations and these are influenced by many ecological factors (both biotic and abiotic). While we have considered a number of these, their combined (interaction) effects on both the emergence and maintenance of diversity remain largely unexplored (although see Buckling *et al.* 2000; Kassen *et al.* 2000; Travisano & Rainey 2000). For example, the effects of bacteriophage on sympatric and allopatric diversity have been examined under a single productivity and disturbance regime and it is likely that the outcomes will be altered at higher and lower levels of both factors. This expectation is supported by one study in which the effect of nutrient specificity on adaptive divergence of *P. fluorescens* populations was examined – subtle alterations had pronounced effects on patterns of diversity and on ecological interactions among types (Kassen *et al.* 2000; Travisano & Rainey 2000). In a recent study, Treves *et al.* (2003) showed that the opportunity for interactions between two competitors in a simple sand-based model system was dependent upon the matric potential of the soil. It is clear that the range of factors likely to impact on patterns of soil diversity is considerable.

Further reason to suspect that our simple interpretation is naïve stems from the fact that we are yet to consider the effect of dispersal and recombination (gene flow) on the emergence and maintenance of diversity. These are of particular relevance: both factors limit the ability of selection to maintain diversity in heterogeneous environments, and both dispersal and recombination are significant factors in soil bacterial communities. The trouble is we have no idea over what scales these operate or the magnitude of their effects. One distinct possibility is that their impacts are sufficiently strong to limit the ability of natural selection to maintain diversity, leading to unstable co-existence. Such a scenario would be amenable to interpretation by neutral models (Hubbell 1997, 2001; Bell 2001) that may prove highly relevant to understanding patterns of abundance and diversity in soils.

As far as studies on the functional significance of diversity are concerned, our results are consistent with a growing stream of work that finds evidence of ecological redundancy and no clear sign that a reduction of diversity adversely affects ecosystem function. However, this interpretation very much depends on the dominance effect – the significance of which in natural communities remains unclear (Huston *et al.* 2000; Loreau *et al.* 2001; Wardle 2001, this volume). But what we find most interesting is not so much the lack of clear causal connection between diversity and function – indeed it is unclear that one should exist, or that there is functional redundancy – but rather the remarkable scope for interactions to evolve quite by accident, among competing genotypes. That these interactions are as likely to be beneficial as detrimental suggests that the interactions themselves, their origin and particularly the scale at which they operate, may be the key to a greater understanding of the functional significance of soil diversity.

Acknowledgements

We thank NERC, NSERC (Canada), BBSRC, The Royal Society and The Wellcome Trust for financial support for aspects of this work.

References

Abrams, P. A. (1987). Alternative models of character displacement and niche shift: 2. Displacement when there is competition for a single resource. *American Naturalist*, **130**, 271–282.

Abramsky, Z. & Rosenzweig, M. L. (1984). Tilman's predicted productivity–diversity relationship shown by desert rodents. *Nature*, **309**, 150–151.

Bell, G. (2001). Neutral macroecology. *Science*, **293**, 2413–2418.

Bohannan, B. J. M., Kerr, B., Jessup, C. M., Hughes, J. B. & Sandvik, G. (2002). Trade-offs and coexistence in microbial microcosms. *Antonie Van Leeuwenhoek International Journal of General and Molecular Microbiology*, **81**, 107–115.

Brockhurst, M. A., Rainey, P. B. & Buckling, A. (2004). The effect of spatial heterogeneity and parasites on the evolution of host diversity. *Proceedings of the Royal Society of London, Series B, Biological Sciences*, **271**, 107–111.

Buckling, A., Kassen, R., Bell, G. & Rainey, P. B. (2000). Disturbance and diversity in experimental microcosms. *Nature*, **408**, 961–964.

Buckling A. & Rainey P. B. (2002a). The role of parasites in sympatric and allopatric host diversification. *Nature*, **420**, 496–499.

Buckling, A. & Rainey, P. B. (2002b). Antagonistic coevolution between a bacterium and a bacteriophage. *Proceedings of the Royal Society, London, Series B, Biological Sciences*, **269**, 931–936.

Chesson, P. (2000). Mechanisms of maintenance of species diversity. *Annual Review of Ecology and Systematics*, **31**, 343–366.

Connell, J. H. (1961). The influence of interspecific competition and other factors on the distribution of the barnacle *Chthamalus stellatus*. *Ecology*, **42**, 710–723.

Connell, J. H. (1978). Diversity in tropical rain forests and coral reefs. *Science*, **199**, 1302–1310.

Curtis, T. P., Sloan, W. T. & Scannell, J. W. (2002). Estimating prokaryotic diversity and its limits. *Proceedings of the National Academy of Sciences, USA*, **99**, 10494–10499.

Czaran, T. L., Hoekstra, R. F. & Pagie, L. (2002). Chemical warfare between microbes promotes biodiversity. *Proceedings of the National Academy of Sciences, USA*, **99**, 786–790.

Darwin, C. (1859). *The Origin of Species*. London: Murray.

Dobzhansky, T. (1951). *Genetics and the Origin of Species*. New York: Columbia University Press.

Dykhuizen, D. E. (1990). Experimental studies of natural selection in bacteria. *Annual Review of Ecology and Systematics*, **21**, 373–398.

Elton, C. S. (1958). *The Ecology of Invasions by Animals and Plants*. London: Methuen.

Grant, P. R. & Grant, B. R. (2002). Adaptive radiation of Darwin's finches. *American Scientist*, **90**, 130–139.

Hardin, G. (1960). The competitive exclusion principle. *Science*, **131**, 1292–1297.

Helling, R. B., Vargas, C. N. & Adams, J. (1987). Evolution of *Escherichia coli* during growth in a constant environment. *Genetics*, **116**, 349–358.

Hodgson, D. J., Rainey, P. B. & Buckling, A. (2002). Demonstration of the direct mechanisms linking community diversity to invasibility. *Proceedings of the Royal Society, Series B, Biological Sciences*, **269**, 2277–2283.

Hopf, F. A., Valone, T. J. & Brown, J. H. (1993). Competition theory and the structure of ecological communities. *Evolutionary Ecology*, **7**, 142–154.

Hubbell, S. P. (1997). A unified theory of biogeography and relative species abundance and its application to tropical rain forests and coral reefs. *Coral Reefs*, **16**, S9–21.

Hubbell, S. P. (2001). *The Unified Neutral Theory of Biodiversity and Biogeography*. Princeton, NJ: Princeton University Press.

Huston, M. A. (1997). Hidden treatments in ecological experiments: re-evaluating the ecosystem function of biodiversity. *Oecologia*, **110**, 449–460.

Huston, M. A., Aarssen, L., Austin, M. P., *et al.* (2000). No consistent effect of plant diversity on productivity. *Science*, **289**, 1255.

Jessup, C. M., Kassen, R., Forde, S. E., *et al.* (2004). Big questions, small worlds: microbial model systems in ecology. *Trends in Ecology and Evolution*, **19**, 189–197.

Kassen, R. (2002). The experimental evolution of specialists, generalists, and the maintenance of diversity. *Journal of Evolutionary Biology*, **15**, 173–190.

Kassen, R., Buckling, A., Bell, G. & Rainey, P. B. (2000). Diversity peaks at intermediate productivity in a laboratory microcosm. *Nature*, **406**, 508–512.

Kassen, R. Llewellyn, M. & Rainey, P. B. (2004). Ecological constraints in a model adaptive radiation. *Nature*, **431**, 984–988.

Kerr, B., Riley, M. A., Feldman, M. W. & Bohannan, B. J. M. (2002). Local dispersal promotes biodiversity in a real-life game of rock–paper–scissors. *Nature*, **418**, 171–174.

Lack, D. (1947). *Darwin's Finches*. Cambridge: Cambridge University Press.

Lenski, R. E. (1988). Experimental studies of pleiotropy and epistasis in *Escherichia coli*: II. Compensation for maladaptive effects associated with resistance to virus T4. *Evolution*, **42**, 433–440.

Lenski, R. E., Mongold, J. A., Sniegowski, P. D., *et al.* (1998). Evolution of competitive fitness in experimental populations of *E. coli*: what makes one genotype a better competitor than another? *Antonie Van Leeuwenhoek International Journal of General and Molecular Microbiology*, **73**, 35–47.

Lenski, R. E. & Travisano, M. (1994). Dynamics of adaptation and diversification: a 10,000-generation experiment with bacterial populations. *Proceedings of the National Academy of Sciences, USA*, **91**, 6808–6814.

Levene, H. (1953). Genetic equilibrium when more than one ecological niche is available. *American Naturalist*, **87**, 331–333.

Levin, B. R. (1972). Coexistence of two asexual strains on a single resource. *Science*, **175**, 1272–1274.

Loreau, M., Naeem, S., Inchaustic, P., *et al.* (2001). Ecology: biodiversity and ecosystem functioning. Current knowledge and future challenges. *Science*, **294**, 804–808.

Maynard Smith, J. (1989). *Evolutionary Genetics*. Oxford: Oxford University Press.

Raff, R. A. (1996). *The Shape of Life: Genes, Development, and the Evolution of Animal Form*. Chicago, IL: University of Chicago Press.

Rainey, P. B., Buckling, A., Kassen, R. & Travisano, M. (2000). The emergence and maintenance of diversity: insights from experimental bacterial populations. *Trends in Ecology and Evolution*, **15**, 243–247.

Rainey, P. B. & Travisano, M. (1998). Adaptive radiation in a heterogeneous environment. *Nature*, **394**, 69–72.

Rosenzweig, R. F., Sharp, R. R., Treves, D. S. & Adams, J. (1994). Microbial evolution in a simple unstructured environment: genetic differentiation in *Escherichia coli*. *Genetics*, **137**, 903–917.

Schluter, D. (1994). Experimental evidence that competition promotes divergence in adaptive radiation. *Science*, **266**, 798–801.

Schluter, D. (2000a). *The Ecology of Adaptive Radiations*. Oxford: Oxford University Press.

Schluter, D. (2000b). Ecological character displacement in adaptive radiation. *American Naturalist*, **157** (suppl.), S4–S16.

Simpson, G. G. (1953). *The Major Features of Evolution*. New York: Columbia University Press.

Spiers, A. J., Kahn, S. G., Bohannon, J., Travisano, M. & Rainey, P. B. (2002). Adaptive divergence in experimental populations of *Pseudomonas fluorescens*: I. Genetic and phenotypic bases of wrinkly spreader fitness. *Genetics*, **161**, 33–46.

Stewart, F. M. & Levin, B. R. (1973). Partitioning of resources and the outcome of interspecific competition: a model and some general considerations. *American Naturalist*, **107**, 171–198.

Tilman, D. (2000). Causes, consequences and ethics of biodiversity. *Nature*, **405**, 208–211.

Tilman, D., Knops, J., Wedin, D., *et al.* (1997). The influence of functional diversity and composition on ecosystem processes. *Science*, **277**, 1300–1302.

Torsvik, V., Daae, F. L., Sandaa, R. A. & Ovreas, L. (1998). Novel techniques for analysing microbial diversity in natural and perturbed environments. *Journal of Biotechnology*, **64**, 53–62.

Torsvik, V., Ovreas, L. & Thingstad, T. F. (2002). Prokaryotic diversity: magnitude, dynamics, and controlling factors. *Science*, **296**, 1064–1066.

Travisano, M. & Rainey, P. B. (2000). Studies of adaptive radiation using model microbial systems. *American Naturalist*, **156**, S35–S44.

Treves, D. S., Manning, S. & Adams, J. (1998). Repeated evolution of an acetate-crossfeeding polymorphism in long-term populations of *Escherichia coli*. *Molecular Biology and Evolution*, **15**, 789–797.

Treves, D. S., Xia, B., Zhou, J. & Tiedje, J. M. (2003). A two-species test of the hypothesis that spatial isolation influences microbial diversity in soil. *Microbial Ecology*, **45**, 20–28.

Wardle, D. (2001). Experimental demonstration that plant diversity reduces invasibility: evidence of a biological mechanism or a consequence of sampling effect? *Oikos*, **95**, 161–170.

Wilson, A. B., Noack-Kunnmann, K. & Meyer, A. (2000). Incipient speciation in sympatric Nicaraguan crater lake cichlid fishes: sexual selection versus ecological diversification. *Proceedings of the Royal Society of London, Series B, Biological Sciences*, **267**, 2133–2141.

CHAPTER SIX

Patterns and determinants of soil biological diversity

RICHARD D. BARDGETT
Lancaster University
GREGOR W. YEATES
Landcare Research
JONATHAN M. ANDERSON
University of Exeter

SUMMARY

1. This chapter examines the vast diversity of organisms that live in the soil and discusses the various factors that regulate its spatial and temporal patterning.
2. There is a dearth of information available on the diversity of soil biota, especially at the species level, but existing data provide little support for the idea that the same forces that regulate patterns of diversity above-ground (i.e. productivity and disturbance) control patterns of biodiversity below-ground, or that regional-scale patterns of soil biodiversity show similar trends to those that occur above-ground.
3. We argue that patterning of soil biodiversity is related primarily to the heterogeneous nature, or patchiness, of the soil environment at different spatial and temporal scales, and that this heterogeneity provides unrivalled potential for niche partitioning, or resource and habitat specialisation, leading to avoidance of competition and hence co-existence of species.
4. We highlight the challenge for soil ecologists to identify the hierarchy of controls on soil biological diversity that operate at different spatial and temporal scales, and to determine the role of spatio-temporal patterning of soil biodiversity as a driver of above-ground community assembly and productivity.

Introduction

The Earth hosts a bewildering diversity of organisms that are distributed in a wide variety of spatial and temporal patterns across, and within, the Earth's ecosystems. Making sense of these complex patterns of diversity, and understanding the dominant forces that control them, has been a major theme of

Biological Diversity and Function in Soils, eds. Richard D. Bardgett, Michael B. Usher and David W. Hopkins. Published by Cambridge University Press. © British Ecological Society 2005.

community ecology (Huston 1994). This is not a simple matter, however, since no single process or theory can explain a phenomenon as complex as biological diversity. Indeed, a range of factors, such as productivity, predation, competition, dispersal and evolutionary history, affect patterns of diversity. The importance of specific mechanisms in regulating the diversity of a particular group of organisms will vary under different sets of environmental conditions that operate on different temporal and spatial scales. The situation is further complicated by a general lack of information on patterns of distribution and abundance of many of the Earth's species, and the fact that many species are yet to be described (Wilson 2002).

Patterns and determinants of diversity have been studied almost exclusively in organisms that live above-ground or in marine ecosystems (Huston 1994; Gaston 2000; Lawton 2000; Mittelbach et al. 2001). In this chapter, we consider patterns and determinants of diversity in soil communities, which remain relatively unexplored. Ecologists are starting to turn their attention to soil communities since it is here where the majority of the Earth's terrestrial species actually dwell (André et al. 1994; Wardle 2002). This surge of interest in soil biological diversity has largely come from the recognition not only that soil organisms regulate major ecosystem processes, such as organic matter turn-over and nutrient mineralisation, but also that they may have key roles in feedbacks between above-ground and below-ground communities and hence influence how ecosystems function (Hooper et al. 2000; van der Putten et al. 2001; Bardgett & Wardle 2003). In this chapter, we first examine what is known about patterns of biodiversity in soil on various spatial and temporal scales. We then consider what factors are likely to act as primary determinants of this diversity in soil on various spatial and temporal scales. As already noted, no single mechanism can explain the complexity of diversity patterns in soil; rather, we seek general rules about some primary determinants of soil biological diversity that operate under certain environmental conditions.

The diversity of soil biota

Biological communities in soil are made up of both microbes and fauna and, together, their diversity is often several orders of magnitude greater than that which occurs above-ground (Heywood 1995). In terms of species number, the bulk of biological diversity in soil is made up of the hundreds or thousands of species of fungi and bacteria. However, there are also numerous animal species that live in soil, including the microfauna (body width less than 0.1 mm; e.g. protozoa and nematodes), the mesofauna (body width 0.1–2.0 mm; e.g. microarthropods and enchytraeids) and the macrofauna (body width more than 2 mm; e.g. earthworms, termites and millipedes). As well as being grouped on the basis of body size, this vast array of animal species can be aggregated into functional groups, based on food choice and life-history parameters; these 'groups' can be then assembled to construct complex soil food webs (de Ruiter *et al.* 1994, 1995; Scheu & Falca 2000). Information on the actual diversity of groups of soil biota, however, is very sparse compared with what is known about above-ground organisms, especially at the species level. Further, information that is known tends to be restricted to a few ecosystem types and a few taxonomic groups of organisms within them. The main reasons for this scarcity of information are the historical neglect of soil organisms by conservation inventories (Lawton *et al.* 1996; Usher this volume); the lack of methodologies that can extract, identify and quantify the diversity of soil biota; the fact that many microbes are actually unculturable (André *et al.* 2002; O'Donnell *et al.* this volume); and the fact that taxonomic work on soil organisms is extremely time consuming, requiring much patience and skill (Bloemers *et al.* 1997).

Despite the above problems, numerous studies illustrate the vast diversity of biological communities that inhabit soil. For example, recent developments in molecular technologies point to the existence of an immense number of bacterial and fungal species in soil (O'Donnell *et al.* this volume), and even traditional culturing approaches suggest that literally hundreds or thousands of microbial species exist in soil (e.g. McLean & Huhta 2002). For the soil fauna, there are some studies that have recorded the diversity of particular groups of fauna in some localities. For example, while a total of 1092 species of ciliate fauna has been recorded globally (Foissner 1997a), the maximum number recorded at any one site is 139 species, in the Manaus, Brazil (Foissner 1997b, 1999a). Substantially more is known about the diversity of nematodes, the most numerous multicellular animals on Earth. Boag and Yeates (1998) reviewed global data on soil nematode diversity and found, at 61.7 species per sample, temperate broadleaf forest had the greatest species richness. More recently, Bloemers *et al.* (1997) recorded a total of 431 species from 24 sites in a Cameroon forest, which is the greatest diversity of nematodes ever recorded. Many species of nematodes also inhabit soils of less fertile ecosystems; an intensive study of a New Zealand sand dune site yielded 44 nematode species under *Ammophila arenaria* (Yeates 1968), while a Scottish sand dune succession, including *A. arenaria* and *Leymus arenarius*, yielded 46 putative species from terrestrial samples (16 to 27 per site) and a further 27 from beach samples (Wall *et al.* 2002). Even in soils of the Antarctic Dry Valleys, which are thought to be the coldest, oldest and driest on Earth, five endemic species of nematodes exist, representing two functional groups (Wall & Virginia 1999).

In general, the diversity of larger fauna is much lower than for microfauna. A review of earthworm species present under various vegetation types around the world by Lee (1985) showed typical diversity to be only 2 to 5 species, with a maximum of 11. However, as with all fauna, the number of species within particular land types or ecosystems varies greatly in space and time, with as many as 13 earthworm species being recorded from intensive sampling of farming systems at Long Ashton in southwest England (Hutcheson *et al.* 2001) or as high as 7 to 9 species in tropical pastures and savannahs in Carimagua, Columbia, such numbers being a subset of the 21 species known in that region (Jiménez et al. 1998). In acid, mor-type soils typical of Boreal forest and heathland, earthworms are virtually absent, being replaced by enchytraeid worms (potworms) as the dominant taxa.

Most studies of Enchytraeidae do not include information on species composition (Didden 1993), but where this is available, such as in Scots pine (Pinus sylvestris) forest in The Netherlands (Didden & de Fluiter 1998) and central Sweden (Lundkvist 1983), some 5 to 6 species are typically found. While enchytraeids typically show low species diversity in northern and temperate coniferous forests, Chalupský (1995) found a total of 23 species across a range of Czech mountain forest soils, and Ştefan (1977) found a total of 35 species in a range of sites in the Cerna Valley, Romania. Many records of microarthropod diversity exist: Behan-Pelletier (1978) recorded a total of 396 mite species across 35 sites in the North American arctic and subarctic; Siepal & van de Bund (1988) found 108 species of microarthropods in 500 cm^2 soil of an unmanaged Dutch grassland; and, some 58 prostigmatid and 59 oribatid species of mites were found in Danish 'poor' pasture (Weis-Fogh 1948). Lower diversities of Collembola have been reported for other grassland systems: 27 species of Collembola were found in tall-grass prairie in Illinois (Brand & Dunn 1998), whereas only 12 species were found in low productivity, acid grassland in Scotland (Cole et al. 2005).

Patterns of soil biodiversity

One of the fascinations of biological diversity is that it is not the same everywhere. There is almost infinite variation in patterns of diversity across and within the Earth's ecosystems. Furthermore, these patterns change constantly with time; the diversity assessed at a particular locality will vary greatly both within and between years, but will also change over longer time scales as a result of processes such as succession and evolutionary change. This tremendous spatial and temporal complexity in diversity, which occurs both above-ground and below-ground, makes it extremely difficult, if not impossible, to evaluate and explain patterns of species richness. However, carefully framed questions about species diversity, which recognise the existence of such complexity, can give insights into the wide range of processes that control diversity on a range of spatial and temporal scales. In this section, we identify some of what is known about patterns of diversity in soils, at global and local scales, and discuss how soil biological diversity might change in particular locations with time.

Latitudinal gradients of soil biological diversity

A host of global patterns of spatial variation in biodiversity have been explored, but perhaps the most prominent and intensively studied are latitudinal gradients of species richness. It has long been recognised that the number of species in most taxonomic groups is lowest in the poles and increases towards the

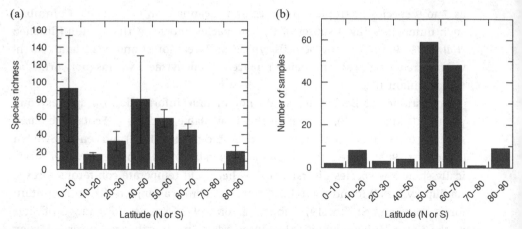

Figure 6.1. Mean species richness (a) and number of samples (b) of nematodes in latitudinal gradients. Derived from a literature review on global nematode diversity by Boag and Yeates (1998).

tropics (Huston 1994). Explaining this pattern of diversity is controversial, and many, often interrelated, mechanisms have been put forward to explain it. These include relationships between species richness and habitat characteristics such as availability of energy and heterogeneity, which increase towards the equator, and greater geographic area and speciation rates in the tropics than near the poles (Gaston 2000). While many other explanations of latitudinal patterns of diversity have been put forward and debated, the issue essentially remains unresolved and is still thought to be one of the biggest intellectual challenges to ecologists today.

Data and theory on latitudinal gradients of taxonomic groups are almost entirely based on organisms that live above-ground and, as noted above, remarkably little is known about patterns of distribution of species that live in soil. Further, no studies have comprehensively assessed the diversity of soil biota across latitudinal gradients using either standardised sampling approaches or similar taxonomic resolution. It is therefore rather dangerous to speculate too much about latitudinal patterns of soil biodiversity when information on species distribution is so scant. These problems are clearly shown by the study of Boag and Yeates (1998); these authors synthesised literature on species richness of soil-inhabiting nematodes and showed that nematode diversity was substantially lower near the poles than in tropical or temperate forests (Fig. 6.1). However, their data also showed that nematode diversity was greater in temperate than tropical forests, suggesting a non-linear relationship between nematode diversity and latitude, which contrasts with what occurs for many organisms above-ground. The problem with this relationship, however, is the gross imbalance in soil sampling in favour of temperate latitudes (80% of samples used were taken from between 40 and 60° N), with very few studies having been done in the tropics or polar regions. Meaningful comparisons of nematode diversity between regions of the world are therefore not really possible, other than at the high end of the latitudinal gradient where the diversity of nematodes drops substantially (Boag & Yeates 1998; Wall & Virginia 1999), as does that of protozoa (Smith 1996) and earthworms (Lavelle *et al.* 1995); presumably, this drop in soil biodiversity towards/near the poles is due to the combined effects of reduced availability of resources and harsh environmental conditions that require special adaptations, especially cold and desiccation tolerance (Bale *et al.* 1997). The only group of soil animals that does appear to follow a clear latitudinal gradient is termites, which clearly increase in diversity towards the tropics (Eggleton & Bignell 1995). Information on latitudinal patterns of microbial diversity is even more scant, and, until recently, we lacked the methods needed to study such patterns.

Whether broad-scale regional patterns of soil biodiversity exist, as for above-ground organisms, is not clear. It was, however, recently suggested by Wardle (2002) that it is unlikely that latitudinal patterns of soil biological diversity will exist, at least across temperate and tropical regions. The rationale for this is largely that many soil organisms, especially the microfauna and microbes, are cosmopolitan, being able to migrate extensively across the globe (Roberts & Cohen 1995; Finlay *et al.* 1999). Furthermore, the abundance of individuals of free-living microbial species is so large that their dispersal is rarely (if ever) restricted by geographical barriers (Finlay 2002). That certain taxonomic groups of soil biota are cosmopolitan, however, has been contested since some free-living ciliates (Foissner 1999b) and heterotrophic soil bacteria (Cho & Tiedje 2000) are known to have restricted geographic distributions. Also, estimates of global soil biodiversity are considered to be very speculative because only a tiny fraction of potential habitats across the globe have been carefully analysed (Foissner 1999b). If soil species are largely cosmopolitan, local patterns of soil biodiversity are likely to be independent of the regional species pool, which contrasts with many above-ground biota whose local species richness is positively related to the regional species pool (Gaston 2000); rather, local factors, such as resource availability, disturbance and habitat heterogeneity, will act as more important determinants of soil biodiversity.

Local patterns of soil biological diversity
When considering local scale diversity, ecologists have traditionally emphasised two factors as key determinants: productivity or resource supply, and consumption or physical disturbance (Fig. 6.2). At relatively local scales, species richness is often found to be unimodally related to productivity, such that peak diversity is at intermediate productivity: the hump-backed model (e.g. Grime 1973; Al-Mufti *et al.* 1977; Grace 1999), with declining diversity at higher levels of

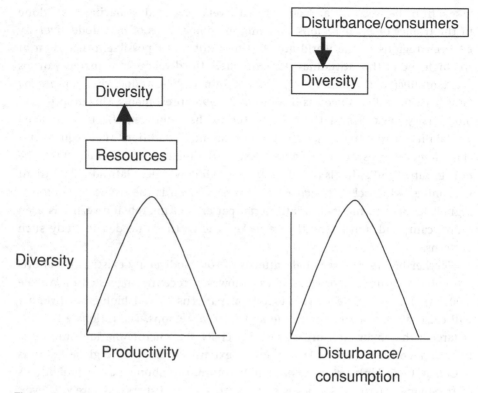

Figure 6.2. Disturbance/consumers versus resource control of species diversity. Both models predict that diversity will be maximal at intermediate levels of disturbance and productivity.

productivity being due to competitive exclusion. Competitive exclusion, however, can be prevented by periodic mortality events that are caused by consumption or physical disturbance (Connell 1978; Huston 1979); these factors likewise display unimodal relationships with diversity, as hypothesised by the 'intermediate disturbance hypothesis' of Connell (1978). These relationships are now embedded in ecological theory, but they do still remain controversial, largely since empirical tests of them are actually quite rare (Mittelbach *et al.* 2001; Sheil & Burslem 2003).

As already noted, there are not enough data on the diversity of soil biota at local scales to allow proper testing of these general ecological theories for the below-ground world. However, from the few data that are available, there is very little support for the notion that soil biodiversity at local scales conforms to either the productivity–diversity or the disturbance–diversity relationship. For example, if we consider the productivity–diversity relationship, it is known that productivity is an important determinant of soil food web structure at the extremely unproductive end of productivity gradients, for example in caves, whereas food chains become longer and more diverse in productive situations with more energy being diverted to higher trophic levels (Moore & de Ruiter, 2000; de Ruiter this volume). However, in a literature synthesis, Wardle (2002) found no data to support the idea that diversity declines along the most favourable portion of the productivity gradient (i.e. that soil organism diversity peaks at intermediate levels of productivity), thereby bringing into question the hump-backed model. This suggests that below-ground communities could differ from their above-ground counterparts, in that soil biodiversity is not so strongly regulated by competition, and competitive exclusion does not occur when resource availability in soil is increased (Wardle 2002). Although certain groups of soil organisms, for example fungi, are clearly regulated by competition (Cooke & Rayner 1984), recent experimental evidence from a grassland manipulation experiment in Scotland supports the idea that competition is not too important in controlling soil biodiversity. Microarthropod diversity did not alter along a gradient of productivity, although the proportion of predators within the community was greater in more productive sites (Cole *et al.* 2005).

There is also little support in the literature for optimisation of local soil animal diversity at intermediate levels of disturbance; on the contrary, disturbances generally lead to dramatic reductions in soil biodiversity. For example, disturbance of soil through tillage has been shown to reduce the abundance and diversity of earthworms (Springett 1992), although the scale of these effects depends on soil type, climate and tillage operation (Chan 2001). Similarly, Wardle (1995) found in a literature synthesis that the diversity of different macrofaunal groups could either be substantially elevated or reduced by tillage, although the diversity of microfauna was little affected. The response of populations of predacious nematodes and earthworms to tillage was found by Yeates *et al.* (1998) to vary with soil type. Other studies have examined gradients of disturbance resulting from the conversion of natural vegetation to agriculture, showing that the diversity of certain faunal groups is reduced as a consequence. For example, the diversity of macrofauna (Lavelle & Pashanasi 1989), termites (Eggleton *et al.* 2002) and nematodes (Bloemers *et al.* 1997) is dramatically reduced by the conversion of primary tropical forest to agriculture (Fig. 6.3); studies in Mexico, Peru and India reveal that earthworm communities of agro-ecosystems have lower species richness and a lower number of native species than do undisturbed ecosystems (Fragoso *et al.* 1997); the diversity of Collembola is higher in native prairie than in prairie that has been influenced by agriculture (Brand & Dunn 1998); and the diversity of soil nematodes has been shown to decrease with agricultural improvement of native pastures in New England Tablelands (Yeates & King 1997).

Measurement of changes in the relative abundance, or evenness, of phospholipids and fatty acids suggests that the broad-scale phenotypic diversity of

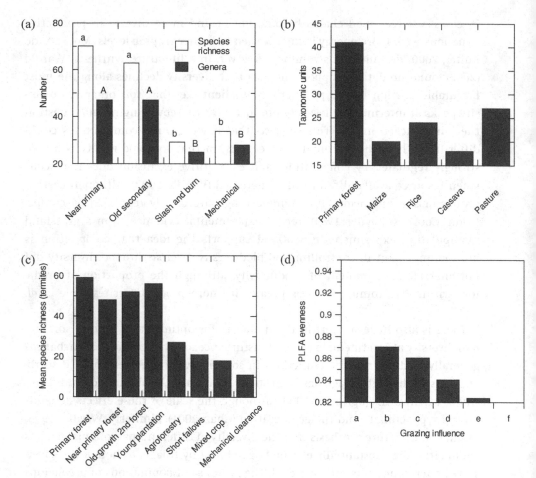

Figure 6.3. Effects of anthropogenic disturbances on the diversity of soil biota. (a) Effects of forest disturbances on the number of species and genera of nematodes in soil of tropical forests, Cameroon. Adapted from Bloemers *et al.* (1997). (b) Effects of land management on the number of taxonomic units of macrofauna in Peruvian Amazonia. Adapted from Lavelle and Pashanasi (1989). (c) Termite diversity across a gradient of anthropogenic disturbance in the humid forest zone of West Africa. Adapted from Eggleton *et al.* (2002). (d) Changes in PLFA evenness index of soil microbial communities along grazing-related disturbance gradient on grasslands in the Snowdonia National Park, UK. Site a, ungrazed control; b, long-term ungrazed; c, short-term ungrazed; d, lightly grazed; e, moderately grazed, and f, heavily grazed. Adapted from Bardgett *et al.* (2001).

microbial communities in soil is also reduced as a consequence of disturbances caused by intensive sheep grazing of semi-natural temperate grasslands (Bardgett *et al.* 2001; Fig. 6.3). While all these studies indicate that disturbances resulting from agricultural intensification can result in reduced soil biodiversity (but see Boag & Yeates 1998), they do not lend support for the notion that biological diversity in soil is optimised at intermediate levels of disturbance. It is important to note, however, that these studies do not explicitly test for the effects of disturbance on soil biological communities, since changes in land management are also associated with alterations in resource availability, and physical and chemical conditions of the soil. In view of this, rigorous experimental testing of the disturbance–diversity relationship is still required to ascertain its importance as a regulator of below-ground diversity.

Temporal patterns of soil biodiversity

Soil communities are not static; they are constantly changing at many different temporal scales, ranging from seasonal changes in the abundance of various groups of organisms, to changes in soil biodiversity over successional time. Temporal patterns of soil biodiversity are also complicated by the fact that many soil organisms can undergo long periods of inactivity when conditions are unfavourable (Joose & Verhoef 1987), thereby allowing them to tolerate periods of harsh soil conditions. The degree to which soil communities change over time is also highly context dependent, varying with a range of factors such as the nature and frequency of disturbance regimes, vegetation change, and changes in soil and climatic conditions. This is perhaps most evident in managed systems, where changes in management can affect soil biota over different time scales. For example, effects of changing management regimes (e.g. cessation of fertiliser application) on soil microbial communities of temperate grasslands in northern England took some nine years to detect, being linked to vegetation change (Smith et al. 2003), whereas in more nutrient-poor grassland, impacts of similar management treatments on microbial communities became evident within only four years, due to rapid changes in soil physico-chemical conditions (Bardgett et al. 1996). Yearly variation in soil biotic communities are also highly evident; for example, over a 23-year period, Yeates et al. (2000) found that various management treatments in a *Pinus radiata* agroforestry trial had little effect on soil nematodes, but there were unexplained, apparently chaotic, changes in the nematode assemblage in the 'control' pasture over this time period. Over a seven-year period, there was also much variation in soil-associated arthropods under various weed management treatments for *Zea mays* and *Asparagus officinale* crops, which appeared to be related to variations in weed biomass, which itself varied from year to year (Wardle et al. 1999). These studies all emphasise the need for careful thought about how soil biodiversity is evaluated over time in field experiments and at specific locations.

Soil biodiversity will also change over time scales of tens or hundreds of years, through processes of succession. A number of studies give insights into the kind of patterns that occur in soil communities over these long time scales. For example, for soil microbes it is evident that during the initial stages of succession – the build-up phase – and towards maximal plant biomass (i.e. community climax), soil microbial communities become increasingly abundant and active (Insam & Haselwandter 1989; Ohtonen *et al.* 1999), and more diverse (Schipper *et al.* 2001) and fungal dominated (over bacteria) in nature (Frankland 1998; Ohtonen *et al.* 1999; Bardgett 2000). Similar shifts in microbial community composition also appear to occur during secondary succession, for example after abandonment of management on agricultural grasslands that typically leads to a shift in the composition of the microbial community towards fungal dominance over bacteria (Bardgett & McAlister 1999; Bardgett *et al.* 2001; Zeller *et al.* 2001). It is not entirely clear why these patterns in microbial communities occur, but they are most likely related to a build-up in the amount and complexity of organic matter, and changes in the quality of resource inputs to soil resulting from vegetation change. They may also be related to changes in the physico-chemical nature of soils, for example a drop in soil pH that commonly occurs in late succession (Bardgett *et al.* 2001).

Little is known about how the faunal community changes during succession, but one might expect that fungal-feeding fauna increasingly dominates it, and that predators become more dominant as more energy becomes available for higher trophic levels. However, a study of different stages of secondary succession (from a wheat field to a beechwood) by Scheu and Schulz (1996) showed that soil invertebrates of similar trophic groups responded very differently to successional changes, suggesting that responses of higher trophic levels to succession are difficult to predict. A study by Kaufmann (2001) of surface-active micro- and macroarthropods in the Central Alpine glacier foreland of the Rotmoostal (Obergurgl, Tyrol, Austria) also indicates that in this group patterns of invertebrate succession are not as might be predicted: although faunal colonisation and succession in Alpine glacier forelands followed a largely predictable pattern, the first colonisers were almost exclusively predators, and herbivores and decomposers appeared later. Similarly, spiders are the earliest colonisers of newly exposed moraine substrates on the glacier foreland of the Midre Lovénbreen at Ny-Ålesund, Spitsbergen, Svalbard (78° N) where their densities are highly correlated with allochthonous inputs of potential prey items, predominantly chironomid midges (Hodkinson *et al.* 2001). In these systems, large allochthonous inputs of insects potentially provide significant quantities of nitrogen and phosphorus to the developing ecosystem from the earliest stages of succession, even before a conspicuous cyanobacterial crust has formed. That spiders entrap nutrients in such a way could be of high importance in early ecosystem development in these extreme environments.

The enigma of soil biodiversity
The 'enigma of soil diversity' posed by Anderson in 1975 still remains: what explains the high diversity of biota in soils and how is it possible for so many species of soil organism to co-exist without competitive exclusion occurring?

Perhaps the most compelling answer to this question lies in the extremely variable nature of the soil environment itself: soil offers an extremely heterogeneous habitat, both spatially and temporally. This results in microhabitat diversity, thereby providing unrivalled potential for niche partitioning, or resource or habitat specialisation, in turn enabling co-existence of species. There are two lines of evidence to support this idea.

First, there are numerous examples in the literature to support the notion that local spatial heterogeneity promotes soil biodiversity. Mixtures of litters, for example, have been shown to support more diverse mite communities than single species litters, due to an increased range of food resources and increased habitat complexity (Gill 1969; Stanton 1979; Hansen & Coleman 1998; Hansen 2000). However, impacts of litter mixing on diversity have been shown to disappear as litters decompose and resources become more morphologically homogeneous (Hansen 2000); mite diversity has been found to be positively related to microhabitat complexity of woodland soil and litter layers (Anderson 1978); species diversity of litter-dwelling gastropods has been shown to increase with increasing plant diversity (Barker & Mayhill 1999); adding even an inert polystyrene material to litters has been shown to increase the size of soil collembolan communities, suggesting the importance of the physical properties of litter mixtures on the associated soil fauna communities (J. Daykin & J. M. Anderson unpublished data); and the extremely high diversity of nematodes in a recent volcanic soil was attributed to unweathered, textural heterogeneity (Yeates 1980).

Second, there is a high degree of specialisation among soil biota. For example, bacteria and fungi have numerous adaptive strategies in terms of spore types, physiological modes of adaptation to resource limitation, sexual and asexual reproductive modes, and eco-physiological tolerance, whereas soil animals show tremendous specialisation in terms of their feeding and life history strategies (Siepel 1996; Scheu & Falca 2000) and their spatial distribution, enabling them to avoid competition. Species also differ markedly in their response to abiotic factors, with reported inter-specific variation in the tolerance of soil animals to soil moisture, food availability and drought (Verhoef & van Selm 1983; Scheu & Schultz 1996; Bongers & Bongers 1998), which may determine the species-specific spatial and temporal distribution of biota in soil. Further, many soil organisms have the ability to undergo periods of inactivity when conditions are unfavourable (e.g. Joose & Verhoef 1987), thereby allowing them to tolerate periods of harsh soil conditions. This is especially the case for microbes that produce spores that are persistent until, for example, resource flushes occur, or fungi that develop extensive mycelial networks that can exploit 'vacant' resource patches that become available in space and time. Together, these multi-dimensional axes of niche partitioning, and hence avoidance of competition, are likely to provide an explanation for the vast diversity of organisms that live in soil.

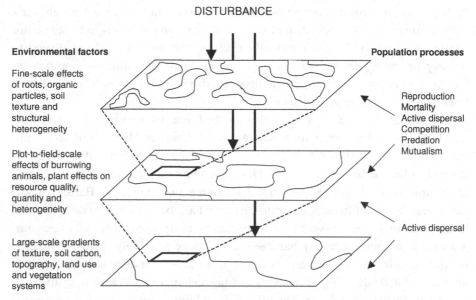

Figure 6.4. Determinants of spatial heterogeneity of soil organisms. The figure shows that spatial heterogeneity in soil organisms is distributed on nested scales, and is shaped by a spatial hierarchy of environmental and biological factors, and disturbance. Adapted from Ettema and Wardle (2002).

It follows, therefore, that spatial and temporal patterns of soil biodiversity are related to variations in the heterogeneity and make-up of the soil microhabitat and, further, that this variation occurs at a range of spatial and temporal scales. Unravelling the nature of these complex relationships therefore presents a key challenge for soil ecologists if we are to understand the factors that regulate patterns of soil biodiversity. This issue was recently explored by Ettema and Wardle (2002), who argued for a more spatially explicit approach to soil ecology not only to identify the factors that drive spatial heterogeneity of populations and activities of soil biota, but also to provide insights into the regulation and maintenance of soil biodiversity. These authors presented evidence to suggest that there is some order to spatial patterning of soil biota; using data from exist-ing studies, they showed that the spatial patterns of soil biodiversity that occur at different scales from millimetres to hectares are not random, as generally thought, but rather that they show predictable patterns with scale-dependent controls (Fig. 6.4). At the landscape scale, for example, patterning of soil biota appears to be related to factors such as soil pattern, topography, tillage and vege-tation patchiness (see Giller *et al.* this volume). In contrast, at the local scale, over ranges of centimetres to metres, patterns of soil biota are often very patchy in their distribution, being structured by plant growth, and varying with plant size and plant spacing. This affects the zone of influence of particular plant species,

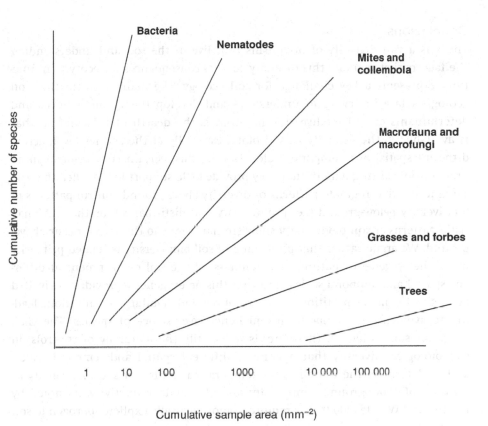

Figure 6.5. Hypothetical species/sample size relationships for soil organisms and vegetation cover affecting soil habitats.

which also determines organic matter quality (Saetre & Bääth 2000; Wardle this volume). At even finer scales, spatial patterns of soil biota relate to resource patches, or hot spots, in the form of buried litter, dead animals, roots and other organic materials, and patterns of rhizosphere carbon flow (Standing *et al.* this volume), or to variations in soil structure that affect the physical dimensions of habitable space in soil (Young *et al.* this volume).

The scales at which habitat heterogeneity impacts on different groups of organisms will further depend on the body size of an organism and the size of its habitat unit, or domain (Fig. 6.5). Spatial and temporal patterns of biodiversity in soil will also relate to intrinsic population processes, for example dispersal, reproduction and competition, and predation, which in turn are influenced by stochastic disturbance events that upset population dynamics of organisms leading to complex spatio-temporal patterns of biodiversity in soil (Ettema and Wardle 2002). This issue, and the geostatistical tools that are available for unravelling order in complex spatio-temporal patterns of biodiversity, is explored in more detail by Ettema and Wardle (2002).

Conclusions

There is a vast diversity of organisms that live in the soil and understanding the factors that regulate this diversity, and its consequences for ecosystem function, represents a key challenge for soil ecology. The main problem that soil ecologists face in trying to understand and develop theory on patterns and determinants of soil biodiversity, however, is the dearth of information that is available on the diversity of soil biota, especially at the species level, across different spatial and temporal scales. Despite this, certain conclusions can be drawn from existing data in that they provide little support for the idea that the same forces that regulate patterns of diversity above-ground control patterns of biodiversity below-ground (i.e. productivity and disturbance), or that regional-scale patterns of soil biodiversity show similar trends to those that occur above-ground. We argue rather that patterning of soil biodiversity is related primarily to the heterogeneous nature, or patchiness, of the soil environment at different spatial and temporal scales, and that this heterogeneity provides unrivalled potential for niche partitioning, or resource and habitat specialisation, leading to avoidance of competition and hence co-existence of species. The challenge for soil ecologists, therefore, is to identify the hierarchy of controls on soil biological diversity that operate at different spatial and temporal scales, and to determine the role of spatio-temporal patterning of soil biodiversity as a driver of above-ground community assembly and productivity. As noted by Ettema and Wardle (2002), this requires a more spatially explicit approach to soil ecology.

Acknowledgements

GWY was supported by the Marsden Fund (administered by the Royal Society of New Zealand).

References

Al-Mufti M. M., Sydes, C. L., Furness, S. B., Grime, J. P. & Bond, S. R. (1977). A quantitative analysis of shoot phenology and dominance in herbaceous vegetation. *Journal of Ecology*, 65, 759–791.

Anderson, J. M. (1975). The enigma of soil animal species diversity. *Progress in Soil Zoology* (Ed. by J. Vanek), pp. 51–58. Prague: Academia.

Anderson, J. M. (1978). Inter- and intra-habitat relationships between woodland Cryptostigmata species diversity and the diversity of soil and litter microhabitats. *Oecologia*, 32, 341–348.

André, H. M., Ducarme, X. & Lebrun, P. (2002). Soil biodiversity: myth, reality or conning? *Oikos*, 96, 3–24.

André, H. M., Noti, M. I. & Lebrun, P. (1994). The soil fauna: the other last biotic frontier. *Biodiversity and Conservation*, 3, 45–56.

Bale, J. S., Hodkinson I. D., Block, W., *et al.* (1997). Life strategies of arctic terrestrial arthropods. *Ecology of Arctic Environments* (Ed. by S. J. Woodin & M. Marquiss), pp. 137–165. Oxford: Blackwell Scientific.

Bardgett, R. D. (2000). Patterns of below-ground primary succession at Glacier Bay, south-east

Alaska. *Bulletin of the British Ecological Society*, **31**, 40–42.

Bardgett, R. D., Hobbs, P. J. & Frostegård, Å. (1996). Changes in fungal : bacterial biomass ratios following reductions in the intensity of management on an upland grassland. *Biology and Fertility of Soils*, **22**, 261–264.

Bardgett, R. D., Jones, A. C., Jones D. L., *et al.* (2001). Soil microbial community patterns related to the history and intensity of grazing in sub-montane ecosystems. *Soil Biology and Biochemistry*, **33**, 1653–1664.

Bardgett, R. D. & McAlister, E. (1999). The measurement of soil fungal : bacterial biomass ratios as an indicator of ecosystem self-regulation in temperate grasslands. *Biology and Fertility of Soils*, **19**, 282–290.

Bardgett, R. D. & Wardle, D. A. (2003). Herbivore mediated linkages between above-ground and below-ground communities. *Ecology*, **84**, 2258–2268.

Barker, G. M. & Mayhill, P. C. (1999). Patterns of diversity and habitat relationships in terrestrial mollusc communities of the Pukemaru Ecological District, northeastern New Zealand. *Journal of Biogeography*, **26**, 215–238.

Behan-Pelletier, V. M. (1978). Diversity, distribution and feeding habits of North American Arctic soil Acari. Unpublished Ph.D. thesis, McGill University.

Bloemers, G. F., Hodda, M., Lambshead, P. J. D., Lawton, J. H. & Wanless, F. R. (1997). The effects of forest disturbance on diversity of tropical soil nematodes. *Oecologia*, **111**, 575–582.

Boag, B. & Yeates, G. W. (1998). Soil nematode biodiversity in terrestrial ecosystems. *Biodiversity and Conservation*, **7**, 617–630.

Bongers, T. & Bongers, M. (1998). Functional diversity of nematodes. *Applied Soil Ecology*, **10**, 239–251.

Brand, R. H. & Dunn, C. P. (1998). Diversity and abundance of springtails (Insecta: Collembola) in native and restored tallgrass prairies. *American Midland Naturalist*, **139**, 235–242.

Chalupský, J. (1995). Long-term study of Enchytraeidae (Oligochaeta) in man-impacted mountain forest soils in the Czech Republic. *Acta Zoologica Fennica*, **196**, 318–320.

Chan, K. Y. (2001). An overview of some tillage impacts on earthworm population abundance and diversity : implications for functioning in soils. *Soil and Tillage Research*, **57**, 179–191.

Cho, J. C. & Tiedje, J. M. (2000). Biogeography and degree of endemicity of fluorescent *Pseudomonas* strains in soil. *Applied and Environmental Microbiology*, **66**, 5448–5456.

Cole, L., Buckland, S. M. & Bardgett, R. D. (2005). Relating soil microarthropod community structure and diversity to soil fertility manipulations in temperate grassland. *Soil Biology and Biochemistry*, in press.

Connell, J. H. (1978). Diversity in tropical rainforests and coral reefs. *Science*, **199**, 1302–1310.

Cooke, R. C. & Rayner A. D. M. (1984). *Ecology of Saprotrophic Fungi*. London: Longman.

De Ruiter, P. C., Neutel, A. N. & Moore, J. C. (1994). Modelling food webs and nutrient cycling in agro-ecosystems. *Trends in Ecology and Evolution*, **9**, 378–383.

De Ruiter, P. C., Neutel, A. N. & Moore, J. C. (1995). Energetics, patterns of interaction strengths, and stability in real ecosystems. *Science*, **269**, 1257–1260.

Didden, W. A. M. (1993). Ecology of terrestrial Enchytraeidae. *Pedobiologia*, **37**, 2–29.

Didden, W. A. M. & de Fluiter, R. (1998). Dynamics and stratification of Enchytraeidae in the organic layer of a Scots pine forest. *Biology and Fertility of Soils*, **26**, 305–312.

Eggleton, P. & Bignell, D. E. (1995). Monitoring the response of tropical insects to changes in the environment: troubles with termites. *Insects in a Changing Environment* (Ed. by R. Harrington & N. E. Stork), pp. 434–497. London: Academic Press.

Eggleton, P., Bignell, D. E., Hauser, S., *et al.* (2002). Termite diversity across an anthropogenic disturbance gradient in the humid forest zone of West Africa. *Agriculture, Ecosystems and Environment*, **90**, 189–202.

Ettema, C. H. & Wardle, D. A. (2002). Spatial soil ecology. *Trends in Ecology and Evolution*, **17**, 177–183.

Finlay, B. J. (2002). Global dispersal of free-living microbial eukaryote species. *Science*, **296**, 1061–1063.

Finlay, B. J., Esteban, G. F., Olmo, J. L. & Tyler, P. A. (1999). Global distribution of free-living microbial species. *Ecography*, **22**, 138–144.

Foissner, W. (1997a). Global soil ciliate (Protozoa, Ciliophora) diversity: a probability-based approach using large sample collections from Africa, Australia and Antarctica. *Biodiversity and Conservation*, **6**, 1627–1638.

Foissner, W. (1997b). Soil ciliates (Protozoa: Ciliophora) from evergreen rain forests of Australia, South America and Costa Rica: diversity and description of new species. *Biology and Fertility of Soils*, **25**, 317–339.

Foissner, W. (1999a). Notes on the soil ciliate biota (Protozoa, Ciliophora) from the Shimba Hills in Kenya (Africa): diversity and descriptions of three new genera and ten new species. *Biodiversity and Conservation*, **8**, 319–389.

Foissner, W. (1999b). Protist diversity: estimates of the near-imponderable. *Protist*, **150**, 363–368.

Fragoso, C., Brown, G. G., Patrón, J. C., *et al.* (1997). Agricultural intensification, soil biodiversity and agroecosystem function in the tropics: the role of earthworms. *Applied Soil Ecology*, **6**, 17–35.

Frankland, J. C. (1998). Fungal succession: unravelling the unpredictable. *Mycological Research*, **102**, 1–15.

Gaston, K. J. (2000). Global patterns in biodiversity. *Nature*, **405**, 220–227.

Grace, J. B. (1999). The factors controlling species density in herbaceous plant communities: an assessment. *Perspectives in Plant Ecology, Evolution and Plant Systematics*, **2**, 1–28.

Grime, J. P. (1973). Control of species diversity in herbaceous vegetation. *Journal of Environmental Management*, **1**, 151–167.

Gill, R. W. (1969). Soil microarthropod abundance following old field litter manipulation. *Ecology*, **50**, 805–816.

Hansen, R. A. (2000). Effects of habitat complexity and composition on a diverse litter microarthropod assemblage. *Ecology*, **81**, 1120–1132.

Hansen, R. A. & Coleman, D. C. (1998). Litter complexity and composition are determinants of the diversity and species composition of oribatid mites (Acari: Oribatida) in litterbags. *Applied Soil Ecology*, **9**, 17–23.

Heywood, V. H. (1995). *Global Biodiversity Assessment*. Cambridge: Cambridge University Press.

Hodkinson, I. D., Coulson, S. J., Harrison, J. & Webb, N. R. (2001). What a wonderful web they weave: spiders, nutrient capture and early ecosystem development in the high Arctic. Some counter-intuitive ideas on community assembly. *Oikos*, **95**, 349–352.

Hooper, D. U., Bignell, D. E., Brown, V. K., *et al.* (2000). Interactions between above-ground and below-ground biodiversity in terrestrial ecosystems: patterns, mechanisms, and feedbacks. *BioScience*, **50**, 1049–1061.

Huston, M. A. (1979). A general model of species diversity. *American Naturalist*, **113**, 81–101.

Huston, M. A. (1994). *Biological Diversity*. Cambridge: Cambridge University Press.

Hutcheson, J. A., Iles, D. R. & Kendall D. A. (2001). Earthworm populations in conventional and integrated farming systems in the LIFE Project (SW England) in 1990–2000. *Annals of Applied Biology*, **139**, 361–372.

Insam, H. & Haselwandter, K. (1989). Metabolic quotient of the soil microflora in relation to plant succession. *Oecologia*, **79**, 174–178.

Jiménez, J. J., Moreno, A. G., Decaëns, T., *et al.* (1998). Earthworm communities in native savannas and man-made pastures of the Eastern Plains of Colombia. *Biology and Fertility of Soils*, **28**, 101–110.

Joose, E. N. G. & Verhoef, H. A. (1987). Developments in ecophysiology research on soil invertebrates. *Advances in Ecological Research*, **16**, 175–248.

Kaufmann, R. (2001). Invertebrate succession on an alpine glacier foreland. *Ecology*, **82**, 2261–2278.

Lavelle, P., Lattaud, C., Trigo, D. & Barois, I. (1995). Mutualism and biodiversity in soils. *Plant and Soil*, **170**, 23–33.

Lavelle, P. & Pashanasi, B. (1989). Soil macrofauna and land management in Peruvian Amazonia (Yurimaguas, Loreto). *Pedobiologia*, **33**, 283–291.

Lawton, J. H. (2000). *Community Ecology in a Changing World*. Oldendorf/Luhe: Ecology Institute.

Lawton, J. H., Bignell, D. E., Bloemers, G. F., Eggleton, P. & Hodda, M. E. (1996). Carbon flux and diversity of nematodes and termites in Cameroon forest soils. *Biodiversity and Conservation*, **5**, 261–273.

Lee, K. E. (1985). *Earthworms: Their Ecology and Relationships with Soils and Land Use*. Sydney: Academic Press.

Lundkvist, H. (1983). Effects of clear-cutting on the enchytraeids in a Scots pine forest soil in central Sweden. *Journal of Applied Ecology*, **20**, 873–885.

McLean, M. A. & Huhta, V. (2002). Microfungal community structure in anthropogenic birch stands in central Finland. *Biology and Fertility of Soils*, **35**, 1–12.

Mittelbach, G. G., Steiner, C. F., Scheiner, S. M., *et al.* (2001). What is the observed relationship between species richness and productivity? *Ecology*, **82**, 2381–2396.

Moore, J. C. & de Ruiter, P. C. (2000). Invertebrates in detrital food webs along gradients of productivity. *Invertebrates as Webmasters in Ecosystems* (Ed. by D. C. Coleman & P. F. Hendrix), pp. 161–184. Oxford: CAB International.

Ohtonen, R., Fritze, H., Pennanen, T., Jumpponen, A. & Trappe, J. (1999). Ecosystem properties and microbial community changes in primary succession on a glacier forefront. *Oecologia*, **119**, 239–246.

Roberts, M. S. & Cohen, F. M. (1995). Recombination and migration rates in natural populations of *Bacillus subtilus* and *Bacillus mojavensis*. *Evolution*, **49**, 1081–1094.

Saetre, P. & Bääth, E. (2000). Spatial variation and patterns of soil microbial community structure in a mixed spruce-birch stand. *Soil Biology and Biochemistry*, **30**, 909–917.

Scheu, S. & Falca, M. (2000). The soil food web of two beech forests (*Fagus sylvatica*) of contrasting humus type: stable isotope analysis of macro- and mesofauna dominated community. *Oecologia*, **123**, 285–296.

Scheu, S. & Schulz, E. (1996). Secondary succession, soil formation and development of a diverse community of oribatids and saprophagous soil macro-invertebrates. *Biological Conservation*, **5**, 235–250.

Schipper, L. A., Degens, B. P., Sparling, G. P. & Duncan, L. C. (2001). Changes in microbial heterotrophic diversity along five plant successional sequences. *Soil Biology and Biochemistry*, **33**, 2093–2104.

Sheil, D. & Burslem, D. F. R. P. (2003). Disturbing hypothesis in tropical forests. *Trends in Ecology and Evolution*, **18**, 18–26.

Siepel, H. (1996). Biodiversity of soil microarthropods: the filtering of species. *Biodiversity and Conservation*, **5**, 251–260.

Siepel, H. & van de Bund, C. F. (1988). The influence of management practices on the microarthropod community of grassland. *Pedobiologia*, **31**, 339–354.

Smith, H. G. (1996). Diversity of Antarctic terrestrial protozoa. *Biodiversity and Conservation*, **5**, 1379–1394.

Smith, R.S, Shiel, R. S., Bardgett, R. D., *et al.* (2003). Diversification management of meadow grassland: plant species diversity and functional traits associated with change in meadow vegetation and soil microbial communities. *Journal of Applied Ecology*, **40**, 51–64.

Springett, J. A. (1992). Distribution of lumbricid earthworms in New Zealand. *Soil Biology and Biochemistry*, **24**, 1377–1381.

Stanton, N. L. (1979). Patterns of species diversity in temperate and tropical litter mites. *Ecology*, **62**, 295–304.

Ştefan, V. (1977). Soil Enchytraeidae from the Cerna Valley. *Fourth Symposium on Soil Biology*, pp. 277–283, Rumanian National Society of Soil Science. Bucureşti: Editura Ceres.

Van der Putten, W. H., Vet, L. E. M., Harvey, J. A. & Wäckers, F. L. (2001). Linking above- and below-ground multitrophic interactions of plants, herbivores, pathogens and their antagonists. *Trends in Ecology and Evolution*, **16**, 547–554.

Verhoef, H. A. & van Selm, A. J. (1983). Distribution and population dynamics of Collembola in relation to soil moisture. *Holarctic Ecology*, **6**, 387–394.

Wall, J. W., Skene, K. R. & Neilsen, R. (2002). Nematode community and trophic structure along a sand dune succession. *Biology and Fertility of Soils*, **35**, 293–301.

Wall, D. H. & Virginia, R. A. (1999). Controls on soil biodiversity: insights from extreme environments. *Applied Soil Ecology*, **13**, 137–150.

Wardle, D. A. (1995). Impacts of disturbance on detritus food webs in agro-ecosystems of contrasting tillage and weed management practices. *Advances in Ecological Research*, **26**, 105–185.

Wardle, D. A. (2002). *Communities and Ecosystems: Linking the Above-ground and Below-ground Components.* Princeton, NJ: Princeton University Press.

Wardle, D. A., Nicholson, K. S., Bonner, K. I. & Yeates, G. W. (1999). Effects of agricultural intensification on soil-associated arthropod population dynamics, community structure, diversity and temporal variability over a seven-year period. *Soil Biology and Biochemistry*, **31**, 1691–1706.

Weis-Fogh, T. (1948). Ecological investigations of mites and collemboles in the soil. Description of some new mites (Acari). *Natura Jutlandica*, **1**, 139–277.

Wilson, E. O. (2002). *The Future of Life*. New York: Alfred A. Knopf.

Yeates, G. W. (1968). An analysis of annual variation of the nematode fauna in dune sand, at Himatangi Beach, New Zealand. *Pedobiologia*, **8**, 173–207.

Yeates, G. W. (1980). Populations of nematode genera in soils under pasture. III. Vertical distribution at eleven sites. *New Zealand Journal of Agricultural Research*, **23**, 117–128.

Yeates, G. W., Hawke, M. F. & Rijske, W. C. (2000). Changes in soil fauna and soil conditions under *Pinus radiata* agroforestry regimes during a 25-year tree rotation. *Biology and Fertility of Soils*, **30**, 391–406.

Yeates, G. W. & King, K. L. (1997). Soil nematodes as indicators of the effect of management on grasslands in the New England Tablelands (NSW): comparison of native and improved grasslands. *Pedobiologia*, **41**, 526–536.

Yeates, G. W., Shepherd, T. G. & Francis, G. S. (1998). Contrasting response to cropping of populations of earthworms and predacious nematodes in four soils. *Soil and Tillage Research*, **48**, 255–264.

Zeller, V., Bardgett, R. D. & Tappeiner, U. (2001). Site and management effects on soil microbial properties of subalpine meadows: a study of land abandonment along a north–south gradient in the European Alps. *Soil Biology and Biochemistry*, **33**, 639–650.

CHAPTER SEVEN

How plant communities influence decomposer communities

DAVID A. WARDLE

Swedish University of Agricultural Sciences, and Landcare Research, New Zealand

SUMMARY

1. The issue of how plant community composition affects decomposer community composition and function is considered, by reviewing recent literature and through the use of two examples.
2. It is apparent from the available literature that plant species identity exerts important effects on soil food webs, and that specific attributes such as the body size distribution of soil animals, and the relative importance of bacterial-based vs. fungal-based energy channels, respond to plant species identity. This has important implications for ecosystem functioning.
3. The first example involves below-ground effects of changes in plant community composition, such as might occur during C4 grass invasion resulting from global warming, in a perennial pasture in New Zealand. The second involves below-ground consequences of changes in vegetation community structure caused by introduced browsing mammals in New Zealand rainforest. Both examples point to above-ground, human-induced changes affecting the composition of the soil food web across several trophic levels, and key ecosystem functions carried out by the soil biota.
4. The issues of how above-ground biodiversity affects below-ground biodiversity, and the nature of reciprocal feedbacks between the above-ground and below-ground biota, are discussed. It is concluded that understanding the nature of above-ground–below-ground feedbacks may offer opportunities for better understanding how ecosystems function and the ecological consequences of global change phenomena.

Introduction

All functional ecosystems consist of explicit producer and decomposer subsystems. Producers fix atmospheric carbon, which is utilised by the decomposer organisms, and the decomposers in turn break down organic matter, which regulates the availability and supply of nutrients required for plant growth. Plant

Biological Diversity and Function in Soils, eds. Richard D. Bardgett, Michael B. Usher and David W. Hopkins. Published by Cambridge University Press. © British Ecological Society 2005.

species differ tremendously in the quantity and quality of resources that they
return to the decomposer subsystem, both in terms of litter from dead shoots
and root material, and substrates that are exuded from the plant-rooting zone.
For this reason, it could be expected for plant species identity to operate as
a major driver of the decomposer subsystem, as well as its functioning. It has
long been recognised that plant species differ in their effects on decomposer-
driven processes such as decomposition, nutrient mineralisation and soil carbon
sequestration (Muller 1884; Handley 1954; Swift et al. 1979), and this topic has
gained considerable attention over the past decade (Hobbie 1992; Lawton 1994).
What remains less well understood is how the decomposers themselves, as well
as decomposer community composition and diversity, respond to plant species
effects. Yet, such information is essential in understanding the responses of
communities and ecosystems to changes in vegetation composition that result
from both natural and human-induced factors (Wardle 2002).

In this chapter an outline is presented as to how plant species affect the
decomposer food web. Two examples are then presented from the work of my
colleagues and myself which investigate how alterations in plant species composition, such as may occur as a result of global change phenomena, affect decomposer communities. The issue of whether plant species diversity as opposed
to species composition is a driver of decomposer organisms is then considered.
Finally, the consequence of plant species effects on decomposers for plant growth
via above-ground–below-ground feedbacks is discussed. The ultimate goal of this
article is critically to evaluate whether and how changes in decomposer communities parallel changes in plant communities.

Soil food web responses to plant species differences

The relative abundance of different components of the soil food web can vary
tremendously among plant species (e.g. Parmelee et al. 1989; Griffiths et al. 1992),
reflecting differences among plant species in both the quantity and quality of
the resources that they produce. There have been few attempts to develop general
principles about how soil food web composition may respond to plant species
identity. However, it could be expected that plant species that differ in terms
of fundamental ecological traits would select for decomposer food webs with
different basic attributes. This first requires recognition that plant species can
be arranged along a fundamental trait axis (cf. Grime et al. 1997). One such axis
involves the spectrum from fast-growing plant species adapted to fertile conditions, to slow-growing species adapted to infertile conditions. In comparison
with slow-growing plants adapted for infertile conditions, plants adapted for
fertile conditions allocate a greater proportion of their carbon to rapid growth
rather than to secondary defence compounds, have a higher specific leaf area,
leaf area ratio and maximum photosynthetic capacity, and produce litter and
foliage with a greater concentration of nitrogen and a lower concentration of

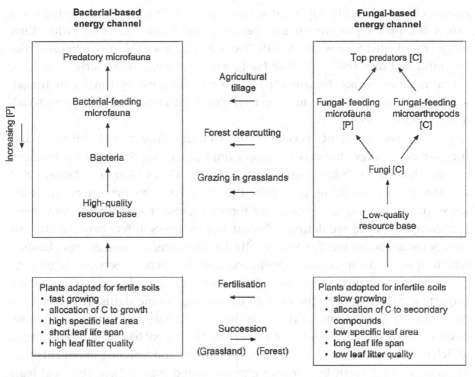

Figure 7.1. The effects of plant species, and extrinsic factors, on the relative roles of the bacterial and fungal energy channels in soil food webs. P, predation regulated (top–down control); C, competition regulated (bottom–up control). Assignment of C and P to trophic groupings follows Wardle (2002).

phenolics and lignin. These differences lead to plant species identity potentially influencing fundamental properties of soil food webs; two examples of these, described below, are the relative importance of bacterial-based vs. fungal-based energy channels, and the relative contribution of animals of differing body sizes.

The soil microfood web (*sensu* Lavelle *et al.* 1995) consists of a fungal-based and a bacterial-based energy channel (Moore & Hunt 1988). Several factors determine the relative dominance of the two channels (Fig. 7.1). The bacterial-based energy channel is favoured by disturbances such as agricultural cultivation (Hendrix *et al.* 1986) and forest clearcutting (Sohlenius 1996), as well as by nitrogen fertilisation (Ettema *et al.* 1999) and grazing of grasslands by livestock (Bardgett *et al.* 1998). This energy channel is also favoured by early successional (vs. late successional) vegetation (Ohtonen *et al.* 1999) and herbaceous rather than woody vegetation (Ingham *et al.* 1989). It is therefore apparent that factors that favour plants with suites of traits that are associated with allocation of carbon to growth rather than defence also favour bacterial-based over fungal-based energy channels. This is consistent with suggestions that plants possessing traits associated with rapid growth may select for bacterial-dominated rather than fungal-dominated food webs (Wardle 2002). Evidence for this also emerges from the mull and mor theory (Muller 1884), with mor soils, characterised by deep, acidic organic surface horizons, promoting domination by fungi and fungal-feeding arthropods, relative to more fertile mull soils which have more bacterial-dominated food webs.

The influence of plant species on the balance between bacterial-based and fungal-based energy channels has major implications for the nature of nutrient cycling that occurs. It has been shown both empirically (Wardle & Yeates 1993) and theoretically (de Ruiter et al. 1995) that the relative importance of top–down forces (i.e. predation) and bottom–up forces (i.e. resource regulation and there-fore competition) in regulating different trophic levels differs between the two energy channels (see Fig. 7.1). Specifically, for the bacterial-based energy channel, which operates as an aquatic system (because interactions between organisms occur in the aqueous component of the soil), predation becomes increasingly important as a trophic regulator with decreasing trophic status. This is consist-ent with the predictions of the hypothesis of Menge and Sutherland (1976), developed for aquatic systems. Meanwhile, the fungal-based energy channel, which involves sessile microorganisms, includes several trophic components that are regulated primarily by resource regulation and competition. This is at least partially consistent with terrestrial-based hypotheses of trophic regulation pro-posed by Hairston et al. (1960) and Oksanen et al. (1981). As a result, a much higher proportion of bacterial productivity and biomass is eaten by their con-sumers than is the case for fungi, and the net result is a more rapid cycling of nutrients taken up by bacteria. This leads to 'fast' cycling of nutrients that results when the bacterial energy channel dominates, as opposed to the 'slow' cycling caused by fungal domination (cf. Coleman et al. 1983). Fast nutrient cycles are likely to benefit rapidly growing plant species adapted for fertile sites.

Soil food webs involve animals that interact with microorganisms over a range of scales. Lavelle et al. (1995) classify soil animals into three groups of generally increasing body size: micropredators (those which directly consume microbes; e.g. nematodes, protozoa, some microarthropods); litter transformers (those that condition litter for microbes as faecal pellets; e.g. micro- and mesoarthropods); and ecosystem engineers (those that create physical structures and support microbial populations in their gut cavities; e.g. termites, earthworms). These animal–microbe interactions become less antagonistic and more mutualistic with increasing scales. Larger soil animals have been shown in microcosm experi-ments to exert positive effects on microbial-driven soil processes, which are of a greater magnitude than can be achieved solely by animals with smaller body sizes (e.g. Coûteaux et al. 1991; Vedder et al. 1996). Plant species identity has the potential to exert important effects on the relative contributions of the three body size groups, although this issue has seldom been addressed. This is especially relevant with regard to earthworms, given their sensitivity to litter quality differences between plant species. For example, co-existing tree species with differing litter qualities support vastly differing earthworm populations under them, contributing to spatial heterogeneity of earthworm populations (Boettcher & Kalisz 1991; Saetre 1998). Further, it is recognised that plant species that give rise to mull soils are often dominated by earthworms whereas those that give rise to mor soils are more likely to be dominated by saprophagous microarthropods. Given the well-known and powerful effects that earthworms have as drivers of bacterial activity and turn-over, selection by plant species for earthworms should in turn promote the 'fast' cycles of nutrients characteristic of bacterial-based soil food webs.

How vegetation change may influence decomposer communities

Although the above discussion provides a conceptual basis for predicting that shifts in plant species composition may cause changes in the basic nature of soil food webs and decomposer communities, there have been surprisingly few attempts to determine how these effects may be manifested in real ecosystems. This issue can be investigated most effectively through experimental or observational studies in which vegetation of differing compositions are compared. One such approach involves experimental manipulation of real ecosystems, for example as is achieved by 'removal experiments' (Díaz et al. 2003) in which different subsets of the above-ground biota are excluded or removed, and the response of the rest of the system is monitored. Two recent examples involving studies conducted by my colleagues and myself are now discussed in this context; both utilise removal experiments to predict how above-ground responses to specific global change phenomena may affect decomposer communities.

Perennial grasslands and plant functional group removals

Global climate change is well known to induce shifts in the functional composition of vegetation (Box 1981, 1988) and reduce (e.g. Harte & Shaw 1995) plant functional groups which produce high-quality litter relative to those which produce poor-quality litter. In grazed perennial grasslands in New Zealand, global warming is expected to cause an expansion of annual C4 grasses to more southerly (cooler) latitudes (Campbell et al. 1999), resulting in a possible displacement of perennial C3 forage species. In pastures in New Zealand's Waikato region, summer-annual C4 grasses facilitate winter-annual C3 grasses and vice versa, suggesting that global warming may promote annual C3 and C4 grasses at the expense of perennial C3 grasses. Since C4 grasses are both generally less productive and produce litter of poorer quality than C3 perennials, this may have implications for decomposer biota, especially those that are bottom–up controlled. In this context, a vegetation removal experiment performed in a grazed perennial grassland in New Zealand's Waikato region (Wardle *et al.* 1999) is now discussed in relation to possible effects of vegetation change (such as may occur through global warming) on the decomposer subsystem. In that study, small plots (20 cm diameter) were sprayed with a non-residual herbicide and all vegetation taken off; these plots were then left for recolonisation by plants over a three-year period. Over this period, different subsets of the flora were continually weeded out of different plots as they emerged, resulting in their effective exclusion. Treatments consisted of no removals, removals of C3 grasses only, removals of annual C3 grasses, removals of C4 grasses, removals of forbs and removals of all plants.

Unsurprisingly, the strongest effects of removals on decomposers were observed for the all plant removal treatment, although different decomposer groups differed in response to this treatment (Fig. 7.2). In the soil microfood web, the basal (microbial) and tertiary (predatory nematode) consumer trophic levels were strongly impaired by this treatment at the end of the experiment while the intermediate trophic level (microbe-feeding nematodes) was not. This is consistent with the prediction that the importance of bottom–up forces differs across trophic levels. The removal of all plants sometimes had particularly strong effects on animals with larger body sizes such as earthworms and collembolans. However, so long as plants were present, the effects of species removals (and therefore which plant functional groups were present) on decomposer trophic groups were generally weak, and there were relatively few instances in which these groups were significantly affected by removal of subsets of the flora relative to the non-removal treatment (even though the removal treatments themselves sometimes differed from each other). When effects were observed, this was usually as a result of adverse effects of C3 grass removal, indicating that the dominant perennial species in this group (*Lolium perenne*) had positive effects on the soil biota that could not be achieved by the remainder of the flora. This may result from either the greater productivity or greater litter quality of this species relative to some of the other species.

When community composition within each of five trophic groups (three decomposer groups: microflora, microbe-feeding nematodes, top predatory nematodes; and two soil herbivorous groups: nematodes and arthropods) was considered across treatments, there were strong effects of plant removals on several taxa within each of these groups, and clear linkages between the community composition of plants and soil trophic groups were observed. These associations were detected over both space and time; community composition within decomposer groups tracked that of the plant community with a time-lag of one to three months. This indicates that even when vegetation change does not alter major groups of soil biota, it could still conceivably exert important consequences on the composition of these groups at higher levels of taxonomic resolution. However, these compositional changes did not translate to differences

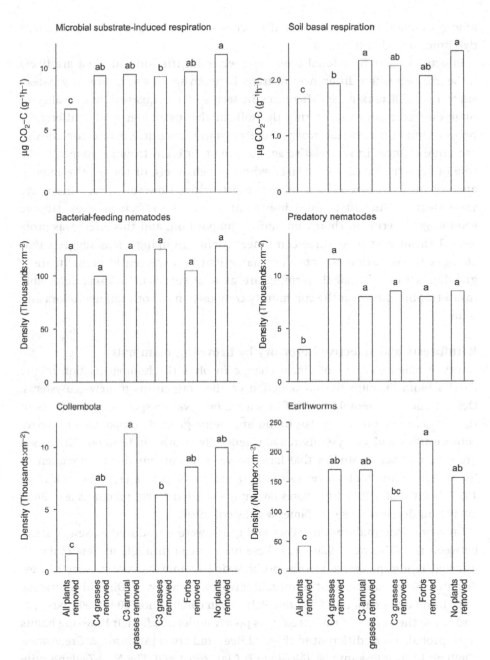

Figure 7.2. Effects of continual removals of different plant functional groups on soil organisms and processes in a three-year 'removal' experiment in a perennial grassland in the Waikato area of New Zealand. All results are for the end of the third year, except for Collembola in which data at the end of the first year are presented. Within each panel, bars topped by the same letter do not differ significantly at $P = 0.05$. Derived from Wardle *et al.* (1999).

among removal treatments for soil processes such as decomposition, nutrient dynamics or soil respiration.

Effects of plant functional group or species identity on soil biota are likely to be greatest when differences in traits between species or groups is greatest, since this will maximise differences in terms of the quantity and quality of plant-derived resources entering the soil. In this particular study, differences between plant functional groups were obviously insufficient to cause consistent large changes in the relative abundance of different trophic groupings, the role of bacterial-based vs. fungal-based energy channels, or the relative importance of different size classes of soil invertebrates. However, these differences were clearly sufficient to exert important bottom–up effects on most trophic groupings in terms of their community composition, and this effect was propagated through at least three consumer trophic groupings. This suggests that changes in vegetation composition that might be expected to occur in these grasslands through global warming are at least sufficient to exert detectable multi-trophic effects on the community composition of both soil microflora and fauna.

Rainforests and selective herbivory by browsing mammals

Another major element of global change involves the homogenisation of the Earth's biota through the introduction of alien organisms to new ecosystems. Despite the considerable body of literature on invasive species, the impacts of these organisms on the above-ground and below-ground components of native communities and ecosystems remain generally poorly understood. There is a growing number of studies that have, however, documented the consequences for specific ecosystem processes of alien species of plants (e.g. Vitousek et al. 1997; Scott et al. 2001), herbivores (McIntosh & Allen 1998), predators (e.g. Boag 2000) and decomposers (e.g. James & Seastedt 1986).

Browsing mammals, notably deer and goats, were introduced to New Zealand between the 1770s and 1920s, and these have spread throughout New Zealand's forested landscape; prior to this, no browsing mammals were present in New Zealand. New Zealand's native megaherbivores, moas (Aves, Dinornithiformes), were hunted to extinction following Polynesian colonisation 800–1000 years ago, and while the impact of their feeding is poorly understood, their browsing habits were probably very different to those of deer and goats (Atkinson & Greenwood 1989) and much less intense (McGlone & Clarkson 1993). The New Zealand rainforest system therefore provides an opportunity, possibly unique, to investigate the effects of the introduction of an entire functional group of alien organisms (forest-dwelling browsing mammals) to an ecosystem from which it was entirely absent. The former New Zealand Forest Service established a large number of fenced exclosure plots (typically 20 × 20 m) throughout New Zealand's rainforests between the 1950s and 1980s to assess the impacts of deer and goats on vegetation. From these exclosures, Wardle *et al.* (2001, 2002) selected 30 exclosure plots with the intention of investigating long-term effects of alien browsers on the decomposer subsystem, across a range of major forest types, range of climatic regimes (subtropical to subalpine) and 12 degrees of latitude.

Plant species that are preferentially eaten by browsing mammals in these forests are those that have large leaves and a high specific leaf area (Wardle *et al.* 2001), and produce foliage with low levels of secondary metabolites and lignin (Forsyth *et al.* 2002), indicating that palatability across species is correlated with fundamental trait axes (Grime *et al.* 1997). Analyses performed for 28 of the 30 exclosure plots (Wardle *et al.* 2002) revealed that those plant species which were most adversely affected by browsing mammals produced litter with lower concentrations of phenolics and lignin than did those promoted by browsing mammals (promotion presumably resulting from unpalatable species gaining a competitive advantage through browser reduction of palatable species). This translated to a significant relationship between the relative response of plant species to browsing in the field and decomposition rates of litter across the range of species sampled in that study (Fig. 7.3). This was due in part to differences among plant functional groups, which showed identical ranking with regard to litter decomposability and vegetation response to browsers (Fig. 7.3). These results extend upon data presented by Grime *et al.* (1996) that suggested that foliage palatability and litter decomposability across plant species are driven by a similar suite of plant traits.

Promotion of poor litter quality plants by browsing mammals could be hypothesised to have adverse effects for decomposer organisms (see Pastor *et al.* 1988). However, assessments of decomposer organisms inside and outside each of the 30 exclosure plots (Wardle *et al.* 2001) yielded a pattern that was only partially consistent with this prediction. With regard to the soil microfood web, there were strong effects of browsing mammals on several trophic groupings (spanning three consumer trophic levels), as well as on the relative importance of the fungal-based vs. bacterial-based energy channels, at several locations. However, these effects were idiosyncratic across sites, with only some showing retardation of decomposers and favouring of the fungal-based energy channel as a result of browsers promoting plant species that produce poor litter quality. However, there was a clear body size effect; all groups of microarthropods (Collembola, and all four orders of the Acari) and soil macrofauna (e.g. Gastropoda, Isopoda, Pseudoscorpionidea, Opiliones, Chilopoda and major families of the Coleoptera) were significantly adversely affected across the entire dataset. However, across the 30 locations the magnitude of this inhibition of soil invertebrates was not significantly correlated with the magnitude of browsing effect on vegetation. Browsing mammal effects on several invertebrate groups across these sites were instead significantly related to a range of other variables including macroclimatic, soil nutrient and tree stand properties. At finer levels of taxonomic

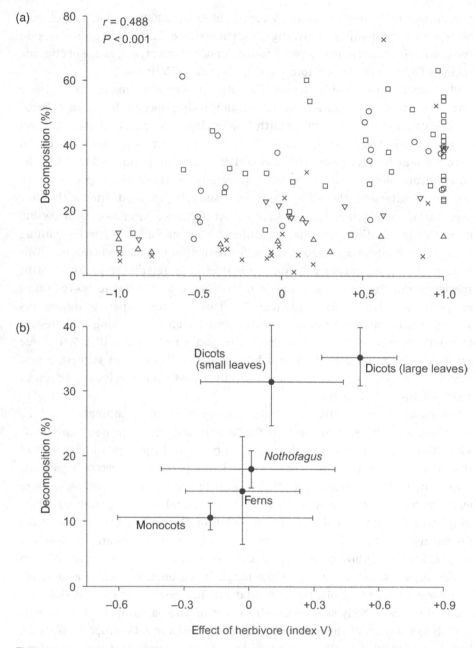

Figure 7.3. Relationships between litter decomposition rates for a range of plant species, and the relative responses of these species to herbivory by browsing mammals, assessed using a series of fenced exclosure plots in natural forests throughout New Zealand. Plant responses to herbivores are measured using the metric 'V' (Wardle 1995, Wardle *et al.* 2001), which ranges from –1 to +1 and becomes increasingly negative and positive as vegetation density in the browse layer is promoted and reduced by herbivores respectively (a value of 0 indicates no effect). The dataset is based on 98 data points

resolution, browsing mammals had important effects at most locations on community composition within the Nematoda, Diplopoda, Gastropoda, Staphylinidae and Coleoptera. Interestingly, across sites the magnitude of these compositional effects did not correlate significantly with the magnitude of browser effects on plant community composition.

This study shows that primary trait axes across plant species may not always be the main factor that drives the decomposer subsystem; although browsers caused predictable effects on vegetation by favouring domination of plant species with particular traits above others (Fig. 7.3), several components of the decomposer food web showed idiosyncratic responses to browsers. This probably results from the fact that herbivores can also affect the decomposer subsystem through other mechanisms operating at both the whole plant and community levels (reviewed by Bardgett *et al.* 1998; Wardle 2002); some of these mechanisms can have positive effects on the decomposer subsystem that could override negative effects of browsers promoting plant species with poor litter quality at some sites. These varied effects have implications for soil processes – Wardle *et al.* (2001) found browsers to have idiosyncratic effects across sites on soil carbon mineralisation and sequestration of carbon and nitrogen in the soil, apparently reflecting their varied effects on the primary decomposers, i.e. the microflora. However, despite the complex nature of results obtained in this study, these results still point to the importance of effects of introductions of browsing mammals to New Zealand's rainforests on most groups of soil biota, and the consequences of this for ecosystem properties.

Above-ground–below-ground biodiversity relationships

Despite the growing interest among researchers in the topic of 'soil biodiversity', there have been few attempts to generate general, theoretically sound principles about the regulation of diversity of soil organisms. For groups of organisms regulated mainly by competition (e.g. plants), it could be expected that highly favourable conditions would reduce diversity relative to intermediate conditions because of the increased role played by competitive exclusion of subordinate

---

Figure 7.3. (*cont.*) (each based on decomposition rates of a given monospecific litter collection, and vegetation density inside vs. outside a fenced exclosure plot for the same plant species and location as that from which the litter was collected), representing 51 plant species and collected from 28 locations. (a) Individual datum points: ×, ferns; △, monocots; □, dicots with leaf lamina width of >6 mm; ○, dicots with leaf lamina width of <6 mm; ▽, *Nothofagus* spp. $R = 0.488$, $P < 0.001$, d.f. $= 96$. (b) Means and 95% confidence intervals for the estimate of the plant group means for data in (a). The five groupings differ significantly among each other at $P < 0.001$ with regard to both decomposition rate and vegetation response to herbivory (one-way ANOVA). From Wardle *et al.* (2002). Reproduced with the permission of the British Ecological Society.

species (Huston 1994). Meanwhile, the diversity of those organisms regulated by top–down control should not necessarily show such a trend (Huston 1994; Wootton 1998). In this light, there is some evidence that most groups of soil organisms are not strongly regulated by competition (Wardle 2002; Bardgett *et al.* this volume). If this is the case, then their diversity need not decline as conditions become more favourable, e.g. through enhanced nutrient availability. Consistent with this prediction, Wardle (2002) found, through the synthesis of the limited body of data available, that the diversity of most groups of soil biota, notably those known not to be regulated by competition, was only neutrally or positively related to measures of soil quality and availability of resources. If this is the case, then plant species that have traits associated with production of high litter quality may be expected to promote diversity within the main groups of the soil biota relative to those which produce poor litter quality. Such a theory remains all but untested, although Schaefer and Schauermann (1990) did find that forests with mull soils supported a higher richness of many soil organism groups than did forests supporting moder soils.

There has been growing interest in the issue of whether plant species richness is an ecological driver in its own right. Evidence for plant diversity promoting the diversity of soil organisms is mixed at best (Hooper *et al.* 2000), although there is evidence that increasing the richness of plant litter mixtures from one to three species can have positive effects on oribatid mite species richness (Hansen & Coleman 1998; Kaneko & Salamanca 1999). This appears to reflect the importance of microhabitat diversity in promoting microarthropod diversity (Anderson 1978). The degree to which plant diversity promotes soil biodiversity depends to a large extent on the level of microhabitat heterogeneity created by plant species richness, an issue which to date remains unexplored. However, it appears unlikely that at higher levels of species richness plant diversity is a major determinant of soil biodiversity, and this is apparent from results presented in both of the examples described in the previous section. First, Wardle *et al.* (1999) found that alteration of plant diversity by removal of subsets of the flora did not cause predictable unidirectional changes in the diversity of microbes, microbe-feeding nematodes, predatory nematodes or soil herbivores, even though the community composition of each of these groups responded to removals. Second, Wardle *et al.* (2001) found that while browsing mammals consistently reduced plant species richness in the forest understorey, taxonomic diversity within the Nematoda, Gastropoda, Diplopoda, Staphylinidae and Coleoptera did not show consistent matching reductions, even though the community composition within each of these groups was altered. These two studies suggest that reductions in plant diversity do not cause predictable changes in the drivers of soil biodiversity, such as microhabitat diversity or favourability of environmental conditions (e.g. soil fertility).

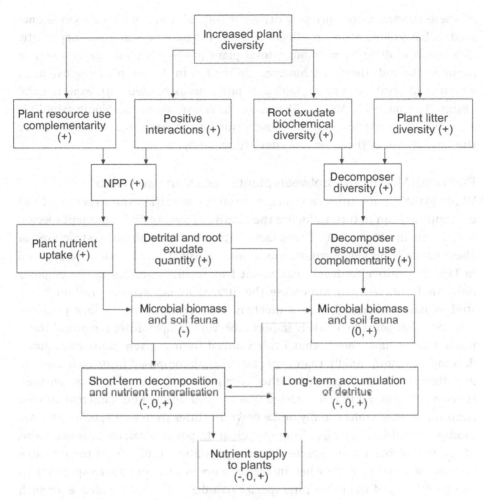

Figure 7.4. Hypothesised mechanisms by which increasing plant species richness may affect decomposer mediated processes. +, 0 and indicate positive, neutral and negative effects, respectively. Reproduced from Wardle and van der Putten (2002) but with correction of errors in the previously published version. Reproduced with permission of Oxford University Press.

There are a number of mechanisms through which plant diversity can theoretically influence abundances or biomasses of soil organisms, the processes that they carry out and, ultimately, the supply of plant-available nutrients from the soil (Fig. 7.4). A growing number of studies have investigated the below-ground consequences of diversity of both live plants and plant litter (reviewed by Schmid et al. 2002; Wardle 2002). The majority of these have detected either neutral or weakly positive effects of plant diversity, and they generally point to the overriding importance of plant species identity (and ultimately the functional traits of these species) as the primary driver of soil biota and processes. Occasional studies have found stronger effects of plant diversity on soil biota, apparently as a result of diversity promoting total plant productivity and hence resource input to the soil; there are, however, difficulties in determining the extent to which such results can be explained by problems associated with experimental design (Huston 1997). Much still remains unknown about the biodiversity linkages that may exist between the above-ground and below-ground subsystems, or the importance of these for ecosystem functioning.

Reciprocal feedbacks between plant species and soil biota

While plant communities can influence soil communities, the structure of soil communities can in turn influence the relative success of different plant species, leading to the likelihood of feedbacks between the community structures of above-ground and below-ground biota. Some possible mechanisms are depicted in Fig. 7.5. Positive feedbacks can result from plants selecting for decomposer biota that preferentially mineralise the litter that they produce. Although few studies have considered such a mechanism, there are fragmentary pieces of evidence that point to its likely importance. For example, litter reciprocal transplant studies have shown that litter sourced from a given plant community decomposes more rapidly than expected when decomposed in its own community than when decomposed in other communities (Hunt et al. 1988). Further, Hansen (1999) found that litter from *Quercus rubra* selected for a microarthropod community that preferentially broke down the litter from that species. Another example, involving mixing of litter of each of the possible pairwise combinations of ten boreal forest plant species, showed that litter from four of these species decomposed more rapidly when in the presence of litter of its own species than when with any of the other nine species (Wardle et al. 2003). Evidence for such feedbacks is also apparent from studies such as those by Setälä and Huhta (1991), which found that seedlings of *Betula pendula* produced leaves with significantly higher concentrations of nutrients when soil fauna were also present; improved quality of litter produced by these seedlings should be expected in turn to promote the soil fauna. If feedbacks such as these are common, then plant species should be able to select for soil biota that are adapted for mineralising nutrients in the patches of soil that they occupy, which could conceivably contribute to maintaining plant species co-existence (Huston & de Angelis 1994; Wardle 2002).

Other positive feedback mechanisms are also possible (Fig. 7.5). Plant species adapted for infertile conditions and which produce poor-quality litter with high levels of phenolics and lignin could conceivably inhibit the mobilisation of nutrients, notably nitrogen. Plants adapted for such habitats are probably highly capable of directly assessing organic nitrogen through effective mycorrhizal

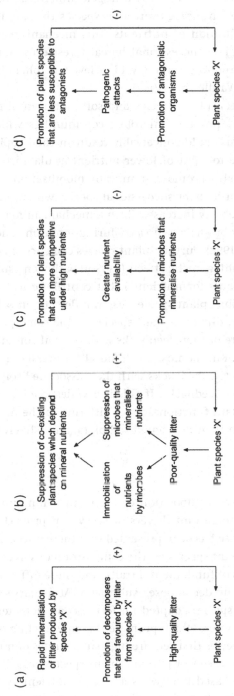

Figure 7.5. Depiction of mechanisms by which a hypothetical plant species 'X' can exhibit either positive or negative feedbacks with the soil community that it promotes. Species X in panel (a) is more competitive than co-existing species in situations of high levels of available mineral nutrients, while species 'X' in panels (b) and (c) are less competitive than co-existing species under high mineral nutrient levels. Panel (d) involves situations in which species 'X' promotes its own antagonists in the soil.

associations and therefore by-passing the nutrient mobilisation pathways driven by saprophytes. If such plants can suppress nitrogen mineralisation then they should derive a competitive advantage over other species that are more dependent upon microbial mobilisation of nutrients. This mechanism may help to explain the domination in late successional boreal forest ecosystems of ericaceous dwarf shrubs that produce extremely high levels of phenolics, such as *Empetrum hermaphroditum* (Wardle *et al.* 1997).

Negative feedbacks between plant species and soil biota can also occur. For example, plant species may promote microbial communities which compete with them for nutrients; this could conceivably result in their replacement by plant species that are more tolerant of lower nutrient availability. Some early successional plant species also promote significant mobilisation of nutrients, leading to their replacement by later successional species which compete more effectively as nutrient availability increases. Such a mechanism may lead to the replacement of *Erica tetralix* by *Molinia caerulea* during fore dune successions in The Netherlands (Berendse 1998). Further, plant species can encourage the build-up of organisms that exert antagonistic effects (e.g. soil pathogens; microbes that effectively compete with plants for nutrients) that contribute to their replacement by other, less susceptible, plant species (e.g. van der Putten & Peters 1997; Packer & Clay 2000). Finally, different plant species in the same plant community may differ in the nature of their feedbacks with the soil communities that they promote. For example, Klironomos (2002) found for grassland species that invasive species showed positive feedbacks with their associated soil biota while rare species showed negative feedbacks. If this is a widespread and consistent pattern, then it points to plant functional types differing in the nature of feedbacks that they exhibit with their associated soil biota, and a likely role for key plant traits in driving this.

Conclusions

Plant species identity and the composition of the plant community are being increasingly recognised as important drivers of ecosystem properties and processes. In this chapter, the case has been presented that these effects may operate through the influences of plant species on the composition of soil communities. Plant species differ in suites of fundamental traits, and these differences in turn have important effects on the decomposer subsystem. At a coarse scale of resolution, fast-growing plant species adapted for fertile conditions tend to select for bacterial-based soil food webs, which promote the rapid cycling of nutrients; this in turn favours plant species that require high supply rates of mineral nutrients from the soil. Conversely, slower growing plant species suited for infertile conditions select for fungal-based food webs and slow nutrient cycling, which favours plant species that can access relatively immobile forms of nutrients, for example through effective mycorrhizal networks. At finer scales of resolution, different plant species can promote differences in community structure within major groups of soil biota, even when the plant species are quite similar to each other. The effects of plant species on soil communities can create important feedbacks that can either favour the persistence of resident species, or encourage replacement of these species by other species, and these feedbacks are likely to serve as important determinants of both plant and soil community structure.

Finally, understanding the nature of above-ground–below-ground feedbacks may offer real opportunities for better understanding global change phenomena. There has been an increasing recognition of the value of plant trait-based approaches for better understanding global change effects in ecosystems (Díaz & Cadibo 1997; Grime 2001). In this chapter, two examples are presented of how global change phenomena may alter both the plant and soil communities through favouring plants with certain traits above others. Ultimately, an improved understanding of how global change affects ecosystems requires a combined above-ground–below-ground approach, because both the producer and decomposer subsystems work in tandem to drive ecosystem properties and processes.

References

Anderson, J. M. (1978). Inter- and intra-habitat relationships between woodland Cryptostigmata species diversity and the diversity of soil and litter microhabitats. *Oecologia*, **32**, 341–348.

Atkinson, I. A. E. & Greenwood, R. M. (1989). Relationships between moas and plants. *New Zealand Journal of Ecology*, **12** (suppl), 67–96.

Bardgett, R. D., Wardle, D. A. & Yeates, G. W. (1998). Linking above-ground and below-ground interactions: how plant responses to foliar herbivory influence soil organisms. *Soil Biology and Biochemistry*, **30**, 1867–1878.

Berendse, F. (1998). Effects of dominant plant species on soils during succession in nutrient poor ecosystems. *Biogeochemistry*, **42**, 73–88.

Boag, B. (2000). The impact of the New Zealand flatworm on earthworms and moles in agricultural land in western Scotland. *Aspects of Applied Biology*, **62**, 79–84.

Boettscher, S. E. & Kalisz, P. J. (1991). Single-tree influence on earthworms in forest soils in eastern Kentucky. *Soil Science Society of America Journal*, **55**, 862–865.

Box, E. O. (1981). *Macroclimate and Plant Form*. The Hague: Junk.

Box, E. O. (1996). Plant functional types and climate at the global scale. *Journal of Vegetation Science*, **7**, 309–320.

Campbell, B. D., Mitchell, N. D. & Field, T. R. O. (1999). Climate profiles of temperate C3 and subtropical C4 species in New Zealand Pastures. *New Zealand Journal of Agricultural Research*, **42**, 223–233.

Coleman, D. C., Reid, C. P. P. & Cole, C. V. (1983). Biological strategies of nutrient cycling in soil systems. *Advances in Ecological Research*, **13**, 1–55.

Coûteaux, M.-M., Mousseau, M., Celerier, M. L. & Bottner, P. (1991). Increased atmospheric CO_2 and litter quality: decomposition of sweet chestnut leaf litter with animal food webs of different complexities. *Oikos*, **61**, 53–64.

De Ruiter, P. C., Neutel, A. M. & Moore, J. C. (1995). Energetics, patterns of interaction strength and stability in real ecosystems. *Science*, **269**, 1257–1260.

Díaz, S. & Cadibo, M. (1997). Plant functional types and ecosystem function in relation to global change. *Journal of Vegetation Science*, **8**, 463–474.

Díaz, S., Chapin, F. S. III, Symstad, A., Wardle, D. A. & Huennecke, L. (2003). Functional diversity revealed through removal experiments. *Trends in Ecology and Evolution*, **18**, 140–146.

Ettema, C. H., Lowrance, R. & Coleman, D. C. (1999). Riparian soil response to surface nitrogen input: temporal changes in denitrification, labile and microbial C and N pools, and bacterial and fungal respiration. *Soil Biology and Biochemistry*, **31**, 1609–1624.

Forsyth, D. M., Coomes, D. A., Nugent, G. & Hall, G. M. J. (2002). The diet and diet preferences of introduced ungulates (order: Artiodactyla) in New Zealand. *New Zealand Journal of Zoology*, **29**, 323–343.

Griffiths, B. S., Welschen, R., van Arendonk, J. J. C. M. & Lambers, H. (1992). The effect of nitrate-nitrogen on bacteria and bacterial-feeding fauna in the rhizosphere of different grass species. *Oecologia*, **91**, 253–259.

Grime, J. P. (2001). *Plant Strategies, Vegetation Processes and Ecosystem Properties*. Chichester: Wiley.

Grime, J. P., Cornelissen, J. H. C., Thompson, K. & Hodgson, J. G. (1996). Evidence of a causal connection between anti-herbivore defence and the decomposition rate of leaves. *Oikos*, **77**, 489–494.

Grime, J. P., Thompson, K., Hunt, R., et al. (1997). Integrated screening validates primary axes of specialization in plants. *Oikos*, **79**, 259–281.

Hairston, N. G., Smith, F. E. & Slobodlin, L. B. (1960). Community structure, population control and competition. *American Naturalist*, **94**, 421–425.

Handley, W. R. C. (1954). *Mull and Mor in Relation to Forest Soils*. London: Her Majesty's Stationery Office.

Hansen, R. A. (1999). Red oak litter promotes a microarthropod functional group that accelerates its decomposition. *Plant and Soil*, **209**, 37–45.

Hansen, R. A. & Coleman, D. C. (1998). Litter complexity and composition are determinants of the diversity and species composition of oribatid mites (Acari: Oribatida) in litterbags. *Applied Soil Ecology*, **9**, 17–23.

Harte, J. & Shaw, R. (1995). Shifting dominance within a montane vegetation community: results of a climate-warming experiment. *Science*, **267**, 876–880.

Hendrix, P. F., Parmelee, R. W., Crossley, D. A., et al. (1986). Detritus food webs in conventional and no-tillage agroecosystems. *BioScience*, **36**, 374–380.

Hobbie, S. E. (1992). Effects of plant species on nutrient cycling. *Trends in Ecology and Evolution*, **7**, 336–339.

Hooper, D. U., Bignell, D. E., Brown, V. K., et al. (2000). Interactions between above- and belowground biodiversity in terrestrial ecosystems: patterns, mechanisms and feedbacks. *BioScience*, **50**, 1049–1061.

Hunt, H. W., Ingham, E. R., Coleman, D. C., Elliott, E. T. & Reid, C. P. P. (1988). Nitrogen limitation of production and decomposition in prairie, mountain meadow and pine forest. *Ecology*, **69**, 1009–1016.

Huston, M. A. (1994). *Biological Diversity. The Coexistence of Species on Changing Landscapes*. Cambridge: Cambridge University Press.

Huston, M. A. (1997). Hidden treatments in ecological experiments: re-evaluating the ecosystem function of biodiversity. *Oecologia*, **110**, 449–460.

Huston, M. A. & de Angelis, D. L. (1994). Competition and coexistence: the effects of resource transport and supply rates. *American Naturalist*, **144**, 854–877.

Ingham, E. R., Coleman, D. C. & Moore, J. C. (1989). An analysis of food web structure and function in a shortgrass prairie, a mountain meadow and a lodgepole pine forest. *Biology and Fertility of Soils*, **8**, 29–37.

James, S. W. & Seastedt, T. R. (1986). Nitrogen mineralization by native and introduced earthworms: effects on big bluestem growth. *Ecology*, **67**, 1094–1097.

Kaneko, N. & Salamanca, E. (1999). Mixed leaf litter effects on decomposition rates and soil arthropod communities in an oak–pine forest stand in Japan. *Ecological Research*, **14**, 131–138.

Klironomos, J. N. (2002). Feedback with soil biota contributes to plant rarity and invasiveness in communities. *Nature*, **417**, 67–70.

Lavelle, P., Lattaud, C., Trigo, D. & Barois, I. (1995). Mutualism and biodiversity in soils. *Plant and Soil*, **170**, 23–33.

Lawton, J. H. (1994). What do species do in ecosystems? *Oikos*, **71**, 367–374.

McGlone, M. & Clarkson, B. D. (1993). Ghost stories: moa, plant defenses and evolution in New Zealand. *Tuatara*, **32**, 1–18.

McIntosh, P. D. & Allen, R. B. (1998). Effect of exclosures on soils, biomass, plant nutrients and vegetation, on unfertilized steeplands, upper Waitaki district, South Island, New Zealand. *New Zealand Journal of Ecology*, **22**, 209–217.

Menge, B. A. & Sutherland, J. P. (1976). Species diversity gradients: synthesis of the roles of predation, competition and spatial heterogeneity. *American Naturalist*, **110**, 351–369.

Moore, J. C. & Hunt, H. W. (1988). Resource compartmentation and the stability of real ecosystems. *Nature*, **333**, 261–263.

Muller, P. E. (1884). Studier over skovjord, som bidrag til skovdyrkningens theori: II. Om muld og mor i egeskove og paa heder. *Tidsskrift for Skovbrug*, **7**, 1–232.

Ohtonen, R., Fritze, H., Pennanen, T., Jumpponen, A. & Trappe, J. (1999). Ecosystem properties and microbial community changes in primary succession on a glacier forefront. *Oecologia*, **119**, 239–246.

Oksanen, L., Fretwell, S., Arruda, J. & Niemelä, P. (1981). Exploitation ecosystems in gradients of primary productivity. *American Naturalist*, **118**, 240–261.

Packer, A. & Clay, K. (2000). Soil pathogens and spatial patterns of seedling mortality in a temperate tree. *Nature*, **404**, 278–281.

Parmelee, R. W., Beare, M. H. & Blair, J. M. (1989). Decomposition and nitrogen dynamics of surface weed residues in no-tillage agroecosystems under drought conditions: influence of resource quality on the decomposer community. *Soil Biology and Biochemistry*, **21**, 97–103.

Pastor, J., Naiman, R. J., Dewey, B. & McInnes, P. (1988). Moose, microbes and the boreal forest. *BioScience*, **38**, 770–777.

Saetre, P. (1998). Decomposition, microbial community structure and earthworm effects along a birch–spruce soil gradient. *Ecology*, **79**, 834–846.

Schaefer, M. & Schauermann, J. (1990). The fauna of beech forests: comparison between a mull and moder soil. *Pedobiologia*, **34**, 299–304.

Schmid, B., Joshi, J. & Schläpfer, F. (2002). Empirical evidence for biodiversity: ecosystem functioning relationships. *The Functional Consequences of Biodiversity: Empirical Progress and Theoretical Extensions* (Ed. by A. P. Kinzig, S. W. Pacala & D. Tilman), pp. 120–150. Princeton: Princeton University Press.

Scott, N. A., Saggar, S. & McIntosh, P. (2001). Biogeochemical impact of *Hieracium* invasion in New Zealand's grazed tussock grasslands: sustainability implications. *Ecological Applications*, **11**, 1311–1322.

Setälä, H. & Huhta, V. (1991). Soil fauna increase *Betula pendula* growth: laboratory experiments with coniferous forest floor. *Ecology*, **72**, 665–671.

Sohlenius, B. (1996). Structure and composition of the nematode fauna in pine forests under the influence of clear-cutting: effects of slash removal and field layer vegetation. *European Journal of Soil Biology*, **32**, 1–14.

Swift, M. J., Heal, O. W. & Anderson, J. M. (1979). *Decomposition in Terrestrial Ecosystems*. Oxford: Blackwell.

Van der Putten, W. H. & Peters, B. A. M. (1997). How soil-borne pathogens may affect plant competition. *Ecology*, **78**, 1785–1795.

Vedder, B., Kampichler, C., Bachmann, G., Bruckner, A. & Kandeler, E. (1996). Impact of faunal complexity on microbial biomass and N turnover in field mesocosms from a spruce forest soil. *Biology and Fertility of Soils*, **22**, 22–30.

Vitousek, P. M., Walker, L. R., Whiteaker, D., Mueller-Dombois, D. & Matson, P. A. (1987). Biological invasion by *Myrica faya* alters ecosystem development in Hawaii. *Science*, **238**, 802–804.

Wardle, D. A. (1995). Impact of disturbance on detritus food-webs in agro-ecosystems of contrasting tillage and weed management practices. *Advances in Ecological Research*, **26**, 105–185.

Wardle, D. A. (2002). *Communities and Ecosystems: Linking the Aboveground and Belowground Components*. Princeton: Princeton University Press.

Wardle, D. A., Barker, G. M., Yeates, G. W., Bonner, K. I. & Ghani, A. (2001). Introduced browsing mammals in natural New Zealand forests: aboveground and belowground consequences. *Ecological Monographs*, **71**, 587–614.

Wardle, D. A., Bonner, K. I. & Barker, G. M. (2002). Linkages between plant litter decomposition, litter quality and vegetation responses to herbivores. *Functional Ecology*, **16**, 585–595.

Wardle, D. A., Bonner, K. I., Barker, G. M., *et al.* (1999). Plant removals in perennial grassland: vegetation dynamics, decomposers, soil biodiversity and ecosystem properties. *Ecological Monographs*, **69**, 535–568.

Wardle, D. A., Nilsson, M. C., Zackrisson, O. & Callet, C. (2003). Determinants of litter mixing effects in a Swedish boreal forest. *Soil Biology and Biochemistry*, **35**, 827–835.

Wardle, D. A. & van der Putten, W. (2002). Biodiversity, ecosystem functioning and aboveground-belowground linkages. *Biodiversity and Ecosystem Functioning* (Ed. by M. Loreau, S. Naeem and P. Inchausti), pp. 155–168. Oxford: Oxford University Press.

Wardle, D. A. & Yeates, G. W. (1993). The dual importance of competition and predation as regulatory forces in terrestrial ecosystems: evidence from decomposer food-webs. *Oecologia*, **93**, 303–306.

Wardle, D. A., Zackrisson, O., Hörnberg, G. & Gallet, C. (1997). Influence of island area on ecosystem properties. *Science*, **277**, 1296–1299.

Wootton, J. T. (1998). Effects of disturbance on species diversity: a multitrophic perspective. *American Naturalist*, **152**, 803–825.

CHAPTER EIGHT

The balance between productivity and food web structure in soil ecosystems

PETER C. DE RUITER
Utrecht University
ANJE-MARGRIET NEUTEL
Utrecht University
JOHN MOORE
University of Northern Colorado

SUMMARY

1. In soil ecosystems, productivity and the structure of community food webs are inextricably interrelated. Central in this interrelationship is that food web stability requires a balance between productivity and food web structure.

2. Model food chains show that productivity must ensure that sufficient energy is available for the subsequent trophic levels, while food chain length should prevent large destabilising population dynamical oscillations. Such large oscillations occur when too high levels of productivity lead to non-pyramidal or inverse pyramidal biomass structures.

3. In real, complex soil food webs, a balance between productivity and food web structure is indicated by the emergence of trophic pyramids in population sizes as well as in feeding rates that translate into stabilising patterns in interaction strengths.

4. Key components in food web structure, i.e. the lengths and weights of trophic interaction loops, provide an ecological as well as mathematical explanation of the observed patterns.

5. Trophic loop analyses also show how the interplay between productivity and food web stability governs and selects pathways in ecosystem development in natural primary succession gradients.

Introduction

Soil harbours a large part of the world's biodiversity organised in highly complex community food webs (Wolters 1997; Griffiths *et al.* 2000). The primary energy sources of soil food webs are formed by various kinds of soil organic matter and root-derived materials. The first trophic level consists of microorganisms,

Biological Diversity and Function in Soils, eds. Richard D. Bardgett, Michael B. Usher and David W. Hopkins. Published by Cambridge University Press. © British Ecological Society 2005.

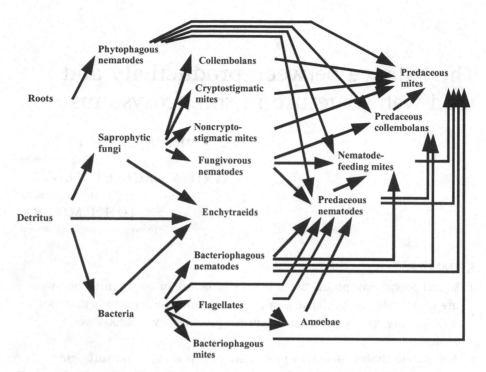

Figure 8.1. Diagram of the below-ground food web from a farming system of the Lovinkhoeve experimental farm de Ruiter *et al.* 1993a). Species are aggregated into functional groups, i.e. based on food choice and life-history parameters (Moore *et al.* 1988). Detritus refers to all dead organic material. Material flows to the detritus pool through death rates and excretion of waste products are not represented in the diagram, but are taken into account in the material flow calculations and stability analyses.

such as bacteria and fungi, grazing on the soil organic matter, and herbivorous nematodes feeding on roots. The microbes are by far the most dominant group of soil organisms, in terms of numbers as well as biomass (Andrén *et al.* 1990; Bloem *et al.* 1994). At the higher trophic levels, soil food webs generally contain a large variety of fauna, like protozoa (amoebae, flagellates, ciliates), nematodes (bacterivores, fungivores, omnivores, herbivores and predators), microarthropods such as mites (bacterivores, fungivores, predators) and collembolans (fungivores and predators), enchytraeids and earthworms (Fig. 8.1).

In the above-ground compartment of ecosystems, biodiversity and ecosystem processes are strongly influenced by intraspecific and interspecific competition and habitat exploitation (Naeem *et al.* 1994; Tilman *et al.* 1996). In soils, however, the relationship between biodiversity and soil processes is thought to be primarily controlled by dynamics and interactions in the soil community food web (Moore & Hunt 1988; de Ruiter *et al.* 1998). Moreover, because of the large amounts of materials that are decomposed and processed in the soil compartment of ecosystems, soil food webs are thought to govern major components in the global cycling of materials, energy and nutrients (Wolters 1997; Griffiths *et al.* 2000).

A key process in ecosystems is productivity. Productivity can be defined as the rate that solar energy is used to produce plant biomass (primary productivity), which in turn provides energy for subsequent trophic levels such as herbivores and predators. In the case of soil ecosystems, productivity can be defined as the rate with which primary energy sources become available to the food web and are consumed and processed by the organisms constituting the soil food web. The relation between productivity and the structure of community food webs has long been a central theme in ecology. It has been hypothesised that community structure depends on ecosystem productivity, in that as productivity increases, communities become more diverse and complex, with longer food chains and a more web-like structure; along with this, communities become more dependent on internal cycling, less dependent on the environment and increasingly stable (Elton 1927; Odum 1971).

The mechanisms that link such trends in ecosystem development have been subject to scientific debate. For example, with respect to *food chain length* it has been argued that higher productivity should give longer food chains, since more energy becomes available to transfer to subsequent trophic levels (Lindeman 1942; Rosenzweig 1971). In contrast to this 'energetic' point of view, the so-called 'dynamic–stability' approach sees food chain length as the result of stability constraints, which depends on the population dynamics of the interacting species (Pimm 1982; Yodzis 1989). However, recent model studies indicate that the effects of productivity and stability constraints are intertwined in that productivity influences the strengths of stability constraints to food web chain length (Moore *et al.* 1993; Moore & de Ruiter 2000). With respect to *food web structure*, effects of productivity and energy flow have been analysed by comparing food web descriptions in terms of: (i) connectedness webs, (ii) energy flow webs and (iii) interaction webs (Paine 1980). *Connectedness webs* describe the topological architecture of food webs in terms of the number of species or trophic groups and the frequency of trophic interactions among the groups. *Energy flow webs* provide quantitative descriptions of food webs expressing the sizes of the energy pools, i.e. population sizes and energy flows, and hence the feeding rates among the trophic groups. *Interaction webs* emphasise the strengths of the interactions among the trophic groups. Interaction strengths may refer to per capita effects of the groups upon one another (May 1972; Pimm & Lawton 1977), or to the outcome of species manipulation experiments (Paine 1980, 1992). Comparisons between energy flow webs and interaction webs have not revealed any clear pattern, which has led to the conclusion that energy flow is not a good indicator of interaction strength. As interaction strengths are considered to be crucial for community stability, it has also been concluded that patterns in energy flow are not good indicators of food web stability (Paine 1980, 1992). Recent studies, however, have emphasised that patterns in the strengths of interactions in real food webs play a crucial role in community stability, while the values of these interaction strengths are directly derived from energetic properties, such as the population sizes, feeding rates, conversion efficiencies and turnover rates of the species (McCann *et al.* 1988; de Ruiter *et al.* 1995). These findings indicate that analysis of the energetics and interaction strengths in food webs may bridge the gap between ecosystem productivity and the structure and stability of food webs.

The present chapter focuses on how productivity relates to food web structure and stability. Central in our approach is that we look at food web structure and ecosystem productivity from the viewpoint that a balance between the two is necessary to ensure community stability. This principle is illustrated by combining results from model food chains and observations on real complex food webs, regarding energy flow rates and interaction strength patterns. Key components in food web structure, i.e. the lengths and weights of trophic interaction loops, will provide an ecological and a mathematical explanation of the observed patterns. Finally, trophic loop analysis will be used to reveal how the interplay between productivity and food web stability governs and selects pathways in ecosystem development in natural primary succession gradients.

Productivity and stability of food chains

The principle of how community stability may depend on the balance between productivity and community structure can be seen from model studies on the determinants of food chain length (Moore *et al.* 1993; Moore & de Ruiter 2000). Consider a food chain of length 3 in which the dynamics of the trophic groups are modelled by Lotka–Volterra differential equations (Fig. 8.2). Productivity levels can be modelled by choosing values for the parameter referring to the specific growth rate (b_i) for the basal species (Fig. 8.2 legend). The results of the food chain model show that the feasibility of the three-level food chain increases with productivity. This indicates the productivity effect on food chain length, in that there should be enough energy for all three trophic levels (Fig. 8.2a). This productivity effect on food chain length is thought to be an especially important determinant in extremely unproductive systems like soils inside caves (Moore & de Ruiter 2000) or in soils in the early vegetation succession stages (Neutel 2001; see Fig 8.6). On the other hand, when productivity is increased to high levels, the dynamics of the populations in the food chain show large oscillations that destabilise the food chain (Moore & de Ruiter 2000). This leads to a 'hump-shaped' curve describing the relationship between productivity and food chain stability, indicating that high levels of productivity create strong stability constraints to food chain length. The model results also show productivity effects on population size distributions (Fig. 8.2b). At low to intermediate productivity levels, population sizes are organised in the form of trophic pyramids with

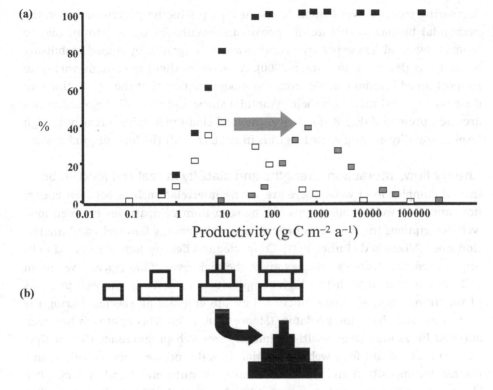

Figure 8.2. The influence of productivity on the length and stability of food chains (Moore *et al.* 1993; Moore & de Ruiter 2000). The food chains are modelled using Lotka–Volterra equations:

$$\dot{X}_i = X_i \left[b_i + \sum_{j=1}^{n} c_{ij} X_j \right].$$

Productivity is defined as $b_i X_i^*$, where X_i^* denotes the equilibrium biomass of the basal primary producers. Productivity levels are modelled by choosing values for b_i for the basal species. (a) Parameters are chosen to be energetically feasible and near values known from literature; the values of productivity are expressed in g m^{-2} a^{-1}. Solid symbols, influence of productivity on food chain feasibility; and open symbols, food chain stability. Feasibility requires that in equilibrium all $X_i^* > 0$ (Moore *et al.* 1993). The decreasing part of the hump-shaped curve of stability is caused by the emergence of large destabilising oscillations (Moore & de Ruiter 2000). (b) The effect of productivity on the organisation of population sizes vs. trophic level. At relatively low productivity, population sizes are organised in the form of trophic pyramids. At high levels of productivity, the pyramidal structure disappears and changes into inverse pyramidal structures. The shaded arrows indicated the effect of the introduction of an extra trophic level on population size distribution and stability (lightly shaded symbols).

decreasing biomass over trophic levels. At high productivity levels, however, the pyramidal biomass structure disappears and eventually turns into an inverse biomass pyramid. These inverse pyramids are accompanied by large destabilising oscillations (Moore & de Ruiter 2000). As soon as there is enough energy to support an additional trophic level, the model predicts that the population size distributions will maintain their pyramidal shape and destabilising oscillations are then prevented (Fig. 8.2b). Hence, the food chain maintains a relatively high level of stability as long as its length is in balance with the level of productivity.

Energy flow, interaction strengths and stability in real soil food webs

In real complex food webs, there are strong interrelationships between energy flow and food web stability. This can be seen from comparisons between food web descriptions in terms of connectedness webs, energy flow webs and interaction webs (Moore & de Ruiter 1991). Connectedness descriptions of soil food webs might become extremely complex given the high level of biological diversity in soils. A way to deal with this high complexity is by defining the web in terms of functional groups, where functional groups embrace all species sharing the same prey and the same predators (Moore *et al.* 1988). This approach has been adopted by various large multi-disciplinary research programmes that analyse the structure of soil food webs in relation to soil processes such as soil organic matter decomposition and the mineralisation of nutrients (Hendrix *et al.* 1986; Hunt *et al.* 1987; Brussaard *et al.* 1988, 1990; Moore *et al.* 1988; Andrén *et al.* 1990; de Ruiter *et al.* 1993b). An example of a *connectedness web* constructed this way shows that complexity in soil food webs can be simplified into a diagram of 18 functional groups (Fig. 8.1). It must be noted that along with simplifying food web structure, the functional group approach may also introduce uncertainties. For many organisms, diets and preferences are not fully known, and particular functional groups, such as collembolans and enchytraeids, might embrace species with different diets and preferences. In some cases, this has led to different grouping of species. For example, some published food webs have treated all protozoa as one group of bacterivorous organisms, while other studies have distinguished flagellates feeding on bacteria from amoebae feeding on bacteria and flagellates (Moore & de Ruiter 1991).

By means of food web modelling, and using available information regarding population sizes, turn-over rates and energy conversion parameters, connectedness webs can be converted into *energy flow webs* (O'Neill 1969; Hunt *et al.* 1987; de Ruiter *et al.* 1993b). Energy flow webs express food web structure in quantitative measures, i.e. population sizes (biomass) and feeding rates. For a series of real soil food webs analysed this way (de Ruiter *et al.* 1998), population sizes as well as feeding rates appear to be organised in the form of trophic pyramids (Fig. 8.3a). Subsequently, from the energy flow webs, it is possible to derive *interaction webs*, emphasising the strengths of the interactions among the functional

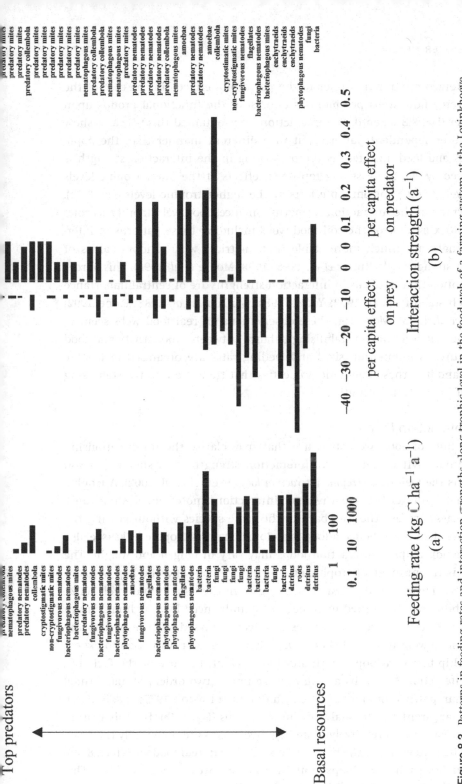

Figure 8.3. Patterns in feeding rates and interaction strengths along trophic level in the food web of a farming system at the Lovinkhoeve experimental farm (The Netherlands), serving as a representative example for seven food webs (de Ruiter et al. 1995, 1998; Moore et al. 1996). (a) Feeding rates (kg C ha^{-1} a^{-1}) derived from population sizes (kg C ha^{-1}), turn-over rates (a^{-1}), energy conversion efficiencies and diets (Hunt et al. 1987; de Ruiter et al. 1993b). (b) Interaction strengths (a^{-1}) derived from feeding rates, population sizes and energy conversion efficiencies. Negative effects are feeding rates divided by the population size of the consumer; positive effects are the production rates (i.e. feeding rates times energy conversion efficiency) divided by the population size of the resource (de Ruiter et al. 1995). Note that the negative effects are much (two orders of magnitude) larger than the positive effects.

groups. Interaction strengths refer to the per capita effects (in the case of the present energy flow webs per biomass effects) of the functional groups upon one another (Fig. 8.3 legend). The interaction webs obtained this way also show a trophic-level dependent pattern, but in a different manner than the population sizes and feeding rates, as the patterning in the interaction strength is characterised by relatively strong top–down effects at the lower trophic levels and relatively strong bottom–up effects at the higher trophic levels (Fig. 8.3b). This patterning in the interaction strengths enhances food web stability, as community matrix models of the soil food webs including these patterns of interaction strength are much more stable than matrices with random values of interaction strengths (de Ruiter *et al.* 1995, 1998; Moore *et al.* 1996). This agrees with the notion that patterns in interaction strengths are of central importance to food web stability (May 1972; Yodzis 1981; McCann *et al.* 1988; Paine 1992; de Ruiter *et al.* 1995). Moreover, these observations on real food webs seem to confirm the principle of the stabilising balance between productivity and food web structure, as population sizes and feeding rates are organised in trophic pyramids, and it is these energetic properties that translate into the stabilising patterns of interaction strengths.

Trophic interaction loops

A key attribute of food web structure that may clarify the interrelationship between productivity, energy flow, interaction strengths and stability in soil food webs is the weight of *trophic interaction loops* (Neutel *et al.* 2002). A trophic interaction loop describes a pathway of interactions (note: not feeding rates) from a species through the web back to the same species without visiting the species more than once; hence a loop is a closed chain of trophic links (see also Fig. 8.5 legend). Trophic interaction loops may vary in length and weight. The *loop length* is the number of trophic groups in the loop, and the *loop weight* is the geometric mean of the interaction strengths in the loop (Neutel *et al.* 2002). System stability can be argued to be negatively influenced by a high maximum loop weight. A series of real food webs has been analysed with respect to the maximum loop weight (Neutel *et al.* 2002). There is an a-priori reason to expect a relationship between loop length and loop weight, because of the fact that negative interaction strengths are on average much (two orders of magnitude) stronger than positive interaction strength (Pimm & Lawton 1977; Yodzis 1981); this is also apparent in the analysed soil food webs (Fig. 8.3b). For this reason, long loops that contain relatively more negative effects are potentially the heaviest ones. The patterning of the interaction strengths in real food webs (Fig. 8.3b), however, makes such long loops contain relatively weak negative links. This means that the maximum of all loop weights will be relatively low and that the level of food web stability will be relatively high compared to food webs with random interaction strengths (Fig. 8.4).

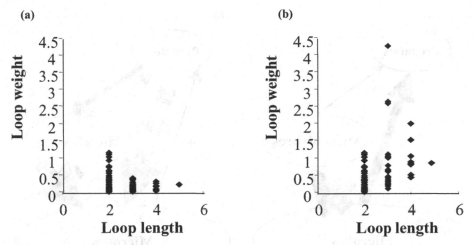

Figure 8.4. Loop length and loop weight in the food web of the shortgrass prairie system in Colorado, serving as a representative example of seven food webs (Neutel *et al.* 2002). (a) Interaction strengths derived from observations (see Fig. 8.3 legend). (b) Interaction strengths randomised through repeatedly exchanging the pairs of interaction strengths, keeping sign structure and predator–prey pairs intact (Yodzis 1981). Long loops with a relatively small weight, those with many positive effects, are not given in the figure, not being relevant for maximum loop weight.

The mechanism behind the aggregation of weak links in long loops is relevant for understanding the effects of productivity on food web structure and stability. Consider a trophic interaction loop of length 3 consisting of an omnivore feeding on two groups on different trophic levels, i.e. on a microbivorous population and a microbial population (Fig. 8.5). Omnivorous loops occur frequently in soil food webs. This omnivorous food chain forms the basis of two interaction loops of length 3 (Fig. 8.5 legend). When population sizes are organised in the form of a trophic biomass pyramid, and when the omnivore is assumed to feed on the two prey types in accordance with their relative availability, the omnivore will feed mainly on the microbes. The negative effect of the omnivore on the microbivores will then be weaker than the effect on the microbes, as the interaction strengths are defined as the feeding rate per biomass of the omnivore (Fig. 8.3 legend). The negative effect between the microbivores and the microbes is defined as the (very large) feeding rate of microbivores on microbes divided by the (relatively large) population size of the microbivores and will therefore be about the same strength as the effect of the omnivore on the microbes. The relatively weak negative effect between omnivores and microbivores will always be part of the loop with the two negative effects, limiting the weight of the loop, i.e. potentially the heaviest one because of the majority of negative links. This mechanism also applies to loops longer than three, especially when the intermediate predators

(a)   **(b)**

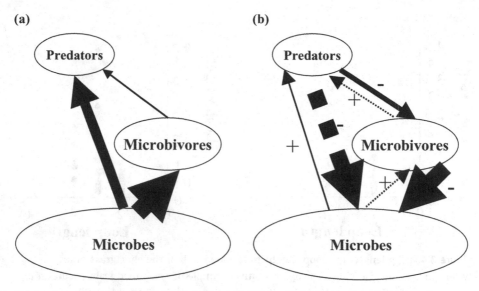

Figure 8.5. Patterns in interaction strengths resulting from biomass pyramids in an omnivorous trophic loop (Neutel *et al.* 2002). (a) Diagram of the relative size of populations and feeding rates. (b) Interaction strengths are partial derivatives of species' growth equations at equilibrium derived from population sizes and feeding rates (Fig. 8.3 legend). The loop of three feeding rates forms the basis of two trophic interaction loops of length 3, i.e. the loop *omnivores–microbivores–microbes* (solid arrows), including two negative and one positive effect, and the loop *omnivores–microbes–microbivores* (dotted arrows), including one negative and two positive effects. The biomass pyramid causes a small negative effect between the omnivores and the microbivores. This means that the long loop with two negative effects contains a relatively small effect, which keeps the weight of this loop low (Neutel *et al.* 2002). This mechanism also applies to loops with a length of longer than 3, especially when the intermediate predators are omnivores as well. The positive effects are much weaker than the negative effects and therefore play only a minor role in maximum loop weight.

in such loops are omnivores as well. Hence, the aggregation of weak links in long loops that enhances stability results from the pyramidal biomass structure of the populations constituting the loop. Moreover, the stronger the pyramidal structure, the lower the maximum loop weight and the higher the level of food web stability (Neutel *et al.* 2002).

Productivity relates to this mechanism in a similar way as in the modelled food chains (Fig. 8.2). Too high levels of productivity may weaken or deteriorate the pyramidal biomass structure. As higher trophic level populations enter the loops as soon as there is enough energy to support a higher trophic level, the pyramidal biomass structures will be maintained, ensuring a relatively low maximum loop weight and a concomitant high level of food web stability. Hence, the patterning in the interaction strengths, as observed in real soil food webs, means that weak links are in the long loops, limiting maximum loop weight and enhancing food web stability. The underlying mechanism is connected to population size distributions in the form of trophic biomass pyramids in omnivorous loops. Productivity influences this mechanism in a similar way as in food chains, in that the length of the omnivorous loops should be in balance with the level of productivity. Such balance ensures enough energy for the higher trophic levels, and prevents the deterioration of the pyramidal biomass structure, which may reduce food web stability.

Productivity and soil food web structure in primary succession gradients

Observations on two real primary succession gradients have been used to see whether the principle of balance plays a role in the co-development of productivity and the structure of the below-ground food web (Neutel *et al.*, in preparation). The two primary succession gradients range from bare soil to forest and are from sandy dune soils; one on the Waddensea island of Schiermonnikoog, The Netherlands (Olff *et al.* 1993), and the other at the Hulshorsterzand (Veluwe), in central Netherlands (Berendse & Elberse 1990). There are many environmental factors that influence succession processes; this study has been carried out to analyse the extent to which the development of the soil community can be explained on the basis of trophic interactions in the soil food web. The development of below-ground food webs is assessed with respect to food web complexity (number of trophic groups and the frequency of interactions), food chain length and food web stability. Stability is investigated by looking at the maximum loop weight and by evaluating the eigenvalues of Jacobian community matrix representations of the observed food webs (Neutel *et al.*, in preparation).

Although there are slight differences between the two succession series, the way that the below-ground food webs have developed are highly similar (Neutel 2001). Along with productivity, the food webs become more complex, with increasing food chain lengths and complexity (Fig. 8.6a–d). By calculating the weights of the loops, it appears that the critical loop, i.e. the heaviest loop of length three or longer (loops of length 2 do not limit food web stability (Hofbauer & Sigmund 1996)), differs between the succession stages. Along with productivity, the critical loop tends to move up higher in the food chain. Introduction of higher trophic level organisms, such as the predatory nematodes in stage 2, are accompanied by changes of the critical loop, i.e. from *bacteria–flagellates–amoebae* in stage 1, to *bacteria–bacterivorous nematodes–predatory nematodes* in stage 2. The mechanism underlying the change in critical loop might be related to the principle of the balance between productivity and food chain length ensuring pyramidal biomass structures and stability (Fig. 8.2). The introduction of predatory nematodes in stage 2 may have prevented large population sizes of the amoebae, the highest trophic level in the critical loop in stage 1,

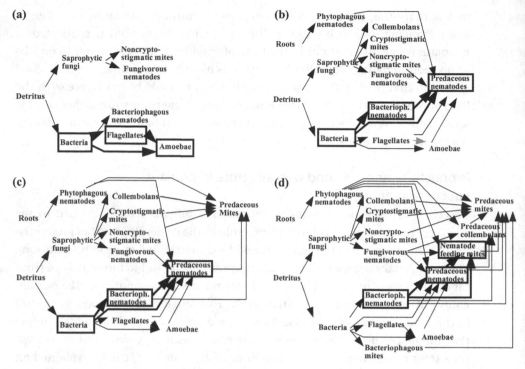

Figure 8.6. Food web structure along primary succession gradients from Schiermonnikoog and Hulshorsterzand (Neutel *et al.*, in preparation). Soil ages of the respective stages are 0, 10, 25 and 100 years in Schiermonnikoog and 0, 5, 15 and 50 years in Hulshorsterzand, approximately. For both series these stages correspond with productivity levels, i.e. for Schiermonnikoog 0.6, 2.6, 12 and 27 (kg C ha^{-1} a^{-1}) and for Hulshorsterzand 0.3, 2.6, 13 and 29 (kg C ha^{-1} a^{-1}). Each diagram was constructed as a representation of four food web replications in a stage, each based on three sample dates: (a) stage 1, (b) stage 2, (c) stage 3, (d) stage 4.

reducing the weight of this loop. The principle of maintaining the balance is also confirmed by the observation that all critical loops along the productivity gradient have trophic pyramidal biomass structures (Fig. 8.7). Moreover, there is a clear relation between the slope of the trophic pyramids (i.e. the relative biomass decrease over trophic levels) in the critical loop, the weight of these loops, and the stability of the food webs (Fig. 8.7).

In summary, productivity and food web structure show a co-development along the primary succession gradients. The increase in productivity is accompanied with higher trophic levels and increasing food web complexity. Moreover, a balance between productivity and food web structure seems to be maintained, ensuring trophic biomass pyramids, low maximum loop weight and concomitant high food web stability.

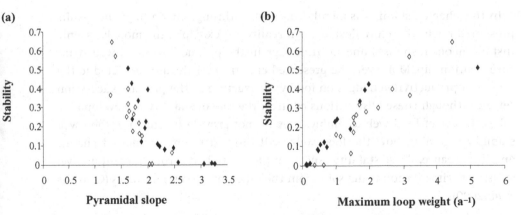

Figure 8.7. Pyramidal slope in the critical loop versus food web stability (Neutel *et al.*, in preparation). Filled diamonds represent the critical loops in the food webs of the Schiermonnikoog gradient; open diamonds represent the food webs of the Hulshorsterzand gradient. Each symbol represents a replicate. The pyramidal slope denotes the decrease over trophic levels of the species' biomass, expressed as a power of 10, i.e. pyramidal slopes of 1 and 2 mean a ten-fold and a hundred-fold decrease in biomass over trophic levels, respectively. The stability measure s (dimensionless) is determined as the value that leads to a minimum level of intraspecific interaction strength required for matrix stability according to $\alpha_{ii} - s \cdot d_i, i - 1, \ldots, n$, where d_i refers to the natural specific death rate (a^{-1}) of species i (Neutel *et al.* 2002).

Conclusions

In soil ecosystems, productivity and the structure of community food webs are inextricably interrelated. Central in this interrelationship is that food web stability requires a balance between productivity and food web structure. In food chains, productivity must ensure that sufficient energy is available for the subsequent trophic levels, while food chain length should prevent large destabilising population dynamical oscillations. Such large oscillations might occur when too high levels of productivity lead to non-pyramidal or inverse pyramidal biomass structures. In real, complex soil food webs, a balance between productivity and food web structure is indicated by the emergence of trophic pyramids in population sizes as well as in feeding rates that translate into stabilising patterns in interaction strengths. These energetic pyramids prevent the occurrence of long heavy loops that reduce food web stability. The principle of balance between productivity and food web structure is indicated to play a role in the co-development of productivity and food web structure along gradients of natural primary succession. In this way, the interplay between productivity and food web stability may govern and select pathways in ecosystem development and hence the ways in which food webs assemble.

In this chapter, stability is merely approached through modelling. The results presented might therefore deviate from reality. For example, the models assume instantaneous responses and no time lags in the productivity–food web structure relationship. Moreover, the presented empirical results are restricted to the effects of productivity changes on food web structure in the primary succession series. Although these observations confirm the role of stability in developmental pathways of food web assembly, they do not provide insight into food web stability regarding how the food webs will respond to environmental change and disturbance. These stability aspects might be studied through manipulative, stress experimentation using soils from the different succession stages (Griffiths *et al.* 2000).

References

Andrén, O., Lindberg, T., Boström, U., *et al.* (1990). Organic carbon and nitrogen flows. *Ecological Bulletin*, **40**, 85–125.

Berendse, F. & Elberse, W. T. (1990). Competition and nutrient availability in heathland and grassland ecosystems. *Perspectives on Plant Competition* (Ed. by D. Tilman), pp. 94–116. New York: Academic Press.

Bloem, J., Lebbink, G., Zwart, K. B., *et al.* (1994). Dynamics of microorganisms, microbivores and nitrogen mineralisation in winter wheat fields under conventional and integrated management. *Agriculture Ecosystems and Environment*, **51**, 129–143.

Brussaard, L., Bouwman, L. A., Geurs, M., Hassink, J. & Zwart, K. B. (1990). Biomass, composition and temporal dynamics of soil organisms of a silt loam soil under conventional and integrated management. *Netherlands Journal of Agricultural Science*, **38**, 283–302.

Brussaard, L., van Veen, J. A., Kooistra, M. J. & Lebbink, G. (1988). The Dutch programme on soil ecology of arable farming systems: I. Objectives, approach and preliminary results. *Ecological Bulletin*, **39**, 35–40.

De Ruiter, P. C., Moore, J. C., Bloem, J., *et al.* (1993a). Simulation of nitrogen dynamics in the belowground food webs of two winter-wheat fields. *Journal of Applied Ecology*, **30**, 95–106.

De Ruiter, P. C., Neutel, A. M. & Moore, J. C. (1995). Energetics, patterns of interaction strengths, and stability in real ecosystems. *Science*, **269**, 1257–1260.

De Ruiter, P. C., Neutel, A. M. & Moore, J. C. (1998). Biodiversity in soil ecosystems: the role of energy flow and community stability. *Applied Soil Ecology*, **10**, 217–228.

De Ruiter, P. C., van Veen, J. A., Moore, J. C., Brussaard, L., & Hunt, H. W. (1993b). Calculation of nitrogen mineralization in soil food webs. *Plant and Soil*, **157**, 263–273.

Elton, C. (1927). *Animal Ecology.* New York: McMillan.

Griffiths, B. S., Ritz, K., Bardgett, R. D., *et al.* (2000). Ecosystem response of pasture soil communities to fumigation-induced microbial diversity reductions: an examination of the biodiversity–ecosystem function relationship. *Oikos*, **90**, 279–294.

Hendrix, P. F., Parmelee, R. W., Crossley, D. A. J., *et al.* (1986). Detritus food webs in conventional and no-tillage agroecosystems. *BioScience*, **36**, 374–380.

Hofbauer, J. and Sigmund, K. (1988). *The Theory of Evolution and Dynamical Systems.* Cambridge: Cambridge University Press.

Hunt, H. W., Coleman, D. C., Ingham, E. R., *et al.* (1987). The detrital food web in a shortgrass prairie. *Biology and Fertility of Soils*, **3**, 57–68.

Lindeman, R. L. (1942). The trophic–dynamic aspect of ecology. *Ecology*, **23**, 399–418.

May, R. M. (1972). Will a large complex system be stable? *Nature*, **238**, 413–414.

McCann, K. S., Hastings, A. & Huxel, G. R. (1988). Weak trophic interactions and the balance of nature. *Nature*, **395**, 794–798.

Moore, J. C. & de Ruiter, P. C. (1991). Temporal and spatial heterogeneity of trophic interactions within belowground food webs. *Agriculture Ecosystems and Environment*, **34**, 371–394.

Moore, J. C. & de Ruiter, P. C. (2000). Invertebrates in detrital food webs along gradients of productivity. *Invertebrates as Webmasters in Ecosystems* (Ed. by P. F. Hendrix), pp. 161–184. New York: CAB International.

Moore, J. C., de Ruiter, P. C. & Hunt, H. W. (1993). Influence of productivity on the stability of real and model ecosystems. *Science*, **261**, 906–908.

Moore, J. C., de Ruiter, P. C., Hunt, H. W., Coleman, D. C., & Freckman, D. W. (1996). Microcosms and soil ecology: critical linkages between field studies and modelling food webs. *Ecology*, **77**, 694–705.

Moore, J. C. & Hunt, H. W. (1988). Resource compartmentation and the stability of real ecosystems. *Nature*, **333**, 261–263.

Moore, J. C., Walter, D. E. & Hunt, H. W. (1988). Arthropod regulation of micro- and mesobiota in belowground food webs. *Annual Review of Entomology*, **33**, 419–439.

Naeem, S., Thompson, L. J., Lawler, S. P., Lawton, J. H. & Woodfin, R. M. (1994). Declining biodiversity can alter the performance of ecosystems. *Nature*, **368**, 734–737.

Neutel, A. M., van der Koppel, J., Berendse, F. & de Ruiter, P. C. (in preparation). Structure and stability in below-ground food webs along gradients of ecosystem development: key loops and key pyramids.

Neutel, A. M., Heesterbeek, J. A. P. & de Ruiter, P. C. (2002). Stability in real food webs: weak links in long loops. *Science*, **296**, 1120–1123.

Odum, E. P. (1971). *Fundamentals of Ecology*, 3rd edition. Philadelphia, PA: Saunders.

Olff, H., Huisman, J. & van Tooren, B. F. (1993). Species dynamics and nutrient accumulation during early primary succession in coastal sand dunes. *Journal of Ecology*, **81**, 693–702.

O'Neill, R. V. (1969). Indirect estimation of energy fluxes in animal food webs. *Journal of Theoretical Biology*, **22**, 284–290.

Paine, R. T. (1980). Food webs: linkage, interaction strength and community infrastructure. *Journal of Animal Ecology*, **49**, 667–685.

Paine, R. T. (1992). Food-web analysis through field measurements of per capita interaction strength. *Nature*, **355**, 73–75.

Pimm, S. L. (1982). *Food Webs*. London: Chapman and Hall.

Pimm, S. L. & Lawton, J. H. (1977). The number of trophic levels in ecological communities. *Nature*, **268**, 329–331.

Rosenzweig, M. L. (1971). Paradox of enrichment: destabilisation of exploitation ecosystems in ecological time. *Science*, **171**, 385–387.

Tilman, D., Wedin, D. & Knops, J. (1996). Productivity and sustainability influenced by biodiversity in grassland ecosystems. *Nature*, **379**, 718–720.

Wolters, V. (1997). *Functional Implications of Biodiversity in Soil*. Luxembuorg: Office for Official Publications of the European Community.

Yodzis, P. (1981). The stability of real ecosystems. *Nature*, **289**, 674–676.

Yodzis, P. (1989). *Introduction to Theoretical Ecology*. New York: Harper and Row.

CHAPTER NINE

Rhizosphere carbon flow: a driver of soil microbial diversity?

D. B. STANDING
University of Aberdeen

J. I. RANGEL CASTRO
University of Aberdeen

J. I. PROSSER
University of Aberdeen

A. MEHARG
University of Aberdeen

K. KILLHAM
University of Aberdeen

SUMMARY

1. Microorganisms play an essential role in modulating the fluxes of organic carbon and nutrients in soil. However, their diversity and functional significance are largely unknown. Recent technical developments in molecular, chemotaxonomic and physiological techniques complement traditional techniques and can now enable us to investigate the linkage between rhizosphere carbon flow, microbial diversity and soil function.

2. Reporter gene systems provide an important method for resolution of rhizosphere carbon flow. Their greatest advantage is that they can be used *in situ* without uncoupling the plant–microbial interaction vital to maintaining both quantity and quality of carbon flow.

3. Any consideration of rhizosphere carbon flow and soil microbial diversity should not only include substrate carbon flow and trophic interactions, but also the role of signal molecules, especially in terms of controlling rhizosphere community structure, diversity and function.

4. Little is known of carbon flow in natural systems. Ecosystem function and, specifically, carbon-cycling pathways can be determined by lipid analysis or nucleic acid stable isotope probing (SIP) using ^{13}C incorporated into microbial biomass. The ability to ascertain which components of the biomass are being enriched in root-derived carbon enables an understanding of how rhizosphere carbon drives microbial diversity.

Biological Diversity and Function in Soils, eds. Richard D. Bardgett, Michael B. Usher and David W. Hopkins.
Published by Cambridge University Press. © British Ecological Society 2005.

5. Changes in 16S rDNA sequence diversity and relative abundance provide indications of which organisms are responding to changing conditions and SIP analysis of mRNA allows for assessment of their activity, and thus may be used to follow changes in the microbial community during rhizosphere development.

Introduction

Despite the fundamental importance of rhizosphere carbon flow in carbon and nutrient cycling (and hence, its importance in ecosystem productivity) and as a key driver of plant–microbe interactions in soil, we remain largely ignorant of the linkage between carbon flow and soil microbial diversity. Most of the information available relates to managed, mainly agricultural systems, where manipulation of rhizosphere biodiversity has been driven by requirements for biocontrol of soil-borne diseases and biofertilisation of crop plants. This manipulation has mainly been through direct use of microbial inocula (typically applied as part of a seed coat; Paau 1988; Turner & Backman 1991). However, it is evident, from reports of how rhizosphere carbon flow can be greatly changed by the presence of microbes (Meharg & Killham 1991), that there is scope to manipulate the influence of the flow of rhizosphere carbon on pathogen–antagonist interactions, since the latter are largely controlled by competition for substrate carbon. Awareness of the potential linkage between rhizosphere carbon flow and soil microbial biodiversity has also been highlighted by investigations of bioremediation where plants provide a rhizosphere population with enhanced capacity relative to non-rhizosphere soil for mineralisation of certain organic pollutants (April & Sims 1990; Walton & Anderson 1990).

Measuring soil microbial diversity

Microorganisms play an essential role in interconverting organic carbon (and associated nutrients) released into the soil, from roots or other sources. The diversity of soil microorganisms is poorly characterised and virtually nothing is known of its significance in terms of soil and ecosystem function in the rhizosphere. However, an array of molecular, chemotaxonomic and physiological techniques that complement the traditional microscopic and selective cultivation techniques now exists for assessing microbial diversity (Fig. 9.1). These techniques are being applied to characterise diversity in soil, particularly in the rhizosphere (Borneman et al. 1996; Borneman & Triplett 1997; Zhou et al. 1997; Duineveld et al. 1998; Felske et al. 1998; Normander & Prosser 2000) and it has been found that rhizosphere carbon can act to reduce microbial diversity by providing a strong selective force for a small number of mainly Gram-negative bacterial groups (Marilley & Aragno 1999). Overall plant nutritional status may also have considerable impact on the quality of root exudation, and Yang and

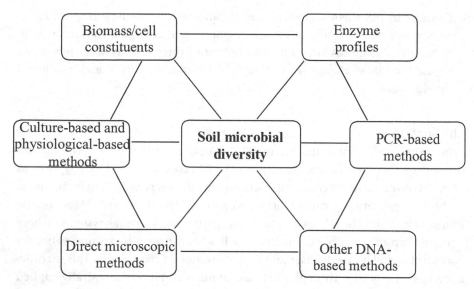

Figure 9.1. Methods available for assessing the diversity of rhizosphere microbial populations. The connectors between the different groups indicate the complexity of methods currently available and that a multi-tiered approach is often necessary in the analysis of soil microbial community structure.

Crowley (2000) demonstrated that differences in plant iron nutritional status were reflected in differences in rhizosphere microbial communities.

Chemotaxonomic, e.g. PLFA and other selective biomarker compounds, and molecular techniques are now routinely used for the analysis of microbial species diversity and community structure in natural rhizosphere populations. These techniques have enabled characterisation of considerable diversity within natural microbial communities, and have led to the discovery of previously uncharacterised microbial groups, e.g. Sait *et al.* (2002). The most common approaches have been based on lipid analysis and on the amplification of 16S rRNA genes amplified from DNA extracted directly from soil samples, with subsequent analysis by cloning and sequencing or using fingerprinting techniques. These approaches avoid the requirement for laboratory cultivation, which has previously been used to study rhizosphere populations and which introduces significant biases. They, therefore, enable analysis of unculturable populations, which are believed to constitute greater than 99% of natural populations (Amann *et al.* 1995; Torsvik *et al.* 1996). For example, in one study of Sourhope rhizosphere soil McCaig *et al.* (1999) showed that more than 25% of sequences in 16S rRNA clone libraries belonged to groups with no cultivated representative, preventing prediction of their physiological and metabolic characteristics, and ecosystem function. Molecular and chemotaxonomic techniques have led to the characterisation and reclassification of established bacterial groups with known ecosystem function, for example nitrifying bacteria (Stephen *et al.* 1996) and methanotrophs (Holmes *et al.* 1999), and of microbial groups with varied function, such as the pseudomonads (Spiers *et al.* 2000). Molecular techniques can be used to assess changes in community structure in response to environmental factors. For example, lipid analysis has been used to assess the broad impact of pH on the diversity of the rhizosphere microbial community in field plots (O'Donnell 2001). Denaturing gradient gel electrophoresis (DGGE) analysis has been used to quantify differences in the soil microbial community in heavy metal polluted soils (Kandeler *et al.* 2000) and quantification can be achieved using competitive PCR (cPCR; Phillips *et al.* 2000). These approaches have been used to investigate changes in diversity during root development. For example, Duineveld *et al.* (1998) found little variation with plant age in DGGE analysis of 16S rDNA PCR products from *Chrysanthemum* rhizospheres. However, Marilley and Aragno (1999) found differences in 16S rRNA gene sequences between rhizosphere soil and root tissue of white clover and ryegrass, while Yang and Crowley (2000) observed different communities along barley roots. Normander and Prosser (2000), also using DGGE analysis of 16S rDNA, found differences between endorhizosphere, rhizoplane and rhizosphere soil of barley plants, with data suggesting that rhizoplane bacteria originated from the surrounding soil, but there was no evidence of a change in bacterial community composition with plant age. The analysis of phospholipid fatty acids (PLFAs) in soils is also increasingly being used to characterise microbial diversity and responses of soil microbial communities to environmental change (e.g. Priha *et al.* 2001). The extraction of PLFAs from soils and subsequent analysis using gas chromatography does not usually reveal the functional or taxonomic identity of the microbial population but can be used to detect change, as demonstrated by O'Donnell (2001). However, there are exceptions where specific PLFAs have been associated with distinct functional groups (e.g. the methanotrophs; see Hanson & Hanson 1996), but such instances are limited.

Measuring rhizosphere carbon flow

The measurement of rhizosphere carbon flow involves methods that range from isotopic techniques (pulse chase and continuous labelling, natural abundance approaches) to the use of reporter genes (Killham & Yeomans, 2001). The isotopic methods are constrained by the extreme difficulty of being able to separate microbial respiration from root respiration. This difficulty does not affect reporter gene methods such as *lux/gfp* marking, where nucleic acid sequences code for easily assayable proteins, and these methods complement the more traditional techniques. These approaches have led to a recognition of the broad selective pressure of this carbon flow during progressive development of the rhizosphere. During the early stages of rhizosphere development when the plant is at seedling stage, carbon flow is dominated by low molecular weight compounds (e.g. glucose) that select for fast-growing soil microorganisms, such as fluorescent pseudomonads, that are readily cultivated on laboratory media. The competitive success of these rhizosphere microorganisms can therefore be explained

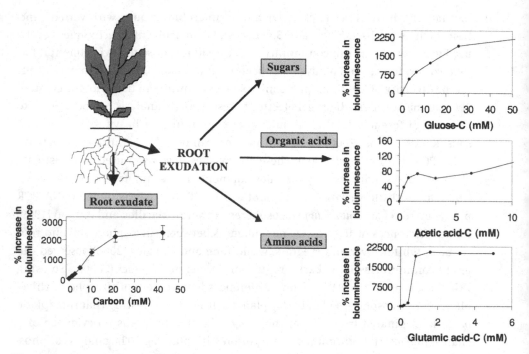

Figure 9.2. Use of a *lux* metabolic reporter system in *Pseudomonas fluorescens* enables quantification of different components of rhizosphere carbon flow as well as characterisation of whole exudate (Yeomans *et al.* 1999).

in terms of comparative Michaelis–Menten kinetics (Killham 1994). Carbon flow then changes considerably, as the plant and rhizosphere mature, and certain molecular constituents of rhizosphere carbon flow provide strong selection for key microbial groups. For example, the ring amino acid proline selects for particular, more specialised pseudomonads (Vilchez *et al.* 2000), while certain phenolics are known to attract *Agrobacterium tumefaciens* (Zhu *et al.* 2000) in the formation of crown gall disease. Furthermore, the role of isoflavones in attracting rhizobia to legume roots has been studied in some detail (e.g. Schell 1993).

Figure 9.2 demonstrates the use of a *lux* reporter gene (the *lux* genes are downstream of a strong, constitutive promoter and hence luminescence is directly linked to metabolic activity) system in the rhizobacterium *Pseudomonas fluorescens* to characterise carbon flow from plant roots. By starving these bacterial reporters prior to use, their light output reports only on the carbon from adjacent roots (i.e. close enough for diffusion of components of root exudates to reach the carbon biosensors). The Michaelis–Menten kinetics (V_{max} and K_m, respectively the maximum reaction velocity of the enzyme and the Michaelis constant, which is the substrate concentration at $1/V_{max}$) of the light output response of the bacterial reporters are highly diagnostic of the exudate components (carbohydrates, amino acids and organic acids) and, in this case, demonstrate that the carbon flow from the young wheat seedling roots is dominated by glucose.

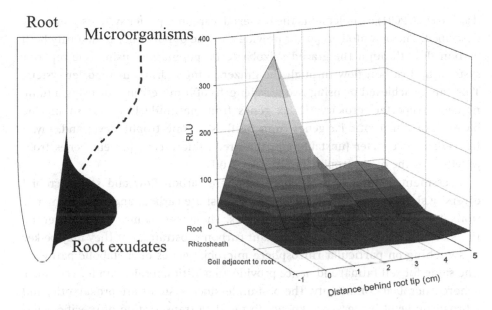

Figure 9.3. Use of *lux* carbon reporting *Pseudomonas fluorescens* for *in situ* resolution of how rhizosphere carbon flow provides the substrate to support the development of a rhizosphere microbial population. Right-hand graph: relative light units (RLUs), with one RLU being equivalent to 1 mV 10 s^{-1}. Left-hand diagram: shows the lag between exudation of low molecular weight substances and the build-up of a rhizosphere microbial community (Adapted from Bowen and Rovira 1991). In this system, roots are first grown in soil and then the roots plus rhizosheath removed to a 35 mm Petri dish and treated with the bacterial reporter prior to luminometric measurements.

The greatest advantage of the bacterial reporter systems for resolving rhizosphere carbon flow is that they can be used *in situ* without any uncoupling of the plant–microbial interaction which is vital to maintaining the integrity of the quantity and quality of the carbon flow itself (Fig. 9.3). This means that carbon flow can be measured through an *in planta* rhizosphere microbial community carrying a reporter gene system. By using different microbes with such systems, the relative carbon sink strength of different rhizosphere communities can begin to be assessed. By using multiple reporters with either different wavelengths of luminescence or a combination of luminescence and fluorescence reporters, it will be possible in future to dissect carbon flow through complex rhizosphere populations and identify key players in terms of competitiveness for rhizosphere carbon substrates at different stages of rhizosphere development.

Of course, rhizosphere carbon flow does not just involve microorganisms. Root-derived carbon also drives interactions with higher trophic levels in the soil. Grazing of the rhizosphere microbial population by protozoa and bacteriophagous nematodes, for example, contributes to rhizosphere carbon flow. Quantifying these aspects has long proved a challenge to soil ecologists. Recent insertion of luminescence-based, reporter genes into these higher organisms

(Lagido *et al.* 2001) complements the bacterial, carbon reporter systems previously described. The nematode reporter provides for the opportunity to quantify both carbon flow through the grazed rhizobacterial population, using one reporter system, and carbon flow through the grazer, using a different reporter system. This can be achieved by using two wavelengths of luminescence or using a luminescence reporter (prokaryotic *lux* genes from naturally luminescent marine bacteria or eukaryotic *luc* genes from fireflies) in one trophic level and, say, a fluorescence reporter (unstable *gfp*, the green fluorescent protein genes from jellyfish) in the other trophic level under study.

Assessment of linkage between rhizosphere carbon flow and soil microbial diversity should not only consider flow of substrate carbon, and trophic interactions, but also the role of signal molecules. These signals may be at concentrations far too low for status as significant carbon substrates, but they may be key in switching on particular rhizosphere microbial genes or metabolic pathways (the same for soil fauna) and hence provide an additional selection for the rhizosphere microbial community. The best-understood systems are prokaryotic, and a broad range of bacteria are known to regulate transcription of specific genes through quorum-sensing systems (Salmond *et al.* 1995). In Gram-negative bacteria, the predominant signal molecules are N-acyl homoserine lactones (AHLs). These compounds readily migrate across cell membranes, with the environmental concentration determined, in part, by the abundance of AHL-producing organisms in that environment and by other factors that include signal diffusional properties and signal degradation, as well as soil properties such as porosity and porewater regime. Above threshold cell densities, AHLs accumulate to concentrations that activate cellular cognate receptors that mediate expression of a wide range of microbial genes (>250 in *Pseudomonas aeruginosa*, Whiteley *et al.* 1999). Importantly, many AHL-regulated genes are involved in bacterial interactions, through the production of antibiotics and chitinases (Stead *et al.* 1996; Wood & Pierson 1996; Raaijmakers *et al.* 1999), and interactions between bacteria and eukaryotic hosts (e.g. pathogenesis). In soil, AHL signalling (evidenced by the presence of AHLs at measurable concentrations) has been found to be particularly prevalent in the rhizosphere (Cha *et al.* 1998; Elasri *et al.* 2001). This is consistent with AHL signalling being important in active interactions between microbial populations. The rhizosphere represents the site where signalling mechanisms will be most important, due to large concentrations of cells and intense competition for resources between populations and plant. A single Gram-negative organism may produce a variety of chemically distinct AHL molecules (Cha *et al.* 1998), each controlling expression of a different array of genes. However, it has also been found that AHLs produced by one Gram-negative species commonly cross-activate expression of AHL-regulated genes in others (Pierson *et al.* 1998). Consequently, there is a strong likelihood of interspecies communication in the rhizosphere. Recently, it has been reported that a number of higher plants produce AHL-like compounds in their root exudates that modify (stimulate or inhibit) AHL-mediated microbial signalling (Teplitski *et al.* 2000; Bauer & Teplitski 2001). Modification by plants (either directly through plant-originated signals or indirectly through changes in microbial substrate exudation) of bacterial AHL signalling in the root environment is likely to have important implications for a broad range of plant–microbe interactions. With the development of bacterial reporter systems for detection of AHL signal molecules (e.g. Winson *et al.* 1998a, b) and the availability of reliable molecular techniques to assess rhizosphere microbial community structure previously described, the opportunity now exists to assess the ecological significance of signalling. This is in terms of controlling the structure of rhizosphere communities and, in particular, the complexity in the way in which rhizosphere carbon flow may drive soil microbial diversity.

Influence of carbon flow on soil microbial diversity

In contrast to our patchy knowledge of the selective role of components of carbon flow in driving soil biodiversity in managed systems, almost nothing is known of natural systems, where carbon flow through the rhizosphere is often from a mixed vegetation community. Kowalchuk *et al.* (2002) report that differences in soil microbial community structure, as a function of above-ground species diversity, may be associated solely with the rhizosphere as they found no association between plant and microbial species diversity in bulk soil. This is critical to our understanding of the linkage between carbon flow and rhizosphere microbial diversity and hence to our understanding of fundamental ecosystem processes. For example, we do not know if microbial populations are resilient to temporal and spatial changes in rhizosphere carbon flow; whether reduced biodiversity or different species composition lead to 'bottle necks' or different pathways for carbon or nutrient flow; or whether biodiversity is related in any way to efficient transformation of organic carbon. These issues are central to our understanding of soil carbon/nutrient cycling and to establishing the link between biological diversity and ecosystem function.

Our lack of knowledge of how rhizosphere carbon flow drives the diversity of the soil microorganisms in natural ecosystems, and of the interrelationships between this diversity and carbon cycling processes, has largely resulted from an inability to dissect the components of soil communities in a way that can be related to their competitive sink strength for carbon. The availability and conjunction of isotopic and molecular approaches provide the potential for a breakthrough in linking diversity and function.

Ecosystem function and, specifically, carbon-cycling pathways can be determined by using ^{13}C to track fixed carbon (from the initial $^{13}CO_2$ pulse labelling) from the plant and into different soil organic carbon compartments. These compartments include a total soil fraction, following removal of roots, soil microbial biomass (chloroform labile) and specific, extractable pools (e.g. lipids and nucleic acids). Soil faunal components can also be included where appropriate.

Two recently developed ^{13}C-based approaches provide information on which components of the microbial populations are using a particular substrate. Both involve application of a ^{13}C pulse, which can be provided directly as a soil microbial substrate or indirectly as ^{13}CO$_2$ to the plant so that the plant fixes the labelled carbon and root carbon flow carries this label through the soil. Non-respired ^{13}C-labelled substrates will be incorporated into key carbon pools of active organisms only and subsequent analysis of these pools enables characterisation of the diversity of active populations. Carbon-13 incorporated into microbial biomass can then be determined by analysis of lipids (PLFA) or nucleic acids by stable isotope probing (SIP; Radajewski *et al.* 2000).

A recent development in the analyses of ^{13}C isotope composition of individual PLFAs in sediments using gas chromatography-isotope ratio-mass spectroscopy (GC-IR-MS; Boschker *et al.* 1998) has enabled the tracing of carbon from labelled organic substrates into individual functional organisms. This approach has relevance to rhizosphere function and has also recently been extended to tracking ^{13}CH$_4$ uptake by soil microorganisms into specific bacterial lipids, enabling characterisation of the organisms responsible for high affinity oxidation of CH$_4$ for the first time (Bull *et al.* 2000). Clearly, the potential for identifying the incorporation of ^{13}C from labelled plants into ^{13}C PLFAs enables the rapid characterisation of *in situ* bacterial and fungal activities, and assessment of which components of these communities are utilising this carbon (Arao 1999). It also allows shifts in the active population to be detected. Reduction of ^{13}C/^{12}C ratios in PLFAs, as microbial biomass is degraded and incorporated into other organisms, indicates if the organisms are primary or secondary consumers. The ability to ascertain which components of the biomass are being enriched in root-derived carbon enables us to understand truly how the carbon economy drives microbial diversity in the rhizosphere.

Stable isotope probing (SIP) has great potential to advance our understanding of rhizosphere ecology as it characterises populations that are active carbon sinks. This is carried out by extraction of nucleic acid and separation of ^{13}C-labelled (heavy) and ^{12}C-labelled (light) nucleic acid pools. Molecular analysis is then carried out using techniques described above (DGGE, cloning, sequencing) to determine the diversity of the active population. While published applications of this method have amplified DNA, methods are now available for analysis of RNA (using real-time PCR). This greatly increases sensitivity and is likely to be necessary for detection of active microbial populations where ^{13}CO$_2$ pulses to vegetation are brief and/or involve lower ^{13}C enrichments.

Rhizosphere carbon derived from recently assimilated root exudates, and carbon derived from root senescence and turn-over, flow through distinct

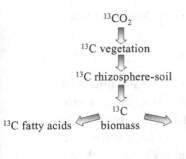

Figure 9.4. A tiered approach to ^{13}C-pulse-chase characterisation of the linkage between rhizosphere carbon flow and soil microbial diversity.

components of the microbial biomass and at different times (Killham & Yeomans 2001). This provides for a number of key hypotheses which can now be tested and which have a considerable bearing on soil microbial ecology:

1. That increasing chemo-diversity, associated with root ageing and senescence, drives an increase in microbial diversity.
2. That fungi and bacteria respond differently to plant-derived carbon of varying complexity, and that there is therefore a shift in the dominant microorganisms (bacteria and fungi) associated with carbon flow from recent assimilate versus more complex carbon flow from root turn-over.
3. That carbon is cycled within the rhizosphere, with secondary consumers utilising the necromass of primary consumers of root-derived carbon.
4. That carbon flow through the different components of the microbial community and its associated driving of rhizosphere diversity are affected by environmental factors such as atmospheric nitrogen deposition to the plant/soil system, and additional factors associated with vegetation type and soil pH.
5. That the resilience of the rhizosphere microbial population to environmental change is related to its diversity.
6. That the residence time of rhizosphere carbon is greater in acid soils than that in corresponding limed soils and soils of naturally higher pH, because of the greater predominance of slower processing fungi and the lower substrate quality of the carbon flow. Furthermore, that more carbon will be sequestered in the short and medium term in more acid soil systems because of the greater substrate assimilation efficiency of the fungi.

The use of ^{13}C pulse chase (tracking ^{13}C incorporated into plant photosynthate through shoots to roots, into the soil, through the microbial biomass and particularly into key pools such as fatty acids and nucleic acids) is now enabling these hypotheses to be tested (Fig. 9.4).

This testing involves tracking ^{13}C over time through the fungal and bacterial biomass using ^{13}C lipid biomarkers and integrating this with molecular analysis of the microbial community. Characterisation of a carbon budget distinguishes the extent to which carbon and energy flow drive soil microbial biodiversity, thereby linking biodiversity to function in natural as well as more managed ecosystems.

Linking soil microbial diversity to function

In addition to the use of ^{13}C to investigate the components of the soil microbial population actively involved in processing carbon from the root, there is now a real opportunity to link observed microbial diversity for soil ecosystems to a range of rhizosphere functions. For example, it has already been mentioned that we do not know whether diversity increases functional resilience to changing environmental conditions. Furthermore, we do not know what portions of the community are active, rather than merely surviving. Changes in 16S rDNA sequence diversity and relative abundance provide indications of which organisms are responding to changing conditions and may be used to follow changes during rhizosphere development. The sensitivity of these techniques can be improved by analysis of rRNA, rather than rDNA. The potential for linking microbial diversity to function can be further extended by application of SIP to mRNA gene probing. In particular, this allows the activity associated with specific genes to be quantified (Bürgmann *et al.* 2003).

Conclusions

This chapter has highlighted the significance of focusing research on the potential linkage between rhizosphere carbon flow and soil microbial diversity. The chapter has also confirmed that we now have the range of methods (molecular, isotopic, chemotaxonomic and physiological) to resolve this linkage and make exciting breakthroughs in terrestrial ecology. So we should soon have the answer as to whether/when and how rhizosphere carbon flow is the key driver of soil microbial diversity.

Acknowledgements

The authors thank BBSRC for their funding through the BIRE initiative and NERC for their funding through the Soil Biodiversity Initiative. We acknowledge Catrin Yeomans for her work on carbon reporting.

References

Amann, R. I., Ludwig, W. & Schleifer, K.-H. (1995). Phylogenetic identification and *in situ* detection of individual microbial cells without cultivation. *Microbiology Reviews*, **59**, 143–169.

April, W. & Sims, R. C. (1990). Evaluation of the use of prairie grasses for stimulating polycyclic aromatic hydrocarbon treatment in soil. *Chemosphere*, **20**, 253–265.

Arao, T. (1999). *In situ* detection of changes in soil bacterial and fungal activities by measuring ^{13}C incorporation into soil phospholipid fatty acids from ^{13}C acetate. *Soil Biology and Biochemistry*, **31**, 1015–1020.

Bauer, W. D. & Teplitski, M. (2001). Can plants manipulate bacterial quorum sensing? *Australian Journal of Plant Physiology*, **28**, 913–921.

Borneman, J., Skroch, P. W., O'Sullivan, K. M., et al. (1996). Molecular microbial diversity of an agricultural soil in Wisconsin. *Applied and Environmental Microbiology*, **62**, 1935–1943.

Borneman, J. & Triplett, E. W. (1997). Molecular microbial diversity in soils from eastern Amazonia: evidence for unusual microorganisms and microbial population shifts associated with deforestation. *Applied and Environmental Microbiology*, **63**, 2647–2653.

Boschker, H. T. S., Nold, S. C., Wellsbury, P., et al. (1998). Direct linking of microbial populations to specific biogeochemical processes by C-13-labelling of biomarkers. *Nature*, **392**, 801–805.

Bowen, G. D. & Rovira, A. D. (1991). The rhizosphere, the hidden half of the hidden half. *Plant Roots: the Hidden Half* (Ed. by Y. Waisel, A. Eshel & U. Kafkafi), pp. 641–669. New York: Marcel Dekker.

Bull, I., Parekh, N., Ineson, P. & Evershed, R. E. (2000). Detection and classification of atmospheric methane oxidizing bacteria in the soil. *Nature*, **405**, 175–178.

Bürgmann, H., Widmer, F., Sigler, W. V. & Zeyer, J. (2003). mRNA extraction and reverse transcription-pcr protocol for detection of *nifH* gene expression by *Azotobacter vinelandii* in soil. *Applied and Environmental Microbiology*, **69**, 1928–1935.

Cha, C., Gao, P., Chen, Y.-C., Shaw, P. D. & Farrand, S. K. (1998). Production of acyl-homoserine lactone quorum-sensing signals by Gram-negative plant-associated bacteria. *Molecular Plant Microbe Interactions*, **11**, 1119–1129.

Duineveld, B. M., Rosado, A. S., van Elsas, J. D. & van Veen, J. A. (1998). Analysis of the dynamics of bacterial communities in the rhizosphere of the *Chrysanthemum* via denaturing gradient gel electrophoresis and substrate utilization patterns. *Applied and Environmental Microbiology*, **64**, 4950–4957.

Elasri, M., Delorme, S., Lemanceau, P., et al. (2001). Acyl-homoserine lactone production is more common among plant-associated *Pseudomonas* spp. than among soilborne *Pseudomonas* spp. *Applied and Environmental Microbiology*, **67**, 1198–1209.

Felske, A., Wolterink, A., van Lis, R. & Akkermans, A. D. L. (1998). Phylogeny of the main bacterial 16S rRNA sequences in Drentse A grassland soils (The Netherlands). *Applied and Environmental Microbiology*, **64**, 871–879.

Hanson, R. S. & Hanson, T. E. (1996). Methanotrophic bacteria. *Microbiology Reviews*, **60**, 439–471.

Holmes, A. J., Roslev P., McDonald I. R., et al. (1999). Characterization of methanotrophic bacterial populations in soil showing atmospheric methane uptake. *Applied and Environmental Microbiology*, **65**, 3312–3318.

Kandeler, E., Tscherko, D., Bruce, K. D., et al. (2000). Structure and function of the soil microbial community in microhabitats of a heavy metal polluted soil. *Biology and Fertility of Soils*, **32**, 390–400.

Killham, K. (1994). *Soil Ecology*, pp. 83–85. Cambridge: Cambridge University Press.

Killham, K. & Yeomans, C. (2001). Rhizosphere carbon flow measurement and implications: from isotopes to reporter genes. *Plant and Soil*, **233**, 91–96.

Kowalchuk, G. A., Buma, D. S., de Boer, W., Klinkhamer, P. G. L. & van Veen, J. A. (2002). Effects of above-ground plant species composition and diversity on the diversity of soil-borne microorganisms. *Antonie van Leeuwenhoek International Journal of General and Molecular Microbiology*, **81**, 509–520.

Lagido, C., Pettitt, J., Porter, A. J. R., Paton, G. I. & Glover, L. A. (2001). Development and application of bioluminescent *Caenorhabditis elegans* as multicellular eukaryotic biosensors. *FEBS Letters*, **493**, 36–39.

Marilley, L. & Aragno, M. (1999). Phylogenetic diversity of bacterial communities differing in degree of proximity of *Lolium perenne* and *Trifolium repens* roots. *Applied Soil Ecology*, **13**, 127–136.

McCaig, A. E., Glover, L. A. & Prosser, J. I. (1999). Molecular analysis of bacterial community structure and diversity in unimproved and improved grass pastures. *Applied and Environmental Microbiology*, **65**, 1721–1730.

Meharg, A. A. & Killham, K. (1991). A novel method of quantifying rhizosphere carbon flow in the presence of soil microflora. *Plant and Soil*, **133**, 111–116.

Normander, B. & Prosser, J. I. (2000). Bacterial origin and community composition in the barley phytosphere as a function of habitat and presowing conditions. *Applied and Environmental Microbiology*, **66**, 4372–4377.

O'Donnell, A. G. (2001). Plants and fertilisers as drivers of change in microbial community structure and function in soil. *Plant and Soil*, **232**, 135–145.

Paau, A. S. (1988). Formulations useful in applying beneficial microorganisms to seeds. *Trends in Biotechnology*, **6**, 276–279.

Phillips, C. J., Paul, E. A. & Prosser, J. I. (2000). Quantitative analysis of ammonia oxidising bacterial using competitive PCR. *FEMS Microbiology Ecology*, **32**, 167–175.

Pierson, E. A., Wood, D. W., Cannon, J. A., Blachere, F. M. & Pierson, L. S. III. (1998). Interpopulation signaling via N-acyl-homoserine lactones among bacteria in the wheat rhizosphere. *Molecular Plant Microbe Interactions*, **11**, 1078–1084.

Priha, O., Grayston, S. J., Hiukka, R., Pennanen, T. & Smolander, A. (2001). Microbial community structure and characteristics of the organic matter in soils under *Pinus sylvestris, Picea abies* and *Betula pendula* at two forest sites. *Biology and Fertility of Soils*, **33**, 17–24.

Raaijmakers, J. M., Bonsall, R. F. & Weller, D. M. (1999). Effect of population density of *Pseudomonas fluorescens* on production of 2,4-diacetylphloroglucinol in the rhizosphere of wheat. *Phytopathology*, **89**, 470–475.

Radajewski, S., Ineson, P., Parekh, N. R. & Murrell, J. C. (2000). Stable-isotope probing as a tool in microbial ecology. *Nature*, **403**, 646–649.

Sait, M., Hugenholtz, P. & Janssen, P. H. (2002). Cultivation of globally distributed soil bacteria from phylogenetic lineages previously only detected in cultivation-independent surveys. *Environmental Microbiology*, **4**, 654–666.

Salmond, G. P. C., Bycroft, B. W., Stewart, G. S. A. B. & Williams, P. (1995). The bacterial 'enigma': cracking the code of cell–cell communication. *Molecular Microbiology*, **16**, 615–624.

Schell, M. A. (1993). Molecular biology of the LysR family of transcriptional regulators. *Annual Reviews in Microbiology*, **47**, 597–626.

Spiers, A. J., Buckling, A. & Rainey, P. B. (2000). The causes of *Pseudomonas* diversity. *Microbiology*, **146**, 2345–2350.

Stead, P., Rudd, B. A. M., Bradshaw, H., Noble, D. & Dawson, M. J. (1996). Induction of phenazine biosynthesis in cultures of *Pseudomonas aeruginosa* by ʟ-N-(3-oxohexanoyl)homoserine lactone. *FEMS Microbiology Letters*, **140**, 15–22.

Stephen, J. R., McCaig, A. E., Smith, Z., Prosser, J. I. & Embley, T. M. (1996). Molecular diversity of soil and marine 16S rRNA gene sequences related to beta-subgroup ammonia-oxidizing bacteria. *Applied and Environmental Microbiology*, **62**, 4147–4154.

Teplitski, M., Robinson, J. B. & Bauer, W. D. (2000). Plants secrete substances that mimic bacterial N-acyl homoserine lactone signal activities and affect population density-dependent behaviours in associated bacteria. *Molecular Plant Microbe Interactions*, **13**, 637–648.

Torsvik, V., Sorheim, R. & Goksoyr, J. (1996). Total bacterial diversity in soil and sediment communities: a review. *Journal of Industrial Microbiology*, **17**, 170–178.

Turner, J. T. & Backman, P. A. (1991). Factors relating to peanut yield increases following *Bacillus subtilis* seed treatment. *Plant Diseases*, **75**, 347–353.

Vilchez, S., Molina, L., Ramos, C. & Ramos, J. L. (2000). Proline catabolism by *Pseudomonas putida*: cloning, characterization, and expression of the *put* genes in the presence of root exudates. *Journal of Bacteriology*, **182**, 91–99.

Walton, B. T. & Anderson, T. A. (1990). Microbial degradation of trichloroethylene in the rhizosphere: potential application to biological remediation of waste sites. *Applied and Environmental Microbiology*, **56**, 1012–1016.

Whiteley, M., Lee, K. M. & Greenberg, E. P. (1999). Identification of genes controlled by quorum sensing in *Pseudomonas aeruginosa*. *Proceedings of the National Academy of Sciences USA*, **96**, 13 904–13 909.

Winson, M. K., Swift, S., Fish, L., *et al.* (1998a). Construction and analysis of *lux* CDABE-based plasmid sensors for investigating N-acyl homoserine lactone-mediated quorum sensing. *FEMS Microbiology Letters*, **163**, 185–192.

Winson, M. K., Swift, S., Hill, P. J., *et al.* (1998b). Engineering the *lux* CDABE genes from *Photorhabdus luminescens* to provide a bioluminescent reporter for constitutive and promoter probe plasmids and mini-Tn5 constructs. *FEMS Microbiology Letters*, **163**, 193–202.

Wood D. W. & Pierson L. S. (1996). The phzI gene of *Pseudomonas aureofaciens* 30–84 is responsible for the production of a diffusible signal required for phenazine antibiotic production. *Gene*, **168**, 49–53.

Yang, C.-H. & Crowley, D. E. (2000). Rhizosphere microbial community structure in relation to root location and plant iron nutritional status. *Applied and Environmental Microbiology*, **66**, 345–351.

Yeomans, C., Porteous, F., Paterson, E., Meharg, A. A. & Killham, K. (1999). Use of *lux* marked rhizobacteria as reporters of rhizosphere C-flow. *FEMS Microbiology Letters*, **176**, 79–83.

Zhou, J. Z., Davey, M. E., Figueras, J. B., *et al.* (1997). Phylogenetic diversity of a bacterial community determined from Siberian tundra soil DNA. *Microbiology*, **143**, 3913–3919.

Zhu, J., Oger, P. M., Schrammeijer, B., *et al.* (2000). The bases of crown gall tumorigenesis. *Journal of Bacteriology*, **182**, 3885–3895.

PART IV

Consequences of soil biodiversity

CHAPTER TEN

Microbial community composition and soil nitrogen cycling: is there really a connection?

JOSHUA P. SCHIMEL
University of California at Santa Barbara
JENNIFER BENNETT
North Carolina State University
NOAH FIERER
University of California at Santa Barbara

SUMMARY

1. In the classical view of nitrogen cycling, the processes that involve nitrogen inputs and outputs (e.g. fixation, denitrification) are physiologically 'narrow' and so should be sensitive to microbial community composition, while internal turn-over (i.e. mineralisation, immobilisation) involves 'aggregate' processes that should be insensitive to microbial community composition.

2. A newly developing view of nitrogen cycling, however, identifies several ways in which mineralisation and immobilisation can be 'disaggregated' into individual components that may be sensitive to microbial community composition. Two of these are extracellular enzyme and microsite phenomena.

3. Exoenzymes are critical in driving decomposition, and hence mineralisation/immobilisation. Different classes of enzymes are produced by different groups of microorganisms. Additionally, the kinetics of exoenzymes may regulate microbial carbon and nitrogen limitation and hence community composition.

4. Microsite phenomena appear to regulate system-level nitrogen cycling (e.g. the occurrence of nitrification in nitrogen-poor soils), yet these effects scale non-linearly to the whole system. Different organisms may live and function in different types of microsites.

5. This new view of the nitrogen cycle provides an intellectual structure for developing research linking microbial populations and the nitrogen cycling processes they carry out.

Biological Diversity and Function in Soils, eds. Richard D. Bardgett, Michael B. Usher and David W. Hopkins.
Published by Cambridge University Press. © British Ecological Society 2005.

Introduction

Since the days of Winogradsky and Beijerink in the late nineteenth century, nitrogen cycling has been at the centre of soil microbiology. Since then, we have largely deciphered the microbial physiology of the important nitrogen cycling processes and have identified some of the important microbial groups involved in them. However, we are just beginning to understand the microbial community ecology of these processes and the interactions between process and population dynamics. Most ecosystem-level models ignore microbial community dynamics, yet still do an adequate job of describing large-scale flows of carbon and nitrogen (Schimel 2001). However, a number of studies suggest that the linkages between microbial processes and community dynamics may have ecosystem-level ramifications (Gulledge *et al.* 1997; Bodelier *et al.* 2000; Cavigelli & Robertson 2000, 2001; Degens *et al.* 2001; Balser *et al.* 2002), suggesting the value of developing this line of study. Understanding the linkages between microbial communities and the processes they mediate is likely to become more important as concern increases about global nitrogen deposition effects on ecosystems. In many cases, increased deposition is creating environmental conditions that have not previously existed, increasing the possibility that 'surprising' microbial responses may have important ecological effects. In this chapter, we will briefly consider what is currently known about linkages between microbial community composition and nitrogen cycling processes, and discuss points useful for developing further research in this area.

Theoretical background: when should microbial community composition matter?

In considering interactions between soil microbial community ecology and nitrogen cycling, there are a few basic concepts that are useful to develop up front. First, current theory suggests that influences of microbial community composition are most likely to be observed for processes that are physiologically or phylogentically 'narrow'. Processes that are physiologically narrow are those that involve a highly defined physiological pathway, such as nitrogen fixation, which is driven by the same enzyme (nitrogenase) in all organisms that fix N_2. Within the framework of narrow processes, it is worth distinguishing those that are phylogenetically diverse from those that are phylogenetically limited. Some narrow physiologies are found in only a phylogenetically limited group of organisms. These organisms are likely to be similar to each other in their overall behaviour but may vary in their specific environmental responses. An example in nitrogen cycling is nitrification; NH_4^+ oxidisers are all relatively closely related and share their overall physiology (aerobic autotrophs) but vary in their response to NH_4^+ concentration, temperature and pH (Laanbroek & Woldendorp 1995; Stark & Firestone 1996). Other narrow physiologies are widely distributed phylogenetically; examples include N_2 fixation and denitrification, where the pathways are distinct, but have spread widely across the microbial world. For example, nitrogenase is almost identical, whether it is found in the Cyanobacteria, the Actinomycetes or the Proteobacteria (Martinez-Romero 2001). In this case, the environmental response may vary dramatically depending on the identity of the organism carrying out the process.

The alternative to 'narrow' processes is, of course, 'broad' processes. It has been argued that such processes should be insensitive to microbial community composition (Schimel 1995). However, two types of processes have been lumped into this classification, and we propose to distinguish them, as they likely differ in terms of their sensitivity to community composition.

The first type of 'broad' processes are those carried out in biochemical pathways that are widely distributed across living organisms, such that almost all organisms carry out the process in the same way. From a process standpoint, it should therefore not matter which organism is actually carrying out the process (Schimel 1995). A classic example of such a process would be glycolysis, the metabolism of glucose, which is remarkably similar in *Escherichia coli* and *Homo sapiens*. The metabolism of simple phenolics appears to follow a similar pattern as well (Sugai & Schimel 1993). This kind of broad physiology should be largely insensitive to community composition at any scale of resolution (Beare *et al.* 1995; Schimel 1995). We propose naming these processes 'universal processes'.

The second type of processes that are considered 'broad' are those that we measure as a single process but are actually an aggregate of multiple distinct physiological processes (Schimel 1995). We propose naming these 'aggregate processes' to distinguish them from those carried out by universal processes. Examples of aggregate processes include microbial respiration and nitrogen mineralisation. We may measure soil CO_2 production as a single process, but it is actually the sum of many respiratory (and possibly fermentative) processes working simultaneously. Such processes are likely to involve a diverse collection of microbes all contributing to the overall process. Thus, when viewed as a whole, an aggregate process should appear insensitive to microbial community composition. However, it is possible that an aggregate process may be governed by a more specific set of narrow processes that are more responsive to community composition. An example of this would be litter decomposition, which may largely be regulated by lignin breakdown, at least in the later stages of decay (Aber *et al.* 1990; Coûteaux *et al.* 1995). Thus, aggregate processes may be quite distinct from universal processes: their observed sensitivity to community composition is a function of the scale at which they are studied.

Consideration of microbial process types in soil nitrogen cycling

This view of the different types of processes provides a useful framework for evaluating interactions among microbial communities and nitrogen cycling, and allows a greater consideration of which microbially mediated nitrogen processes are likely to be sensitive to changes in microbial community composition.

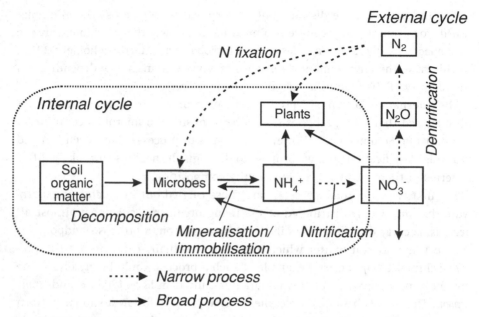

Figure 10.1. A 'classical' view of the nitrogen cycle, identifying internal vs. external nitrogen cycling processes and broad vs. narrow processes.

For example, the 'classical model' of soil nitrogen cycling (Fig. 10.1; Schimel & Bennett 2004) highlights several points where microbial community composition is potentially able to affect process dynamics: the narrow physiologies of nitrogen fixation, nitrification and denitrification. It is worth noting that these processes are primarily those regulating the import and export of nitrogen from the ecosystem. While substantial amounts of nitrification may be involved in the internal recycling of nitrogen-limited ecosystems (Stark & Hart 1997), nitrification still acts as a 'gatekeeper' process in many ways, connecting the internal cycling to the loss pathways of denitrification and leaching. Case studies exist for each of these processes showing how changes in microbial community composition may alter the dynamics of the nitrogen transformations.

The classic study showing how community differences may influence denitrification rates is the work by Cavigelli and Robertson (2000, 2001). In this pair of papers, they documented that denitrification varied substantially between an agricultural field and an untilled successional field. Major differences existed between the sites in the sensitivity of denitrification to O_2 and pH, which were attributed to different populations of denitrifiers in the sites. These same sites show strong differences in their response to wetting history, with the agricultural field showing dramatic changes in the products of denitrification (N_2 vs. N_2O) in response to varying moisture, while the successional site shows limited change in products under parallel conditions (Bergsma *et al.* 2002). There are also case studies showing at least moderate influence of community composition on process dynamics for nitrification. Stark and Firestone (1996) showed that nitrifier communities varied functionally between an open grassland and an adjacent oak woodland. The temperature sensitivities of the communities were different, with the temperature responses alone causing nitrification potentials to vary by as much as 20–25% between the different soils. Distinct nitrifier communities may also vary somewhat in their kinetic responses to NH_4^+ concentrations (Laanbroek & Woldendorp 1995; Stark & Firestone 1996). Similar examples can be shown for nitrogen fixation, where different strains vary in their efficacy at fixing nitrogen (Alexander 1985; Harrison *et al.* 1988; Benson & Silvester 1993).

Thus, it is becoming increasingly clear that the inputs and outputs of nitrogen from an ecosystem may be partially controlled by microbial community composition. However, except in rare cases (ecosystems dominated by nitrogen-fixing plants for example), nitrogen inputs and outputs are a relatively small fraction of the total amount of nitrogen cycling through an ecosystem. In the context of the classical model of nitrogen cycling, the bulk of nitrogen processing remains focused within the 'internal' cycle of mineralisation/immobilisation (Hart *et al.* 1994; Chapin *et al.* 2002).

According to the classical view of nitrogen cycling, mineralisation is an aggregate process, as defined above. Nitrogen is mineralised by many specific physiological processes involving a wide variety of nitrogen-containing compounds. Immobilisation, on the other hand, can be considered either a universal or an aggregate process. It is a universal process because nitrogen assimilation follows the same set of pathways in almost all organisms (Moat & Foster 1995). However, it can be seen as an aggregate process because nitrogen immobilisation can be coupled to the metabolism of any nitrogen-poor organic substrate. From either perspective, the classical view of nitrogen cycling leaves little room for sensitivity to microbial community composition. For a paper discussing the links between microbial community composition and nitrogen cycling that would be a frustrating conclusion, even if it were accurate.

An alternative perspective: changing the view of the soil nitrogen cycle

Over the last decade, however, our view of nitrogen cycling has changed dramatically from the classical view presented in Fig. 10.1. There have been a number of important developments in our thinking about the soil nitrogen cycle that alter our perception of nitrogen mineralisation/immobilisation as an aggregate process. We have been breaking down overall mineralisation/immobilisation into smaller scale processes that may be physiologically narrow and thus more sensitive to the effects of microbial community composition. We argue that there are two critical changes in our thinking that allow us to recognise separate components within the aggregate process of nitrogen mineralisation. These are an increased recognition of the role of (a) extracellular enzyme processes in soil nitrogen cycling, and (b) microsite dynamics in regulating large-scale nitrogen cycling phenomena. Research in these areas suggests that whole-soil nitrogen cycling phenomena are a non-linear integration of narrow processes, and that only through understanding how those narrow processes scale-up to the whole plant–soil system can we fully understand the nature of soil nitrogen cycling and the linkages between soil, microbial and plant processes.

In parallel with our improving understanding of the mechanisms involved in nitrogen cycling is our recognition of important factors that may actively modify the population-process linkages underlying them. One important factor in modifying nitrogen cycling is that nitrogen deposition and fertilisation may affect both communities and nitrogen cycling processes in ways that are counter-intuitive. A second factor is the specific role of roots in manipulating both substrate availability and microbial communities directly. In both of these cases, some of those observed effects may be predictable given a classical perspective on nitrogen cycling, while other effects may only be predictable or understandable when we break down the aggregate process of mineralisation into its individual components.

In the rest of this chapter, we will briefly discuss these aspects of nitrogen cycling and their relationships to microbial community composition. We do not intend this to be in any way a comprehensive review of these areas; a number of papers address some of the points we are covering, but not necessarily within the specific framework of trying to link microbial community composition and nitrogen cycling processes. Because of the dearth of research specifically addressing this topic, we draw few substantive conclusions, and our primary objective is to develop an intellectual framework for advancing research in this area.

Extracellular enzyme processes

While the existence and function of extracellular enzymes in a soil context has been known for some time (Dick & Tabatabai 1993; Sinsabaugh 1994), we are only now truly coming to terms with the full implications of their role in overall carbon and nitrogen cycling (Sinsabaugh 1994; Sinsabaugh *et al.* 1994; Schimel & Weintraub 2003; Schimel & Bennett 2004). Incorporating exoenzyme processes into the nitrogen cycle breaks the soil organic matter pool into two fundamentally distinct pools: insoluble macromolecular material and dissolved organic nitrogen compounds. Thus, this change separates the single mineralisation step into two processes that are under different controls. Depolymerisation is carried out by exoenzymes and so is under 'biochemical' control, while the use of the monomers produced is under direct organismal, or 'biological', control.

There are two important aspects of the role of exoenzymes that need to be considered when attempting to link soil nitrogen cycling and microbial communities. The first is that breakdown of polymeric organic matter is carried out by several exoenzymes, each of which may be synthesised by a limited suite of microorganisms. Thus, enzyme processes become potentially narrow physiologies and sensitive to shifts in microbial community composition (Waldrop et al. 2000). The second important aspect is that incorporating exoenzymes into kinetic models of soil organic matter dynamics changes our perspectives on the regulation of microbial growth and carbon vs. nitrogen limitation, and thus of the factors regulating microbial community composition. If certain portions of a microbial community are strongly nitrogen limited, elevating nitrogen may accelerate their growth, shifting community composition and the processes carried out by that community.

Enzyme processes as narrow physiologies

There are at least three critical enzyme systems that regulate soil carbon and nitrogen cycling: protease, polysaccharide-degrading enzymes (cellulases, xylanases) and phenol oxidases. Proteases are produced very widely by microorganisms and protease activity in soil may be important in regulating nitrogen mineralisation (Zaman et al. 1999). While cellulose and phenol oxidases are not directly involved in nitrogen metabolism, they are critical in driving carbon cycling, and thus must be considered in evaluating mineralisation and immobilisation dynamics. Cellulose breakdown is a critical step in litter decomposition. While cellulase is produced by a diverse array of bacteria and fungi (Bhat & Bhat 1997; Bayer et al. 2001), it is not produced by all microbes and thus there is the potential for cellulose metabolism to be a function of community composition. Polyphenol-degrading enzymes may be particularly important in nitrogen cycling, since they are involved in degrading lignin and tannin (Gamble et al. 1996), two important polyphenols that can bind to polysaccharides and proteins respectively, potentially regulating the flow of carbon and nitrogen in soil.

Lignin forms chemical complexes with cellulose and hemicellulose in plant cell walls, making those materials inaccessible to direct microbial breakdown. Thus lignin breakdown may regulate overall decomposition and nitrogen mineralisation from plant structural material, which comprises the bulk of older litter and organic soil materials. For this reason, cellulose breakdown is constrained by the rate of lignin metabolism (Aber et al. 1990; Coûteaux et al. 1995; Berg 2000). For example, in year-long laboratory incubations of Alaskan tundra organic soils, as much as one-third of the total carbon was respired but the proportions of lignin and cellulose were unchanged (Weintraub & Schimel 2003). Variation in the ability of different microorganisms to degrade lignin therefore has the potential to substantially alter the rate of organic matter breakdown and the associated mineralisation/immobilisation dynamics.

One important aspect of the behaviour of lignolytic microorganisms in the context of microbial community composition and nitrogen cycling is their variable sensitivity to nitrogen (Keyser *et al.* 1978). For example, some white rot fungi, such as *Phanaerochaete chrysosporium*, are highly sensitive to NH_4^+ concentrations, with lignolytic enzyme synthesis repressed by elevated levels of NH_4^+ (Keyser *et al.* 1978; Gold & Alic 1993). However, other white rot fungi are either insensitive to NH_4^+ or are actually stimulated by it (Périé & Gold 1991; Kaal *et al.* 1993). Thus, the complex interaction of community composition and nitrogen availability creates a strong possibility for variation in lignin metabolism and corresponding variation in the rates of decomposition and nitrogen mineralisation. Resolving these interactions requires answering some specific questions, such as: which organisms actually produce these enzymes? Under what conditions? How does the environmental control of enzyme synthesis vary with microbial community composition?

Tannins, both hydrolysable and condensed, are primarily involved in directly complexing proteins, making them unavailable for microbial attack (Bhat *et al.* 1998; Bradley *et al.* 2000) and this process may be important in making nitrogen unavailable in nitrogen-poor ecosystems (Fierer *et al.* 2001). Thus, tannin-degrading enzymes (tannase and other phenol oxidases) may be immediately involved in nitrogen cycling and mineralisation. The ability to break down tannin–protein complexes is highly variable among different microorganisms (Bhat *et al.* 1998). While the ability to degrade tannin–protein complexes has been demonstrated in some saprophytic fungi and ericaceaous mycorrhizae, ectomycorrhizal fungi that have been tested show only limited ability to degrade these complexes (Bending & Read 1996, 1997). Additionally, tannins are toxic to some microorganisms, but sensitivity differs between microbes and specific tannins (Scalbert 1991; Field & Lettinga 1992; Schulz *et al.* 1992). Whole communities, in fact, may vary in their sensitivity to tannins. For example, Fierer *et al.* (2001) found that small molecular weight condensed tannins from Alaskan balsam poplar were used as substrates by microbes in a balsam poplar stand but were toxic to litter microbes in a nearby alder stand. The available data therefore suggest that there are strong variations at the community level in the narrow physiologies associated with tannin tolerance and polyphenol breakdown. Hence, the distribution of microbes that produce tannin-degrading enzymes may be an important factor in regulating nitrogen mineralisation.

Exoenzymes and microbial nitrogen limitation

It is generally accepted in ecology that the composition of a biotic community is regulated by the available resources and, in particular, by the nature of the limiting resource. Since microbes with differing physiological characteristics (e.g. fungi vs. bacteria) have widely different nitrogen requirements, varying nitrogen availability may select for microbial communities with varying function. Thus, understanding both the composition and the functioning of soil communities probably requires understanding the nature of resource limitation and the extent to which different components of the community may be limited by different resources, including nitrogen. There is some evidence of this; for example competition among basidiomycetes and ascomycetes in decomposing wood (Fog 1988).

Since the early part of the last century, the dogma has been that soil microbes are generally carbon, rather than nitrogen limited, as argued by Waksman (1932): 'The fact that the addition of available nitrogen, phosphorus, and potassium did not bring about any appreciable increase in the evolution of CO_2 points definitely to the fact that nitrogen is not a limiting factor in the activities of microorganisms in peat but that the available carbon compounds are.' However, increasingly, evidence is accumulating that microbes may frequently be nitrogen limited (Nadelhoffer et al. 1984; Jackson et al. 1989; Schimel & Firestone 1989; Giblin et al. 1991; Polglase et al. 1992; Hart et al. 1994; Hart & Stark 1997; Wagener & Schimel 1998; Chen & Stark 2000). The commonly observed lack of a CO_2 response to added nitrogen often appears contradictory to responses in nitrogen metabolism, and suggests that these results may, in fact, be the result of shifted carbon and nitrogen metabolism rather than a lack of response to nitrogen. Furthermore, the fundamental logic used by Waksman (1932) has recently been challenged (Schimel & Weintraub 2003). Microbes will process and respire simple carbon compounds through 'overflow metabolism' (Tempest & Neijssel 1992), even if they lack the nutrients to produce new biomass (Schimel & Weintraub 2003). Thus, the tenet that microbial activity is limited by carbon may be over simplified. It therefore becomes important to understand how exoenzyme dynamics regulate carbon flow and the extent of microbial carbon vs. nitrogen limitation and, through this, microbial community composition and function.

A theoretical model by Schimel and Weintraub (2003) explored the implications for microbial carbon and nitrogen limitation of incorporating exoenzymes in the decomposition and soil carbon flow system. While traditional soil organic matter (SOM) models all use kinetics that are first order on substrate, fundamental chemistry argues that the concentration of the catalyst must be a part of the rate expression. In a variable environment like soil, it is likely that catalyst concentration is not constant, and may need to be considered in biogeochemical models (Schimel 2001). Incorporating enzyme concentration into an SOM model requires considering the investment in enzyme production, which can be substantial (Umikalsom et al. 1997; Romero et al. 1999). When carbon flow to enzymes is incorporated into a model that examines decomposition and microbial growth, several conclusions arose that have implications for this chapter:

1. Microbial carbon limitation is regulated by the return on investment in exoenzymes – only if this value is zero or less will microbes be carbon limited. As long as microbes receive more carbon and energy back from

decomposition than they 'spent' synthesising the enzymes responsible, they will be able to grow and produce more enzymes.

2. Enzyme kinetics must be non-linear, with a decreasing rate of activity with increasing enzyme concentration. Linear kinetics make the system unstable, since a constant, positive return on investment leads to runaway enzyme synthesis and decomposition rates. Non-linear kinetics, however, ultimately lead to carbon-limited microbes even in the presence of potentially available carbon.

3. Because the kinetics of the exoenzyme system stabilise carbon flow, microbial growth may be limited by nitrogen without altering the overall flow of carbon in the system. Carbon that is not used to support growth is still consumed through one of several 'overflow metabolism' processes (Tempest & Neijssel 1992).

The role of exoenzymes in regulating microbial carbon vs. nitrogen limitation and the fate of carbon flowing through the soil system has not been well developed in experimental studies (Schimel & Weintraub 2003). In evaluating microbe–nitrogen interactions, we need to focus better on microbial growth and turn-over to understand the nature and extent of nutrient limitation and how this then alters community composition and function (Hart & Stark 1997).

Microsite phenomena

The second major development that helps us better understand the role of specific microbial community composition in nitrogen cycling comes from our growing understanding of microsite phenomena in soil.

The soil is not a homogeneous medium, rather it is the most physically heterogeneous environment for life on the planet. We often analyse soil communities as an aggregate entity and are amazed by the number of species of bacteria that can be detected in even one-tenth of a gram of soil. Yet to say that the microflora of soil is diverse is analogous to saying that the plant flora of Europe is diverse. Likely, much of that diversity results from there being a plethora of communities within the larger area. Just as Europe spans from the tundra high in the Alps to the arid shrublands of Spain, a gram of soil incorporates environments as varied as a fresh fragment of decomposing leaf and mineral-bound humics in the interior of a microaggregate. While microsite phenomena have been recognised for many years, the full implications of microsite dynamics in linking microbial community composition and function have not.

Only starting in the late 1980s have researchers attempted meaningfully to analyse how soil microstructure really regulates ecosystem processes (Ladd *et al.* 1993). Early examples include work on denitrification, such as that of Parkin (1987) who showed that 85% of the total denitrification in a soil core occurred in a single fragment of decomposing plant material. Microsites have also been used to explain how microbial nitrate consumption can occur in the presence of available NH_4^+ in soil (Jackson *et al.* 1989; Rice & Tiedje 1989). More recently, however, work has strongly suggested the importance of microsite phenomena in regulating mineralisation/immobilisation dynamics (Chen & Stark 2000; Schimel & Bennett 2004).

Relatively nitrogen-rich and nitrogen-poor microsites exist in soil, producing zones where mineralisation and immobilisation occur separately but simultaneously (e.g. Jackson *et al.* 1989; Davidson *et al.* 1990; Chen & Stark 2000). These sites are linked by diffusion of available nitrogen, alternatively as organic nitrogen, NH_4^+ or NO_3^- (Jingguo & Bakken 1997; Chen & Stark 2000). Schimel and Bennett (2004) developed these ideas to explore how these phenomena could explain the observed differences in nitrogen cycling across a wide nitrogen-availability gradient (Fig. 10.2). When the primary substrates in soil are very nitrogen poor, microbes will be nitrogen limited, even in microsites of relatively high nitrogen availability. These organisms will take up organic nitrogen compounds and will largely retain the nitrogen contained within them. Other organisms (heterotrophs in nitrogen-poor sites and roots) will depend on organic nitrogen forms that 'escape' and diffuse out from the nitrogen-rich sites. As nitrogen availability increases, net nitrogen mineralisation will begin to occur in the nitrogen-rich sites, and NH_4^+ or NO_3^- will diffuse away from those sites to be competed for between plant roots and heterotrophs. As substrate quality increases further, microbes will become generally nitrogen saturated, and net mineralisation and nitrification will ultimately dominate the ecosystem. At the extreme, such as in agricultural systems, the nitrogen cycle appears to become dominated by NO_3^-. Although some of the critical processes appear to be happening at micro scales, these phenomena do not scale-up as simply a linear combination to regulate whole-system nitrogen cycling; hence the occurrence and importance of NO_3^- in apparently nitrogen-limited ecosystems (e.g. Stark & Hart 1997).

An appreciation of microsite phenomena in ecosystem nitrogen cycling frames a new perspective on possible links between community composition and nitrogen cycling processes. Simply due to space limitation, community composition in individual microsites is likely to be constrained. For this reason, individual nitrogen cycling processes may be carried out by a small subset of the entire microbial community present in the bulk soil, effectively 'narrowing' what we would assume to be a relatively broad process. Additionally, recognising the importance of microsites highlights the potential importance of interactions between species and functional groups in regulating process dynamics (Wheatley *et al.* 2001). This microsite hypothesis raises a number of related questions that we are only beginning to answer (Blackwood & Paul 2003): How do different types of microbes respond to different types of microsites? Do

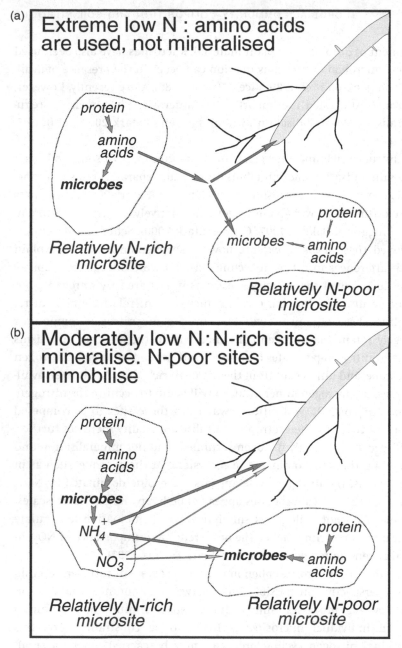

Figure 10.2. A view of nitrogen cycling that incorporates both extracellular enzyme processes and microsite processes. The two panels represent two cases from along a nitrogen-availability gradient and demonstrate how the nature of nitrogen cycling may appear to change as overall ecosystem nitrogen availability changes, even though the underlying microbial and enzyme processes remain the same. The width of the arrows indicates the relative magnitude of the flows. In (b), diffusion of amino acids has been left out purely for clarity – it still likely occurs. It should be noted that mycorrhizal fungi may be the agents of depolymerisation (by producing extracellular enzymes) in nitrogen-rich sites and thus the microbes take up amino acids directly, making that nitrogen available to plants. Derived from Schimel and Bennett (2004). Reproduced with permission of the Ecological Society of America.

you find the same collection of organisms in different types of soil microsites? How dynamic in time are different types of microsites? How does the dynamic nature of microsites affect the demands for growth and colonisation ability of organisms that may be specialised for particular microsite types? Are nitrogen-rich vs. nitrogen-poor microsites created entirely by the nature of substrates or do different microbial communities produce areas that have a nitrogen surplus or a nitrogen deficit? Only research that attempts to link microsite conditions with microbial communities and processes will be able to answer these questions and perhaps facilitate scaling microsite processes up to the ecosystem level.

Effects of elevated nitrogen on ecosystem processes

As mentioned above, there is growing concern over the effects of elevated nitrogen on ecosystem processes, including nitrogen cycling. The observed effects of nutrient additions on soil processes are highly variable. For example, nitrogen additions can have positive, negative or minimal effects on the decomposition of soil organic matter, depending on the specific soil type examined (Fog 1988; Berg & Matzner 1997). The particular response of a soil process to elevated nitrogen may partly be a function of how the nitrogen additions affect the composition of soil microbial communities and their physiologies.

Previous studies have shown that nitrogen additions can alter the composition of soil microbial communities (Bardgett et al. 1999; McCaig et al. 1999; Sarathchandra et al. 2001). In all likelihood, increasing nitrogen inputs alter microbial community dynamics by selecting for different groups of microbes with distinct physiologies. We already discussed one particular example of this in terms of the different nitrogen responses of lignolytic organisms. Others may exist as well.

Overall, the variability in the observed effects of elevated nitrogen on soil processes probably results from a complicated mix of specific effects in soil. This is highlighted by the work of Neff et al. (2002) who showed that increasing nitrogen deposition to soil accelerated the breakdown of labile, light-fraction carbon, while inhibiting the breakdown of recalcitrant, heavy-fraction carbon. Thus, although there was no net effect on soil respiration, they observed substantial alterations of soil carbon and nitrogen cycling processes. They were not able to define clearly the specific mechanisms responsible for the observed results. However, several possible mechanisms could explain the reduced mineralisation of recalcitrant organic matter (OM): (1) nitrogen complexation increases the stabilisation of recalcitrant OM; (2) substrate switching – with adequate nitrogen, microbes may reduce dependence on recalcitrant, low carbon:nitrogen ratio material as a nitrogen source; (3) nitrogen addition may cause a shift in the microbial community, reducing the numbers of organisms capable of degrading recalcitrant OM; or (4) nitrogen additions repress the production of

enzymes required to mineralise recalcitrant soil carbon. Since lignin biomarkers in the light fraction disappeared rapidly, the often-postulated inhibition of
lignolytic activity by added nitrogen (see Fog 1988) would not be a plausible
explanation in this specific case study.

Thus, elevated nitrogen concentrations have the potential to alter critical processes directly, such as the mineralisation of recalcitrant soil carbon, but the
actual overall response may be hard to predict due to the interactive effects of
nitrogen addition on both microbial community composition and physiology.

Effects of plant roots on microbial communities and nitrogen cycling

Plant roots and their exudates play complex roles in regulating all factors that
affect soil communities and nitrogen cycling. Many of these effects have been
evaluated in recent reviews (Cheng 1999; Dakora & Phillips 2002; Kuzyakov 2002).
There is little new to add to those syntheses, and so we wish merely to highlight
that root interactions are critical in understanding how community–process
interactions actually function in soil. Plant roots alter the structure of microbial
communities in the rhizosphere (Jaeger *et al.* 1999; Kuzyakov 2002), and thereby
alter soil carbon and nitrogen cycling including mineralisation/immobilisation
reactions (Cheng 1999; Dakora & Phillips 2002). The mechanisms for root effects
on soil nitrogen cycling may be complex and are likely to involve the direct
manipulation of carbon vs. nitrogen availability, microbial community composition and microsite structure. As a result of these multiple influences, roots may
either stimulate or inhibit nitrogen mineralisation depending on the specific
balance of multiple variables (Cheng 1999; Kuzyakov 2002).

Conclusions

Our changing understanding of soil nitrogen cycling and how it may couple with
carbon cycling processes provides a new perspective and framework with which
to assess nitrogen cycling–microbe interactions. This new understanding should
allow us to explore the topic with a greater level of sophistication than we could
by applying the 'traditional' view of the nitrogen cycle. As long as we view nitrogen mineralisation as a simple aggregate process driven simply by OM turn-over
and carbon:nitrogen ratios, we will be blind to the specific roles of microbial
community composition in regulating nitrogen cycling. An advanced perspective
on nitrogen cycling, one that considers both the full chemistry of decomposition
and the spatial structure of soil, can break nitrogen cycling processes down into
individual narrow processes, increasing the likelihood that we will be able to
understand when and how nitrogen cycling processes are mediated by microbial
community composition. Through this, we will develop a better understanding
of ecosystem-level nitrogen cycling and how it will respond to novel conditions
and anthropogenic disturbances.

References

Aber, J. D., Melillo, J. M. & McClaugherty, C. A. (1990). Predicting long-term patterns of mass loss, nitrogen dynamics, and soil organic matter formation from initial fine litter chemistry in temperate forest ecosystems. *Canadian Journal of Botany*, **68**, 2201–2208.

Alexander, M. (1985). Ecological constraints on nitrogen fixation in agricultural ecosystems. *Advances in Microbial Ecology*, **8**, 163–183.

Balser, T., Kinzig, A. & Firestone, M. (2002). The functional consequences of biodiversity. *The Functional Consequences of Biodiversity* (Ed. by A. Kinzig, S. Pacala & D. Tilman), pp. 265–293. Princeton, NJ: Princeton University Press.

Bardgett, R. D., Mawdsley, J. L., Edwards, S., et al. (1999). Plant species and nitrogen effects on soil biological properties of temperate upland grasslands. *Functional Ecology*, **13**, 650–660.

Bayer, E. A., Shoham, Y. & Lamed, R. (2001). Cellulose-decomposing bacteria and their enzyme systems. *The Prokaryotes: An Evolving Electronic Resource for the Microbiological Community*, 3rd edition, release 3.7, November 2, 2001 (Ed. by M. Dworkin, et al.). New York: Springer-Verlag, www.prokaryotes.com.

Beare, M. H., Coleman, D. C. & Crossley, D. A. (1995). A hierarchical approach to evaluating the significance of soil biodiversity to biogeochemical cycling. *Plant and Soil*, **170**, 5–22.

Bending, G. D. & Read, D. J. (1996). Nitrogen mobilization from protein–polyphenol complex by ericoid and ectomycorrhizal fungi. *Soil Biology and Biochemistry*, **28**, 1603–1612.

Bending, G. D. & Read, D. J. (1997). Lignin and soluble phenolic degradation by ectomycorrhizal and ericoid mycorrhizal fungi. *Mycological Research*, **101**, 1348–1354.

Benson, D. R. & Silvester, W. B. (1993). Biology of Frankia strains, actinomycete symbionts of actinorhizal plants. *Microbiological Reviews*, **57**, 293–319.

Berg, B. (2000). Litter decomposition and organic matter turnover in northern forest soils. *Forest Ecology and Management*, **133**, 13–22.

Berg, B. & Matzner, E. (1997). Effect of N deposition on plant litter and soil organic matter in forest systems. *Ecological Reviews*, **5**, 1–25.

Bergsma, T. T., Robertson, G. P. & Ostrom, N. E. (2002). Influence of soil moisture and land use history on denitrification end-products. *Journal of Environmental Quality*, **31**, 711–717.

Bhat, M. K. & Bhat, S. (1997). Cellulose degrading enzymes and their potential industrial applications. *Biotechnology Advances*, **15**, 583–620.

Bhat, T. K., Sing, B. & Sharma, O. P. (1998). Microbial degradation of tannins: a current perspective. *Biodegradation*, **9**, 343–357.

Blackwood, C. B. & Paul, E. A. (2003). Eubacterial community structure and population size within the soil light fraction, rhizosphere, and heavy fraction of several agricultural systems. *Soil Biology and Biochemistry*, **35**, 1245–1255.

Bodelier, P. L. E., Roslev, P., Henckel, T. & Frenzel, P. (2000). Stimulation by ammonium-based fertilizers of methane oxidation in soil around rice roots. *Nature*, **403**, 421–424.

Bradley, R. L., Titus, B. D. & Preston, C. P. (2000). Changes to mineral N cycling and microbial communities in black spruce humus after additions of $(NH_4)_2SO_4$ and condensed tannins extracted from *Kalmia angustifolia* and balsam fir. *Soil Biology and Biochemistry*, **32**, 1227–1240.

Cavigelli, M. A. & Robertson, G. P. (2000). The functional significance of denitrifier community composition in a terrestrial ecosystem. *Ecology*, **81**, 1402–1414.

Cavigelli, M. A. & Robertson, G. P. (2001). Role of denitrifier diversity in rates of nitrous oxide consumption in a terrestrial ecosystem. *Soil Biology and Biochemistry*, **33**, 297–310.

Chapin, F. S., III., Matson, P. & Mooney, H. (2002). *Principles of Terrestrial Ecosystem Ecology*. New York: Springer-Verlag.

Chen, J. & Stark, J. (2000). Plant species effects and carbon and nitrogen cycling in a sagebrush–crested wheatgrass soil. *Soil Biology and Biochemistry*, **32**, 47–57.

Cheng, W. (1999). Rhizosphere feedbacks in elevated CO_2. *Tree Physiology*, **19**, 313–320.

Coûteaux, M. M., Bottner, P. & Berg, B. (1995). Litter decomposition, climate and litter quality. *Trends in Ecology and Evolution*, **10**, 63–66.

Dakora F. D. & Phillips, D. A. (2002). Root exudates as mediators of mineral acquisition in low-nutrient environments. *Plant and Soil*, **245**, 35–47.

Davidson, E. A., Stark, J. M. & Firestone, M. K. (1990). Microbial production and consumption of nitrate in an annual grassland. *Ecology*, **71**, 1968–1975.

Degens, B. P., Schipper, L. A., Sparling, G. P. & Duncan, L. C. (2001). Is the microbial community in a soil with reduced catabolic diversity less resistant to stress or disturbance? *Soil Biology and Biochemistry*, **33**, 1143–1153.

Dick, W. A. & Tabatabai, M. A. (1993). Significance and potential uses of soil enzymes. *Soil Microbial Ecology* (Ed. by F. B. Metting, Jr.), pp. 95–127. New York: Marcel Dekker.

Field, J. A. & Lettinga, G. (1992). Toxicity of tannin compounds to microorganisms. *Plant Polyphenols* (Ed. by R. W. Hemingway & P. E. Laks), pp. 673–692. New York: Plenum Press.

Fierer, N., Schimel, J. P., Cates, R. G. & Zou, J. (2001). The influence of balsam poplar tannin fractions on carbon and nitrogen dynamics in Alaskan taiga floodplain soils. *Soil Biology and Biochemistry*, **33**, 1827–1839.

Fog, K. (1988). The effect of added nitrogen on the rate of decomposition of organic matter. *Biological Review*, **63**, 433–462.

Gamble, G. R., Akin, D. E., Makkar, H. P. S. & Becker, K. (1996). Biological degradation of tannins in *Sericea lespedeza* (*Lespedeza cuneata*) by the white rot fungi *Ceriporiopsis subvermispora* and *Cyathus stercoreus* analyzed by solid-state ^{13}C nuclear magnetic resonance spectroscopy. *Applied and Environmental Microbiology*, **62**, 3600–3604.

Giblin, A. E., Nadelhoffer, K. J., Shaver, G. R., Laundre, J. A. & McKerrow, A. J. (1991). Biogeochemical diversity along a riverside toposequence in arctic Alaska. *Ecological Monographs*, **61**, 415–435.

Gold, M. H. & Alic, M. (1993). Molecular biology of the lignin-degrading Basidiomycete *Phanerochaete chrysosporium*. *Microbiological Reviews*, **57**, 605–622.

Gulledge, J. M., Doyle, A. P. & Schimel, J. P. (1997). Different NH_4^+-inhibition patterns of soil CH_4 consumption: a result of distinct CH_4 oxidizer populations across sites? *Soil Biology and Biochemistry*, **29**, 13–21.

Harrison, S. P., Jones, D. G., Schunmann, P. H. D., Forster, J. W. & Young, J. P. W. (1988). Variation in *Rhizobium leguminosarum* Biovar *trifolii* sym plasmids and the association with effectiveness of nitrogen-fixation. *Journal of General Microbiology*, **134**, 2721–2730.

Hart, S. C., Nason, G. E., Myrold, D. D. & Perry, D. A. (1994). Dynamics of gross nitrogen transformations in an old-growth forest: the carbon connection. *Ecology*, **75**, 880–891.

Hart, S. C. & Stark, J. M. (1997). Nitrogen limitation of the microbial biomass in an old-growth forest soil. *Ecoscience*, **4**, 91–98.

Jackson, L. E., Schimel, J. P. & Firestone, M. K. (1989). Short-term partitioning of ammonium and nitrate between plants and microbes in an annual grassland. *Soil Biology and Biochemistry*, **21**, 409–415.

Jaeger, C. I., Lindow, S., Miller, W., Clark, E. & Firestone, M. K. (1999). Mapping of sugar and amino acid availability in soil around roots with bacterial sensors of sucrose and tryptophan. *Applied Environmental Microbiology*, 65, 2685–2690.

Jingguo, W. & Bakken, L. (1997). Competition for nitrogen during decomposition of plant residues in soil: effect of spatial placement of N-rich and N-poor plant residues. *Soil Biology and Biochemistry*, 29, 153–162.

Kaal, E. E. J., DeJong, E. & Field, J. A. (1993). Stimulation of ligninolytic peroxidase-activity by nitrogen nutrients in the white-rot fungus *Bjerkandera sp.* strain BOS55. *Applied and Environmental Microbiology*, 59, 4031–4036.

Keyser, P., Kirk, T. K. & Zeikus, J. G. (1978). Ligninolytic enzyme system of *Phanerochaete chrysosporium*: synthesized in absence of lignin in response to nitrogen starvation. *Journal of Bacteriology*, 135, 790–797.

Kuzyakov, Y. (2002). Review: factors affecting rhizosphere priming effects. *Journal of Plant Nutrition and Soil Science*, 165, 382–396.

Laanbroek, H. J. & Woldendorp, J. W. (1995). Activity of chemolithotrophic nitrifying bacteria under stress in natural soils. *Advances in Microbial Ecology*, 14, 275–304.

Ladd, J. N., Foster, R. C. & Skjemstad, J. O. (1993). Soil structure: carbon and nitrogen metabolism. *Geoderma*, 56, 401–434.

Martinez-Romero, E. (2001). Dinitrogen-fixing prokaryotes. *The Prokaryotes: An Evolving Electronic Resource for the Microbiological Community*, 3rd edition, release 3.7, November 2, 2001 (Ed. by M. Dworkin, et al.) New York: Springer-Verlag, www.prokaryotes.com.

McCaig, A. E., Phillips, C. J., Stephen, J. R., et al. (1999). Nitrogen cycling and community structure of proteobacterial beta-subgroup ammonia-oxidizing bacteria within polluted marine fish farm sediments. *Applied and Environmental Microbiology*, 65, 213–220.

Moat, A. G. & Foster, J. W. (1995). *Microbial Physiology*, 3rd edition. New York: Wiley-Liss.

Nadelhoffer, K. J., Aber, J. D. & Melillo, J. M. (1984). Seasonal patterns of ammonium and nitrate uptake in nine temperate forest ecosystems. *Plant and Soil*, 80, 321–335.

Neff, J. C., Townsend, A. R., Gleixner, G., et al. (2002). Variable effects of nitrogen additions on the stability and turnover of soil carbon. *Nature*, 419, 915–917.

Parkin, T. B. (1987). Soil microsites as a source of denitrification variability. *Soil Science Society of America Journal*, 51, 1194–1199.

Périé, F. H. & Gold, M. H. (1991). Manganese regulation of manganese peroxidase expression and lignin degradation by the white rot fungus *Dichomitus-squalens*. *Applied and Environmental Microbiology*, 57, 2240–2245.

Polglase, P. J., Attiwill, P. M. & Adams, M. A. (1992). Nitrogen and phosphorus cycling in relation to stand age of *Eucalyptus regnans* F. Muell: II. N mineralization and nitrification. *Plant and Soil*, 142, 167–176.

Rice, C. W. & Tiedje, J. M. (1989). Regulation of nitrate assimilation by ammonium in soils and in isolated soil-microorganisms. *Soil Biology and Biochemistry*, 21, 597–602.

Romero, M. D., Aguado, J., González, L. & Ladero, M. (1999). Cellulase production by *Neurospora crassa* on wheat straw. *Enzyme and Microbial Technology*, 25, 244–250.

Sarathchandra, S. U., Ghani, A., Yeates, G. W., Burch, G. & Cox, N. R. (2001). Effect of nitrogen and phosphate fertilisers on microbial and nematode diversity in pasture soils. *Soil Biology and Biochemistry*, 33, 953–964.

Scalbert, A. (1991). Antimicrobial properties of tannins. *Phytochemistry*, 30, 3875–3883.

Schimel, J. (1995). Ecosystem consequences of microbial diversity and community structure. *Arctic and Alpine Biodiversity: Patterns, Causes and Ecosystem Consequences* (Ed. by F. S. Chapin & C. Korner), pp. 239–254. Berlin: Springer-Verlag.

Schimel, J. P. (2001). Biogeochemical models: implicit vs. explicit microbiology. *Global Biogeochemical Cycles in the Climate System* (Ed. by E. D. Schulze, S. P. Harrison, M. Heimann, et al.), pp. 177–183. San Diego, CA: Academic Press.

Schimel, J. P. & Bennett, J. (2004). Nitrogen mineralization: challenge of a changing paradigm. *Ecology*, **85**, 591–602.

Schimel, J. P. & Firestone, M. K. (1989). Inorganic nitrogen incorporation by coniferous forest floor material. *Soil Biology and Biochemistry*, **21**, 41–46.

Schimel, J. P. & Weintraub, M. N. (2003). Implications of exoenzyme activity on microbial carbon and nitrogen limitation in soil: a theoretical model. *Soil Biology and Biochemistry*, **35**, 549–563.

Schulz, J. C., Hunter, M. D. & Appel, H. M. (1992). Antimicrobial activity of polyphenols mediates plant–herbivore interactions. *Plant Polyphenols* (Ed. by R. W. Hemingway & P. E. Laks), pp. 621–637. New York: Plenum Press.

Sinsabaugh, R. L. (1994). Enzymic analysis of microbial pattern and process. *Biology and Fertility of Soils*, **17**, 69–74.

Sinsabaugh, R. L., Moorhead, D. L. & Linkins, A. E. (1994). The enzymic basis of plant litter decomposition: emergence of an ecological process. *Applied Soil Ecology*, **1**, 97–111.

Stark, J. M. & Firestone, M. K. (1996). Kinetic characteristics of ammonium-oxidizer communities in a California oak woodland–annual grassland. *Soil Biology and Biochemistry*, **28**, 1307–1317.

Stark, J. M. & Hart, S. C. (1997). High rates of nitrification and nitrate turnover in undisturbed coniferous forests. *Nature*, **385**, 61–64.

Sugai, S. F. & Schimel, J. P. (1993). Decomposition and biomass incorporation of ^{14}C-labeled glucose and phenolics in taiga forest floor: effect of substrate quality, successional state, and season. *Soil Biology and Biochemistry*, **25**, 1379–1389.

Tempest, D. W. & Neijssel, O. M. (1992). Physiological and energetic aspects of bacterial metabolite overproduction. *FEMS Microbiology Letters*, **100**, 169–176.

Umikalsom, M. S., Ariff, A. B., Shamsuddin, Z. H., *et al.* (1997). Production of cellulase by a wild strain of *Chaetomium globosum* using delignified oil palm empty-fruit-bunch fibre as substrate. *Applied Microbiology and Biotechnology*, **47**, 590–595.

Wagener, S. M. & Schimel, J. P. (1998). Stratification of soil ecological processes: a study of the birch forest floor in the Alaskan taiga. *Oikos*, **81**, 63–74.

Waksman, S. (1932). *Principles of Soil Microbiology*, 2nd edition. Baltimore, Md: Williams and Wilkins.

Waldrop, M. P., Balser, T. C. & Firestone, M. K. (2000). Linking microbial community composition to function in a tropical soil. *Soil Biology and Biochemistry*, **32**, 1837–1846.

Weintraub, M. N. & Schimel, J. P. (2003). Interactions between carbon and nitrogen mineralization and soil organic matter chemistry in Arctic tundra soils. *Ecosystems*, **6**, 129–143.

Wheatley, R. E., Ritz, K., Crabb, D. & Caul, S. (2001). Temporal variations in potential nitrification dynamics in soil related to differences in rates and types of carbon and nitrogen inputs. *Soil Biology and Biochemistry*, **33**, 2135–2144.

Zaman, M., Di, H. J., Cameron, K. C. & Frampton, C. M. (1999). Gross nitrogen mineralization and nitrification rates and their relationships to enzyme activities and the soil microbial biomass in soils treated with dairy shed effluent and ammonium fertilizer at different water potentials. *Biology and Fertility of Soils*, **29**, 178–186.

CHAPTER ELEVEN

Biodiversity of saprotrophic fungi in relation to their function: do fungi obey the rules?

CLARE H. ROBINSON
King's College, University of London
E. JANIE PRYCE MILLER
King's College, University of London
and Centre for Ecology and Hydrology Lancaster
LEWIS J. DEACON
King's College, University of London
and Centre for Ecology and Hydrology Lancaster

SUMMARY

1. The mycelia of fungal communities in soil and plant litter are strongly structured by soil horizon and resource availability. Resource quality is important in determining species composition and a certain degree of 'host' specificity exists. In soil fungal communities, a few taxa occur much more frequently than the large number of rare ones. The taxa detected are highly dependent on the techniques used. Therefore, it is necessary to cross-reference the information obtained from different methods.
2. Fungal communities in soil and plant litter are enormously diverse taxonomically, with possibly hundreds of species present in a particular soil horizon. There is still much work to be carried out at the local scale to detect the mycelia of fungi and identify them, together with estimating fungal species richness. Without these initial taxonomic studies, it is impossible subsequently to relate mycelial location and function to species diversity.
3. Scattered data exist about functional diversity of fungi in soil and plant litter, and there is still far to go before specific fungal decomposer functions are satisfactorily described, especially in the natural environment. Again, a combination of methods is needed. The results of functional tests, especially for 'key' species, should be related to community structure.
4. Are all the possibly hundreds of fungal species present on decomposing plant litter necessary to maintain decomposition rates? There is some evidence for functional redundancy because frequently isolated species have

Biological Diversity and Function in Soils, eds. Richard D. Bardgett, Michael B. Usher and David W. Hopkins.
Published by Cambridge University Press. © British Ecological Society 2005.

been found to have the same specific enzyme capabilities for decomposition as occasional ones. The idiosyncratic hypothesis may also be supported. The existence of 'keystone' species, on which the maintenance of whole ecosystems may rely, suggests that decomposition rate is dependent more on fungal species composition, and its functional repertoire, rather than on simply richness alone.

Introduction

Saprotrophic (decomposer) fungi often underpin nutrient cycling in soil and plant litter in terrestrial ecosystems but are sensitive to disturbance, pollution and environmental change. Current thinking about fungal diversity has focused on global taxonomic richness (e.g. Hawksworth 1991, 2001; May 1991), whereas the community structure and the detailed location of specific mycelia of saprotrophic fungi in soil are inadequately characterised despite over a century of research into the nature of these fungi. The lack of information on these two factors has hampered investigation of fungal function, and more information is needed to understand the relationship between taxonomic richness of saprotrophic soil fungi and the functional processes they carry out. The species richness of saprotrophic fungi associated with plant litter is often extremely high and the question immediately arises: are all these species necessary to maintain decomposition rates?

   The first part of the chapter is a synthesis of studies from different ecosystems of community structure of saprotrophic fungi in soil and plant litter, highlighting work where specific mycelia have been located in relation to resources. Second, functional diversity of decomposer fungi is examined. Subsequently, data on community structure of fungi in soil and plant litter are related to studies of fungal function to determine whether there is a predictive relationship between species richness and resource decomposition. The advantages of new techniques in characterising fungal diversity and function are discussed throughout the review and, finally, future challenges are identified.

Community structure and detailed location of specific mycelia
Definitions

The following paragraph is the definition of a community taken from Cooke and Rayner (1984).

A community . . . is the biotic component of an ecosystem. Implicit in this is that it comprises taxonomically diverse organisms, and has its own distinctive structure, activities and laws, including a unique internal economy which depends on relationships between the organisms that constitute it. Studies of communities should therefore be multidisciplinary, but the inevitable tendency has been to consider only those components which are relevant to any particular ecological discipline.

A fungal assemblage is a collection of different species in the same (micro-) environment whose activities, unlike in a community, may not be interactive (after Cooke & Rayner 1984). A resource is defined as any identifiable component of detritus (Swift *et al.* 1979). For the purposes of this chapter, it is worth noting the occurrence of two fungal groupings, the basidiomycetes and the microfungi. Basidiomycetes are members of the phylum Basidiomycota, the diagnostic character of which is the presence of a typically macroscopic fruit-body, the basidiome, bearing sexually produced basidiospores. Enzymes produced by basidiomycete mycelium present in soil, plant litter or wood are often capable of decomposing lignocellulose. Microfungi have microscopic fruit-bodies, and this artificial grouping of convenience includes fungi in soils and plant litter which produce large numbers of spores asexually. Enzymes from their mycelium are usually unable to degrade lignin.

What is meant by structure and which methods have been used?

The identity of species, location and spatial arrangement of mycelia, resource relationships, and involvement with other organisms are all of significance in the structure of fungal communities or assemblages in soil.

Ideally, to describe the structure of a fungal community completely it would be necessary to identify each species *in situ* and have separate biomass values for each of them. It would also be essential to map where these quantities of the species are within the soil profile, or in a leaf, and to show their association with particular resources. By direct observation alone this may be impossible because of the similarity in appearance of mycelia of different fungal species. It is also difficult to separate fungi from the complex medium of soil or litter, and it is demanding to differentiate living from dead mycelium by eye. It is important, but extremely difficult, to examine a fungal community *in situ*.

Using traditional techniques, the closest one can come to quantifying the abundance of single fungal species in soil or litter is to bring into the laboratory specific fractions of the substratum (e.g. soil particles of a particular size or from a specific location in the profile) which are serially washed to remove 'contaminant' surface spores (after Harley & Waid 1955) and plated into defined media (e.g. Robinson *et al.* 1994). Estimates of percentage frequency of occurrence can be obtained by relating the number of observations, or isolations, of each species to unit amounts of the substratum. For example, presence and absence data obtained from Warcup soil plates (by this method, a soil crumb is distributed as a thin layer within the culture medium; Warcup 1957) are usually expressed as the percentage number of plates on which the fungus has grown. An example of the type of results obtained is shown in Fig. 11.1. This technique of isolation on defined media has the obvious problem that fungi which cannot

(a)

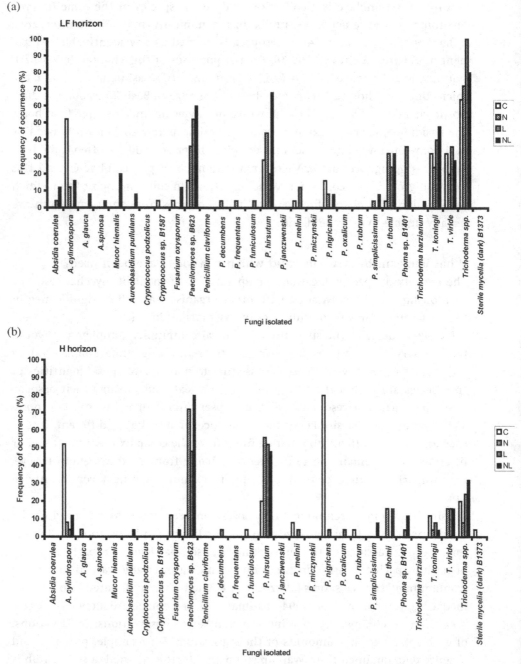

Figure 11.1. Mean ($n = 5$) percentage frequency of occurrence of fungi isolated from soil samples collected in July 2001 from an upland grassland at the NERC Soil Biodiversity Programme field-site, Sourhope, Roxburghshire, UK, for: (a) litter-fermentation (LF) horizons, (b) humus (H) horizons and (c) mineral (A) horizons. C, control; N, nitrogen; L, lime; NL, nitrogen plus lime. After Pryce Miller (2002).

(c)

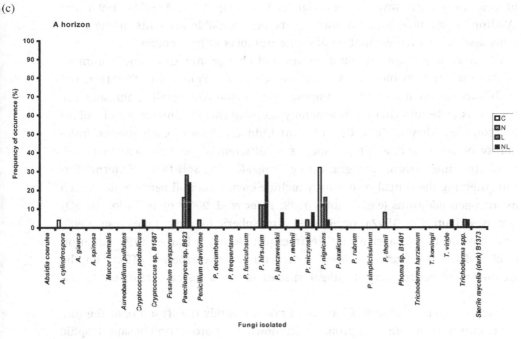

Figure 11.1. (cont.)

grow (e.g. AM fungi) will not be isolated. Much has also been written about problems of competition between fungi within the plated particle affecting fungal outgrowth (e.g. Bååth 1988).

Isolation frequencies from Warcup plates that favour mycelial growth, rather than spore germination, are preferable to dilution plate counts of fungal colonies. As dilution plate colonies arise from hyphal fragments and spores, they are meaningless in terms of either abundance or biomass of a species. These colony counts are useful only in limited circumstances, for example, in a comparative study of different terrestrial ecosystems, numbers of fungal colony-forming units (CFUs) in the upper layer of soil ranged from 0.005×10^4 in tundra to 100×10^4 in *Betula* woodland in Japan (Kjøller & Struwe 1982). Heavily sporulating species of decomposer microfungi (e.g. *Penicillium*, *Trichoderma*) are typically isolated by such dilution plate methods. Numerical surveys of macroscopic fungal fruit-bodies also give unreliable data on the frequency of species occurrence, unless there is repeated sampling both within a season and also over a period of several years to offset the bias arising from the ephemeral nature of many species and the vagaries of fruiting. A further complication is that although the occurrence of fruit-bodies does indicate with certainty the presence of a mycelium in the substratum, the lack of fruit-bodies does not necessarily reflect the absence of mycelia. Indeed, using molecular techniques, fruiting species of ectomycorrhizas were found to constitute merely 20–30%

of ectomycorrhizal mycelial abundance (Dahlberg 2001). Arnolds (1995) and Watling (1995) have both produced extremely readable accounts of the problems associated with estimating absolute numbers of fruit-bodies.

The structure of fungal communities will change over time, and a number of these substratum successions have been described. Frankland (1992) reviewed such successions involving decomposer (and ectomycorrhizal) fungi, and the mechanisms behind them. It is generally accepted that the interactions involved are complex. They include the inherent individualities of each species, availability of space, availability of species of differential performance, dispersal, combative interactions and grazing by fungal-feeding soil fauna. External factors affecting the fungal community include environmental perturbations such as nitrogen additions (e.g. Arnolds 1989; Peter *et al.* 2001) or pollution by SO_2 (e.g. Newsham *et al.* 1992a) or elevated atmospheric CO_2 (e.g. Jones *et al.* 1998), or extensification of agricultural regimes (e.g. Bowen & Harper 1989; Robinson *et al.* 1994). Most work has been carried out on perturbation effects on the frequency of occurrence of fungal species or, to a lesser extent, on fungal biomass.

Many different approaches have been taken to study the structure of the fungal community in soils. It is probably unknown what proportion of saprotrophic fungi is unculturable, but to circumvent the inability to quantify non-culturable fungi, traditional culture-based methods are being reinforced by molecular methods that study either nucleic acids or some other cellular components. Of these molecular options, most studies currently use either phospholipid fatty acids (PLFA; e.g. Frostegård & Bååth 1996) or ribosomal sequence analysis. PLFA provides a broad-scale level of detail and has advantages in that it examines the whole microbial community (i.e. bacteria and fungi), and does so in a linear manner, although PLFA cannot be used to give species identity and it has not been used to show where fungal mycelia are located in soil at the fine scale. Most ribosomal studies involve the (non-linear) PCR amplification of DNA. Because of non-linearity, there are often difficulties in extrapolating back to the starting concentrations of ribosomal templates extracted from soils. When analysing fungal community structure in soil, care must be taken at various stages not to replace the biases inherent in culture-based protocols with biases specific to molecular studies. Such stages include DNA extraction (see Krsek & Wellington (1999) for an assessment of DNA extraction strategies) and PCR amplification.

The resulting view of fungal community structure is affected greatly by the methodology used and the sample under investigation. Some studies have generated clone libraries of 18S rRNA genes (18S rDNA) after extracting the DNA directly from soil (e.g. Valinsky *et al.* 2002; Anderson *et al.* 2003) or roots (Vandenkoornhuyse *et al.* 2002). Data from these studies showed the specific fungal communities, produced by selection of particular primers, contained a relatively high proportion of basidiomycetes (Valinsky et al. 2002; Anderson et al. 2003) and were highly diverse (Vandenkoornhuyse et al. 2002).

Three further molecular approaches have been devised to characterise the structure of the community. Profiling methods such as denaturing gradient gel electrophoresis (DGGE; Muyzer & Smalla 1998; Muyzer 1999), temperature gradient gel electrophoresis (TGGE; Muyzer & Smalla 1998; Muyzer 1999) and terminal-restriction fragment length polymorphism (T-RFLP; Liu et al. 1997; Marsh 1999), developed originally for bacterial communities, provide means of assessing the structure of the fungal community in soil.

Two notable early attempts to circumvent culture techniques to examine fungal communities have been made in samples from the field, one in roots of *Ammophila arenaria* (Kowalchuk et al. 1997) and the other in soil from the rhizosphere of *Triticum aestivum* (Smit et al. 1999). Both studies used DNA amplified by primers for 18S rDNA gene sequences in DGGE analyses. The molecular data in the study of Kowalchuk et al. (1997) revealed fungal types not detected in previous culture-based surveys, although this is understandable to some extent as the roots would contain arbuscular mycorrhizal (AM) fungi which are completely unculturable, as well as root-pathogenic fungi. Both studies cited the incompleteness of existing genetic databases and the taxonomic resolution of the 18S rDNA as short-comings in their work. Generally, when comparing the fungi detected in natural samples by culturing with those from DGGE analysis, different groups of fungi have been detected by the two methods (e.g. Borneman & Hartin 2000; Vainio & Hantula 2000).

The technique T-RFLP has been used to assess fungal species richness in soils under different atmospheric CO_2 concentrations (Klamer et al. 2002), in various soil horizons (Dickie et al. 2002) and in sand contaminated with petroleum (Lord et al. 2002). In our study of saprotrophic fungi in a temperate, acidic grassland at Sourhope, Roxburghshire, UK, we attempted to identify, using T-RFLP, the niches (i.e. the nature of soil and litter particles) in which the mycelium of 'key' fungal species occurred in soil (Pryce Miller 2002). 'Key' species were chosen because they were isolated frequently and because of their potential enzymatic capabilities. In model systems, T-RFLP profiles of four of the 'key' fungi, namely *Cladosporium cladosporioides*, *Fusarium oxysporum*, *Penicillium hirsutum* and *Trichoderma koningii*, were successfully visualised for soil 'spiked' with added mycelium. The T-RFLP profile generated did differentiate between the fungal mycelium in pairs and for all four species *en masse* (Fig. 11.2). Unfortunately, T-RFLP analysis on field soil samples failed at the stage of amplification of soil DNA extracts, even though amplification of PCR products from native soil and litter extracts had been successful previously.

In attempting to characterise the structure of a particular fungal community as above, it is essential to cross-reference the information obtained from several different techniques, which each may have inherent biases.

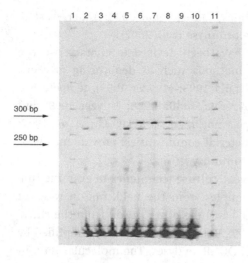

Figure 11.2. T-RFLP profile of the internally transcribed spacer (ITS) region of soil 'spiked' with pairs and all of the four chosen fungi. Lane 1: Microstep 15a (IR700; Microzone Ltd., Haywards Heath, UK); lane 2: *Trichoderma koningii* and *Cladosporium cladosporioides*, lane 3: *T. koningii* and *Penicillium hirsutum*, lane 4: *T. koningii* and *Fusarium oxysporum*, lane 5: *C. cladosporioides* and *F. oxysporum*, lane 6: *C. cladosporioides* and *P. hirsutum*, lane 7: *F. oxysporum* and *P. hirsutum*, lane 8: *T. koningii*, *C. cladosporioides*, *F. oxysporum* and *P. hirsutum*, lane 9: *T. koningii*, *C. cladosporioides*, *F. oxysporum* and *P. hirsutum* (positive control where all four fungal mycelia mixed underwent DNA extraction not in the presence of soil), lane 10: negative control, lane 11: Microstep 15a (IR700). After Pryce Miller (2002).

What exactly has been found?

Findings by horizon

Contributing a great deal towards the 'complete' picture of the structure of a fungal assemblage, fungal community structure, in relation to resources for two broad species' groupings, rather than for individual species, was detailed in a painstaking study by Frankland (1982). The distribution of living and total fungal biomass in a deciduous woodland soil, estimated from hyphal length, with hyphae with cell contents assumed to be living, and classified as belonging to basidiomycetes or microfungi according to the presence or absence of clamp connections, is shown in Table 11.1. A relatively low quantity of non-basidiomycete mycelium (kg ha^{-1}) occurred in the lowest soil horizons. However, this reflected the sheer bulk of the subsoil. Fungal mycelium of all types was most concentrated (g g^{-1} substrate) in the thin organic L and (Oh + Ah) surface horizons, and basidiomycete mycelium, excluding as far as possible that of mycorrhizal and pathogenic species on and in living roots, was almost confined to this area of the profile, dominating the mycoflora during the decomposition of the cell walls of plant debris. Basidiomycetes, therefore, can form a significant proportion of the microbial biomass of a woodland soil, but their ecological importance in biomass terms becomes much more obvious if account is taken of the large quantities of fungal mycelium in dead wood and dead roots (Table 11.1). In mixed deciduous woodland at Meathop Wood, Cumbria, UK (54° 12′ N, 2° 51′ W), these substrates 'were often packed with mycelium, and many dead tree branches and roots contained virtually a "pure culture" of a basidiomycete, such as *Stereum hirsutum* or *Armillaria mellea*' (Frankland 1982).

A further example of characterisation of the structure of fungal assemblages, in this instance in relation to different experimental perturbations applied and

Table 11.1. *Comparison of the distribution of the biomass of basidiomycetes with that of other microbial decomposers in the floor of a temperate deciduous woodland with mull humus (Meathop Wood, Cumbria, UK). Mycelial biomass was estimated from measurements of fungal hyphal length by direct microscopic observation.*

Substrate or horizon	Basidiomycetes (kg ha^{-1} dry wt.)		Other fungi (kg ha^{-1} dry wt.)		Bacteria and actinomycetes (kg ha^{-1} dry wt.)	
	Living	Total	Living	Total	Living	Total
Woody debris	30.5	216.9	7.3	34.7	2.6	601.6
L	3.1	8.7	0.5	4.1		
(Oh + Ah)	8.9	31.7	3.4	12.9	37.3	8 433.3
A	<1.0	<1.0	26.4	97.5		
B	<1.0	<1.0	31.4	155.6		
Dead roots	228.0	1 628.5	65.1	325.7	8.0	1 851.1
Total	271.5	1 886.8	134.1	630.5	47.9	10 866.0

Source: Frankland, J. C. (1982) In *Decomposer Basidiomycetes: Their Biology and Ecology* (Ed. by J. C. Frankland *et al.*), pp. 241–261. Cambridge: Cambridge University Press.

by soil depth, is shown in Fig. 11.1. The relative abundance of fungal species was determined in soils sampled from four different treatments applied to temperate, acidic grassland at Sourhope, Roxburghshire, UK (Pryce Miller 2002). In the third year of treatment with nitrogen additions as NH_4NO_3 at 12 g of N m^{-2} a^{-1} and lime as $CaCO_3$ at 600 g m^{-2} a^{-1}, all treatments showed a greater fungal diversity (richness and abundance) than the control plots when comparing the same horizons, and there was a decline in frequency of occurrence and species richness in the mineral A soil horizon compared with the LF and H horizons (Fig. 11.1). As often found in studies of soil fungi, in our Sourhope study a few taxa occurred much more frequently than the large number of rare taxa (Fig. 11.1).

By isolation in the laboratory of fungi from leaf and litter samples of defined ages from *Dryas integrifolia*, and from soil from 0 to 5 and from 10 to 15 cm depths, collected from a raised beach in the Canadian high arctic, Widden and Parkinson (1979) were able to show the structure of the fungal community in relation to resources at various depths. Visser and Parkinson (1975) also provided as complete a picture of the fungal community as possible by identifying species in live leaves, the L, F_1, F_2 and humus horizons, plus measuring total fungal biomass in each. Most biomass was located in the humus horizon (196 g wet weight m^{-2}) and least in the L horizon (0.91 g m^{-2}).

Fluorescent monoclonal antibodies, produced to mycelia of particular fungi (Dewey *et al.* 1997), have been applied in defined environmental samples to pinpoint the location of specific mycelia in relation to resources. In such autoecological studies it is possible to map the structure of one, or a few, species and to estimate their biomass. Using such a technique, *Mycena galopus* mycelium was found to be concentrated in the F_1 horizon of a *Picea sitchensis* plantation within and beyond the area of arcs of basidiomes observed around some of the trees. Biomass of *M. galopus* increased outwards from the tree bole to a maximum at the position of fruiting (Frankland *et al.* 1995). Unfortunately, this technique is unlikely to be applied to characterise the structure of the whole fungal community, as usually at least tens of fungal species are present, each requiring the time-consuming production of species-specific antibodies.

Findings by different size soil particles

A fungal community could be structured according to soil particle size, as well as by soil horizon, or age class of plant litter, as outlined above. Soil washing, to remove surface spores, of a humus horizon from a coniferous forest soil was carried out on sieves by Bååth (1988), resulting in various particle size fractions. Organic soil particles from the different fractions were plated on carboxymethyl-cellulose agar medium. The mean colonisation frequencies (isolates plated per particle) of the different particle sizes were 0.71 in the 50–80 µm fraction, 0.96 in the 80–100 µm fraction, 1.26 in the 100–125 µm fraction, 1.47 in the 125–180 µm fraction and 1.84 in the 180–250 µm fraction, showing that there were fewer fungi associated with smaller particles.

Approximately the same species were isolated from all size fractions and the species diversity values were also similar in all fractions. However, the abundance of different species differed between particles of different sizes, and the larger the difference in size of the particle, the greater the difference in abundance of the isolated species. The fungal assemblage isolated from the smaller particles was characterised by higher abundances of slow-growing fungi (e.g. *Oidiodendron echinulatum*) compared with those from the larger particles. Those species that were preferentially isolated from larger particles were all fast-growing species (e.g. *Penicillium spinulosum*). These results suggest that, rather than this fungal assemblage being structured by particle size, slow-growers were prevented from growing out from larger organic particles because more than one, perhaps faster-growing, fungus was present. Thus, Bååth's results show the use of smaller particles, where mostly not more than one isolate per particle was present, would overcome problems with fungal isolates interacting on the agar media. His findings also illustrate the difficulty in accurately characterising the structure of the fungal community, and that this assemblage from coniferous soil appears not to be structured by organic particle size.

'Is everything everywhere?' and host specificity

Saprotrophic fungal species in soil are probably largely cosmopolitan (Fenchel *et al.* 1997; Finlay *et al.* 1997), and it is the *structure* of the decomposer fungal community, specifically the balance of dominant and rare species and the location of the mycelia, which differs between ecosystems (e.g. Swift 1976; Frankland 1998; Robinson *et al.* 1998). Evidence that 'everything is not everywhere' comes from the fact that the structure and composition of fungal assemblages, especially in mycorrhizal fungi, is affected by host specificity. For example, Newton and Haigh (1998) examined host specificity in ectomycorrhizal fungi within the UK, and found that 233 species of ectomycorrhizal fungi, out of a total of 577 for which host information was collected, appeared to be specific to a single host plant species.

'Host' specificity can also occur between decomposer fungi and their substrata. Although comprehensive analysis of the fungi decaying plant debris will yield many hundreds of species, an initial partitioning of this community can be readily made on the basis of specificity for different types of plant litter, even though the nature of this specialisation is unclear (Swift 1976). Such fungi have been termed resource specific. Many microfungi, however, are found on a wide variety of plant tissues, and some overlap between fungal species on different types of resources is common, resulting in the term resource non-specific fungi. An example of this phenomenon has been found by Robinson *et al.* (1994) who followed the frequency of occurrence of fungal species in stem internodes or leaves of *Triticum aestivum* buried in an arable soil at 10 cm depth for 32 weeks. *Cladorrhinum foecundissimum* was a resource-non-specific fungus found in similar frequencies on both leaves and internodes, whereas *Epicoccum nigrum* was an example of a resource-specific species, found significantly more frequently on leaves. Similarly, within living tree branches, host and 'substratum' specificity have been demonstrated when *Hypoxylon fragiforme* and *H. nummularium* were restricted to *Fagus sylvatica* and, at the finer scale, to pockets of decay probably because transient activity occurred as the sapwood was not saturated with water (Chapela & Boddy 1988). Thus, the decomposer fungal community is structured by 'host'-specific fungi, and species that are more widespread, or plurivorous (Ellis & Ellis 1997). As well as host identity, the age and condition of the substrata (i.e. soil particle quality rather than size, as outlined above) can be important in determining which fungi are present in a community.

In summary to '*What exactly has been found?*', it appears that the decomposer fungal community is structured strongly by soil horizon and resource availability (e.g. dead wood and roots in Table 11.1). Resource quality is also important, and it is apparent that a certain degree of 'host' specificity (or host exclusivity, Zhou & Hyde 2001) exists. The community is structured less by the size of organic particles, according to the work of Bååth (1988) outlined above. Even though there is a clear correspondence between species composition in the soil microfungal community and vegetation type on a world-wide basis (Christensen 1989), it is the balance of dominant and rare species and the location of mycelia which differs between ecosystems.

Taxonomic diversity and its interpretation

Fungal communities are enormously diverse. For example, 250 species of fungi were isolated from a European agricultural soil (Kjøller & Struwe 1982). Even so, typically for microfungi in relatively undisturbed soils, a small number of species are isolated more frequently than others under particular conditions. At the two sites which have the best total fungal inventories in the world, Esher Common, Surrey, UK, and Slapton Ley Nature Reserve, Devon, UK, 2900 and 2500 species have been recorded, respectively (Hawksworth 2001). There is still much work to be carried out at the local scale on detecting and identifying the mycelia of micro- and macrofungi, together with estimating fungal species richness, even in the UK. These gaps in knowledge are even wider elsewhere in the world, for example in the tropics (e.g. Kaul 2002). Without detailed taxonomic studies of sites, it is impossible subsequently to relate mycelial location and function to species diversity.

Diversity indices

It is possible to calculate indices to estimate the species diversity of a sampled habitat, and both the Shannon–Weiner and the Brillouin diversity indices (Pielou 1975) have been used for fungal assemblages (e.g. Kjøller & Struwe 1982; Durrall & Parkinson 1991; Robinson *et al.* 1994; Donnison *et al.* 2000). Kjøller and Struwe (1982), in their comprehensive review of the occurrence and activity of microfungi in soil and litter, showed that for sites worldwide, the maximum values of the diversity indices were reached in the uppermost F or H layers of soil, whereas dead leaves and mineral soil generally showed lower diversity. Similarly, Pryce Miller (2002) found that diversity (abundance and richness) of microfungi isolated, using modified Czapek Dox agar medium, decreased with soil depth in a temperate, acidic agricultural grassland in the UK. However, in marked contrast, when using soil extract agar medium, the A horizon showed an exceptionally high species richness of fungi compared with the uppermost LF and H horizons of the profile (Pryce Miller 2002).

Researchers have used molecular profiling methodologies to generate diversity indices (e.g. Sigler & Turco 2002). In such studies, the index is often not intended to describe 'true' diversity, but to provide a numerical indicator to compare changes in the structure of the most abundant members of the community.

Diversity in macroecology

The wider ecological community has moved beyond simple evaluations of species richness to assess the importance of functional diversity (see next section below) in ecosystem processes (Zak & Visser 1996). The suggestion is that, in particular communities, some species are 'redundant', i.e. they could be lost with little effect on the structure and functioning of the whole community (Gitay *et al.* 1996). At present, it is largely unclear whether such redundancy occurs in fungal communities.

Function

What is meant by function?

As early as 1982, Kjøller and Struwe stated that a full understanding of the role of fungi in an ecosystem is not reached through independent observations on numbers, biomass, lists of species or physiological groups, but only through combined investigations where the relative occurrence of the different groups of fungi is linked to their function, that is, activity or capacity for substrate utilisation. Over 20 years later, the need for insight into the functional role of fungi in ecosystems is still great, partly because of the inadequate description of fungal assemblages in relation to resources. Some successful attempts, however, have been made to match organisms and their activities together (e.g. Flanagan & Scarborough 1974; Boddy 1986; Newsham *et al.* 1992a, b, c), but this type of project is very time-consuming, involving several years' work for many people. Table 11.2 lists some of the functions of fungi in ecosystems.

How is function assessed and what has been found?

Given the diversity of fungi within soil, it is unsurprising that there is a similarly diverse set of metabolic potentials within soil fungal communities. Not all of this potential, however, is expressed at any one time (White 1995). Measuring the metabolic activity of fungi is therefore another challenge, one made more complex by heterogeneity at the fine scale in soil and litter. Current understanding of the metabolic activities of fungi in natural environments derives in large part from in vitro assessments of fungal physiology.

Decomposition and nutrient cycling

The broad functions of fungal mycelium in soil and litter are decomposition and nutrient cycling. Decay rates (usually estimated from changes in weight, tensile strength and chemical composition) of individual litter components or pure substrates in the field have been studied extensively. However, these methods measure activities of a decomposer community as a whole, and attempts to partition functions between specific groups are difficult. Frankland *et al.* (1990) stated that the principal value of decay rates from field samples lies in

Table 11.2. *Functions of fungi in ecosystems.*

Physiological and metabolic
  Decomposition of organic matter: volatilisation of C, H and O; fragmentation
  Elemental release and mineralisation of N, P, K, S and other ions
  Elemental storage: immobilisation of elements
  Accumulation of toxic materials
  Synthesis of humic materials

Ecological
  Facilitation of energy exchange between above-ground and below-ground systems
  Promotion and alteration of niche development
  Regulation of successional trajectory and velocity

Mediative and integrative
  Facilitation of transport of essential elements and water from soil to plant roots
  Facilitation of plant-to-plant movement of essential elements and carbohydrates
  Regulation of water and ion movement through plants
  Regulation of photosynthetic rate of primary producers
  Regulation of C allocation below ground
  Increased survivability of seedlings
  Protection from root pathogens
  Modification of soil permeability and promotion of aggregation
  Modification of soil ion exchange and water-holding capacities
  Detoxification of soil (degradation, volatilisation or sequestration)
  Participation in saprotrophic food chains
  Instigation of parasitic and mutualistic symbioses
  Production of environmental biochemicals (antibiotics, enzymes and
    immunosuppressants)

Source: Miller, S. L. (1995) *Canadian Journal of Botany*, **73**, S50–S57.

the clues they can give to the functions of the predominant decomposer fungi, especially when observations are backed up by laboratory tests on isolated species. For example, Hering (1967), using pure cultures growing on γ-irradiated *Quercus* leaves, showed that lignin and hemicelluloses were most actively decomposed by *Mycena galopus* and *Collybia peronata*; cellulose by *M. galopus* and *Polyscytalum fecundissimum*. In a similar set of laboratory experiments on relatively realistic substrata, nutrient release, rather than decomposition, was followed from γ-irradiated internodes of *Triticum aestivum* inoculated with single and mixed species (Robinson *et al.* 1993b). Release of Na^+, K^+ and NH_4^+–N was similar from all combinations, but Ca^{2+}, Mg^{2+} and PO_4^{3-}–P release depended on species. These results suggest the fungi were functionally 'equivalent' for the monovalent cations above, but species' identity was important for the other ions analysed.

  Spatial relocation of nutrients (e.g. nitrogen and phosphorus) and water by fungi are ecosystem functions likely to be of utmost importance. Such functions have been illustrated, usually in realistic substrata in the laboratory, by labelling compounds translocated by fungal mycelia that form specialised linear organs, cords and rhizomorphs (reviewed by Boddy 1999). Two studies are especially noteworthy: one by Gray et al. (1995), which demonstrated relocation of caesium in situ by Schizophyllum commune, and the second by Lindahl et al. (1999), who conducted an elegant demonstration in microcosms of phosphorus transfer from saprotrophic to mycorrhizal mycelium, thus 'short-cutting' conventional pathways of nutrient cycling. Many more studies of the latter type are necessary, where interactions between different types of microorganism that share the same microsite are studied.

Potential decomposer activity using specific substrates

Fungal function is often tested in artificial media, which may not be good approximations to more natural substrata, but are useful to try and understand the potential roles of fungal isolates within a community. A wide range of tests exist to show utilisation of a complex carbon source in agar media; for example, Robinson et al. (1993a) showed that, of the isolates tested, only basidiomycetes were able to clear lignin agar medium, whereas Chaetomium globosum and four basidiomycetes cleared cellulose agar medium. However, it is worthwhile remembering that the ability to decompose lignin and cellulose in vitro is not necessarily related to the abilities of the organism to decompose plant tissues (Swift 1976).

Biomass increase of single species in liquid culture on various carbon and nitrogen sources can be measured as indications of the use of these compounds (reviewed by Paterson & Bridge 1994). The ultimate extension of the type of substrate tests outlined above is the commercially available BIOLOG plate (Garland & Mills 1991) used to assess functional differences among soil and water bacteria based on their patterns of utilisation of 96 substrates. The substrates include carbohydrates, carboxylic acids, amino acids, amines and amides. A thorough critique of BIOLOG plates to assess the functional diversity of microorganisms in environmental samples has been provided by Preston-Mafham et al. (2002). Dobranic and Zak (1999) have developed this method (FungiLog) to examine fungal functional diversity. Ideally, one should know which of the substrates are present in the field and their relative abundance. Studies including more ecologically relevant substrates, for example root exudates, have been developed for microbes by Hodge et al. (1998). Fluorogenic enzyme substrates have been used to quantify specific components of fungal chitinase and cellulase activities in situ in soil (Miller et al. 1998).

Enzyme activity

Biochemical tests for a wide range of extracellular enzymes exist, but most involve addition of a substrate so, although useful in laboratory studies, they

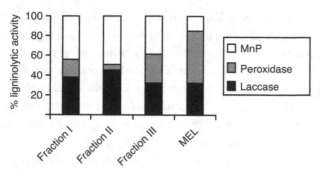

Figure 11.3. Relative contributions of three types of ligninolytic enzyme (manganese peroxidase (MnP), a further peroxidase and laccase) produced by *Mycena galopus* to the activity of fractions I, II and III from *Picea sitchensis* needle litter extracts and from malt extract liquid cultures (MEL). Fractions I and II were sequential water extractions from F_1 horizon *P. sitchensis* needle litter incubated with *M. galopus* at 25 °C for four weeks; fraction III was the final extraction with 0.1 mol sodium phosphate buffer (pH 6.5).

measure potential rather than actual activity, as do the majority of the tests in the previous section. Examples of the type of 'potential activity' results obtained are shown in Fig. 11.3. Using guaiacol as a substrate, the contribution of three enzymes, a manganese-dependent peroxidase, a further peroxidase, and a laccase, produced by *Mycena galopus* in malt extract liquid medium and in needle litter extracts was assayed (Ghosh *et al.* 2003).

The range of enzymes produced from a single fungus can be great: for example, 18 types of extracellular enzyme from *Agaricus bisporus* (the commercial white mushroom) have been detected in compost cultures (Wood 1998). Enzyme activity has also been measured using specific functional gene sequences. For example, Lamar *et al.* (1995) quantitatively assessed two lignin peroxidase mRNA transcripts from *Phanerochaete chrysosporium* in soil.

Isotopes of carbon and nitrogen

Stable and radioactive isotopes of carbon and stable isotopes of nitrogen are being used to understand the decomposer and nutrient cycling functions of fungi in natural environments. At the fine scale, Radajewski *et al.* (2000) demonstrated how bacteria that are actively involved in specific metabolic processes could be identified through the acquisition of a specific ^{13}C label from growing on a ^{13}C-labelled substrate, and similar studies could be performed with decomposer fungi and relevant substrates.

'Stable isotopes have the potential to help clarify the role of fungi in ecosystem processes' according to Hobbie *et al.* (1999). Differences in carbon and nitrogen isotope ratios (δ^{13}C and δ^{15}N values) among fruit-bodies of mycorrhizal and saprotrophic fungi may provide a way of determining trophic strategies (or 'function' at the broad scale) for a broad spectrum of fungi whose strategies have been a matter of speculation only (Hobbie *et al.* 2001). Mycorrhizal fruit-bodies have been found to be consistently enriched in $\delta^{15}N$ and depleted in $\delta^{13}C$ compared with those of saprotrophic fungi (Hobbie et al. 1999, 2001). Subsequently, however, Emmerton et al. (2001) and Hobbie et al. (2002) stated that these results should be interpreted with caution. Emmerton et al. (2001) showed that two species of ecto-mycorrhizal fungi, Paxillus involutus and Leccinum scabrum, exhibited significant net fractionation in favour of ^{15}N when grown on inorganic and organic nitro-gen sources of predetermined nitrogen isotope combination, suggesting that any attempts to use the tissue ^{15}N abundance as a means of identifying the sub-strates being exploited by mycorrhizal fungi are likely to be unrealistic. Hobbie et al. (2002) have used ^{14}C to provide additional insight into possible mycorrhizal status, showing that the isotopic composition of fruit-bodies of known mycor-rhizal fungi resembled that of current years' coniferous needles or atmospheric CO_2 and indicated an average age of 0–2 years for incorporated carbon, whereas those of saprotrophic genera were composed of carbon at least 10 years old.

The broad-scale 'function' of arbuscular mycorrhizal fungi in ecosystems has also been examined using ^{14}C and ^{13}C labelling. Such studies have been carried out at the UK Soil Biodiversity Sourhope grassland site (Johnson et al. 2002a, b), and show that AM mycelia provide a rapid and important pathway of carbon flux to the soil and atmosphere.

In summary, as seen from the scattered information above, there is still far to go before fungal function is satisfactorily described and analysed. Model studies using mixtures of decomposer (and mycorrhizal) fungi in relatively realistic sub-strata may help to explain the succession of functionally different fungi during decomposition. mRNA has potential as an in-situ indication of enzyme activity. It is necessary to relate results of functional tests to community structure, which should be possible for 'key' species, if not for the whole community.

Relationship between species richness and resource decomposition

Several models (Fig. 11.4a–d; Gaston & Spicer 1998) have been proposed to describe the relationship between species richness and ecosystem stability or function, all developed from studies in plant and animal ecology. Fungi have been neglected in this debate, even though this group is one of the most diverse in the world with estimates of around 1.5 million species of which 74 000 are currently known (Hawksworth 2001).

As stated previously, saprotrophic fungi are organisms that usually underpin nutrient cycling in ecosystems, but they are sensitive to disturbance, pollution and environmental change. The possible impact of environmental change on fungal diversity could influence ecosystem function via decomposition, so it is important to understand the degree of dependence of decay rates on the number of fungal species present. It is well known that the species richness of decomposer fungi associated with plant litter is often extremely high, with for example the petioles of one species (Pteridium aquilinum) alone being colonised by up to 177 fungal species (Frankland 1966), and the question immediately arises

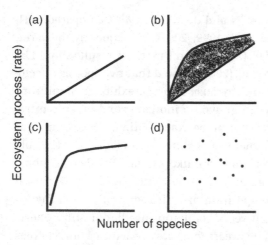

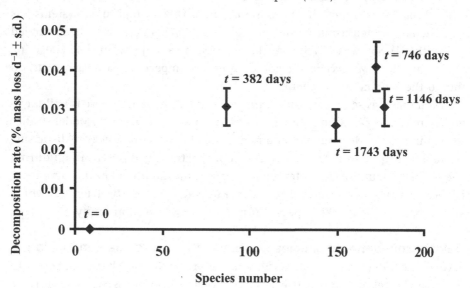

Number of species

Figure 11.4. Examples of hypotheses relating species richness and ecosystem function: (a) diversity–stability, which predicts there is a direct positive relationship between species number and ecosystem function; (b) rivet, where all species have a function although a few may be lost without notice since they are redundant, until beyond a particular threshold species' reduction will result in ecosystem collapse; (c) species redundancy, which predicts that beyond a certain critical diversity species are functionally redundant; and (d) idiosyncratic, in which diversity changes ecosystem function but not in a predictable way. After Gaston and Spicer (1998).

Figure 11.5. Data for numbers of fungal species isolated from washed particles of petioles of *Pteridium aqulinum* and decomposition rates (mean percentage mass loss $d^{-1} \pm$ s.d.) of these petioles over a five-year period. After Frankland (1966).

whether all these species are essential to maintain decomposition rates. The relationship between species richness and decomposition rate apparently has not been tested for decomposer fungi, even though theoretical arguments about this subject have been proposed for soil organisms more generally (Wardle & Giller 1996).

From the only long-term field study which contains both decomposition rates plus 'complete' species richness data at corresponding time intervals (Fig. 11.5; Frankland 1966), the general relationship most closely followed is apparently the diversity–stability hypothesis, although the increasing (species richness) and decreasing (mass) time series offer plenty of potential to find a correlation with nothing to do with causation (D. Elston personal communication). The observed change in species richness implies that the species composition of decomposer fungal communities varies as decomposition proceeds. This 'succession' involves the addition of new species to the community at specific times during the process of decomposition and the extinction of others (Frankland 1998), and partly relates to the type of carbon and nutrient substrates available in the litter at a particular time. In the example in Fig. 11.5 (Frankland 1966), the first three time points of the graph exhibit a direct positive relationship between increasing species richness and decomposition rate, consistent with the diversity–stability hypothesis (Fig. 11.4a). However, at the two later time points, the decomposition rate declines even though there are approximately the same numbers of species as at the third time point. One reason for the decline in rate at the later sampling times (Fig. 11.5) could be that the substratum contains proportionately more recalcitrant substrates as decomposition proceeds.

More studies and data are necessary before any conclusions can be drawn about the form of the relationship shown in Fig. 11.5. It is widely believed that many decomposer species are functionally redundant (Andrèn et al. 1995). This would mean that members of groups (guilds) involved in specific functions, such as cellulose or lignin decomposition, could be lost providing those functions remain represented in the community (Fig. 11.4c). There is good evidence that redundancy should operate in nature because, in the laboratory in a wide variety of studies (e.g. Gochenaur 1984; and our own work with grassland fungi, (Fig. 11.6)), frequently isolated species have been found to have the same specific enzymatic capabilities for decomposition as occasional ones, suggesting the capacity for interspecific competition is high. It is likely, however, that actual species composition rather than richness alone is important, for example, there is some evidence for 'keystone' species in the study in Fig. 11.5, as one basidiomycete fungus, Mycena galopus, was responsible for 35% weight loss after one year (Frankland 1966).

Other researchers have proposed the alternative view that communities of decomposer fungi are relatively disordered and stochastic (Swift 1985), perhaps arguing for the idiosyncratic hypothesis. Constant changes in physical and biotic conditions (e.g. constant grazing by Colembolla and Acarina) may mean that there is no time for the competitive exclusion principle to operate to conclusion, leaving the fungal community in disequilibrium, and explaining the paradox of how an assemblage of species apparently requiring the same resources could co-exist (Swift 1985).

There are two potential problems with such studies of decomposition and fungal diversity: confounding effects of pieces of plant litter becoming smaller and more recalcitrant as decomposition continues, and it is difficult to measure

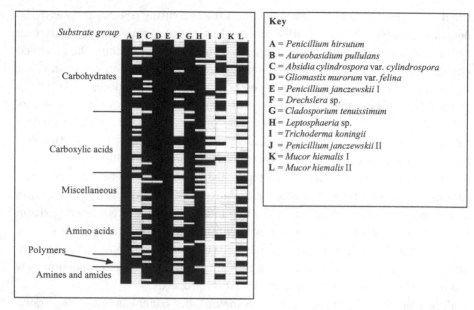

Figure 11.6. Twelve fungal isolates (A to L), from soil and standing-dead litter from an upland grassland at Sourhope, were inoculated in pure culture onto BIOLOG plates. Isolates A and B, C and D, E and F, and G and H were paired species of matched ability from primary potential function tests on solid media. Each pair was made up of one abundantly isolated species (A, C, E and G) and one occasionally isolated species (B, D, F and H). I, J, K and L are 'functionally unique' isolates. Black horizontal bars indicate positive utilisation of a substrate. After Deacon (2003).

the 'complete' species richness of the vegetative growth stage of the fungal community. However, the data in Fig. 11.5 are exemplary in comparison with other studies because fungi were enumerated over a five-year period. Molecular methods could be important in characterising this 'complete' fungal species richness but, to date, such measurements have not been combined with decomposition rates.

In conclusion, using the best dataset available for decomposer fungi, it is difficult to draw conclusions about the relationship between species richness and decomposition rate. Aspects of redundancy and stochasticity and the existence of 'keystone' species are reported elsewhere in the literature. Even though the relationship is likely to depend on species composition (and by implication the species' functional repertoire), the various hypotheses (Fig. 11.4a–d) are useful frameworks to characterise the relationship between species richness and decomposition rate. Although these hypotheses may not fit particularly well, they do highlight new areas for research and the necessity for conservation strategies, especially as decomposition rate appears to decline with species richness (Fig. 11.5).

Conclusions

There is much to be done in order to characterise fungal communities *in situ* in the natural environment. We should know more about the distribution of mycelia of particular species, with which resources they are located, what the characteristics of that resource are, and what specific decomposer functions are being performed in the overall ecosystem processes of decomposition and nutrient cycling by the abundant and occasional species colonising their respective resources. Priority has rightly been given by field mycologists to recording the vast diversity of species and to saving some from extinction (Frankland 1998). Although ecologists know that there are large numbers of species of decomposer fungi in soil, they need not be overwhelmed by the wealth of taxonomic diversity. It is possible to concentrate on a few 'star performers' ('keystone' species?) on which the maintenance of entire ecosystems may depend (Frankland 1998), plus other species chosen for particular reasons. Relatively realistic substrata can be used as a bridge between entirely artificial media and all the complexities of soil in the field. The number of techniques to determine structure and function of the saprotrophic fungal community is increasing, and one can only echo the sentiments of Bruns (1995) that now is an exciting time to work on fungal ecology. More than ever, to characterise fungal community structure and function it is necessary to cross-reference the information obtained from different methods (Frankland et al. 1990).

Regarding the relationship of species richness to ecosystem function, it is unknown whether fungi obey any of the theories commonly reported by plant and animal ecologists. There is some evidence of functional redundancy because common species have the same function as occasional ones in vitro, although it is impossible to know whether 'functions' of particular species in the field are the same as measured in pure culture in the laboratory. Also, researchers are unlikely ever to have exhaustive knowledge of the functional roles played by fungi in the natural environment, or to have a full suite of tests to measure them. Possible support for the idiosyncratic hypothesis is suggested by the stochastic nature of fungal communities in response to continued disturbance. However, for the ecosystem process of decomposition rate, it is very simplistic to assume that the relationship with fungal species richness is the only important one. Fungal species composition (e.g. 'keystone' species), and thus their functional repertoire, is important, and other soil organisms (bacteria, soil animals) are involved, as are abiotic variables such as microclimate and resource quality (Swift et al. 1979).

One future challenge is to quantify further the effects of interspecific interactions of saprotrophic fungi, including with mycorrhizas, on ecosystem processes such as decomposition and nutrient cycling. Interactions of fungi with other groups of soil organisms, such as bacteria, Collembola, Acarina and Lumbricidae, may be important in ecosystem carbon and nitrogen fluxes, and it is in this research area where more multi-disciplinary research programmes are needed.

Acknowledgements

Our work in this review was funded by the Natural Environment Research Council's Soil Biodiversity and Ecosystem Function thematic programme, which provided CASE studentships to EJPM and LJD plus an associated grant to CHR, Juliet C. Frankland and Brian W. Bainbridge. We are very grateful to Pam Self, Nisha R. Parekh and Jai David for scientific support, and it is a pleasure to thank Kenneth D. Bruce and Peter D. Moore for, respectively, kind and constructive discussions about analysing fungal communities by molecular techniques and the ecological importance of commonness and rarity.

References

Anderson, I. C., Campbell, C. D. & Prosser, J. I. (2003). Potential bias of fungal 18S rDNA and internal transcribed spacer polymerase chain reaction primers for estimating fungal biodiversity in soil. *Environmental Microbiology*, 5, 36–47.

Andrèn O., Clarholm, M. & Bengtsson, J. (1995). Biodiversity and species redundancy among litter decomposers. *The Significance and Regulation of Soil Biodiversity* (Ed. by H. P. Collins, G. P. Robertson & M. J. Klug), pp. 141–151. Dordrecht: Kluwer.

Arnolds, E. (1989). The influence of increased fertilisation on the macrofungi of a sheep meadow in Drenthe, the Netherlands. *Opera Botanica*, 100, 7–21.

Arnolds, E. (1995). Problems in measurements of species diversity of macrofungi. *Microbial Diversity and Ecosystem Function* (Ed. by D. Allsopp, R. R. Colwell & D. L. Hawksworth), pp. 337–353. Wallingford: CAB International.

Bååth, E. (1988). A critical examination of the soil washing technique with special reference to the size of the soil particles. *Canadian Journal of Botany*, 66, 1566–1569.

Boddy, L. (1986). Water and decomposition processes in terrestrial ecosystems. *Water, Fungi and Plants* (Ed. by P. G. Ayres & L. Boddy), pp. 375–398. Cambridge: Cambridge University Press.

Boddy, L. (1999). Saprotrophic cord-forming fungi: meeting the challenge of heterogeneous environments. *Mycologia*, 91, 13–32.

Borneman, J. & Hartin, R. J. (2000). PCR primers that amplify fungal rRNA genes from environmental samples. *Applied and Environmental Microbiology*, 66, 4356–4360.

Bowen, R. M. & Harper, S. H. T. (1989). Fungal populations on wheat straw decomposing in arable soils. *Mycological Research*, 93, 47–54.

Bruns, T. D. (1995). Thoughts on the processes that maintain local species diversity of ectomycorrhizal fungi. *Plant and Soil*, 170, 63–73.

Chapela, I. H. & Boddy, L. (1988). Fungal colonization of attached beech branches: II. Spatial and temporal organization of communities arising from latent invaders in bark and functional sapwood, under different moisture regimes. *New Phytologist*, 110, 47–57.

Christensen, M. (1989). A view of fungal ecology. *Mycologia*, 81, 1–19.

Cooke, R. C. & Rayner, A. D. M. (1984). *Ecology of Saprotrophic Fungi*. London: Longman.

Dahlberg, A. (2001). Community ecology of ectomycorrhizal fungi: an advancing interdisciplinary field. *New Phytologist*, 150, 555–562.

Deacon, L. J. (2003). Functional biodiversity of grassland saprotrophic fungi. Unpublished Ph.D. thesis, King's College, University of London.

Dewey, F. M., Thornton, C. R. & Gilligan, C. A. (1997). Use of monoclonal antibodies to detect, quantify and visualise fungi in soils. *Advances in Botanical Research*, **24**, 275–308.

Dickie, I. A., Xu, B. & Koide, R. T. (2002). Vertical niche differentiation of ectomycorrhizal hyphae in soil as shown by T-RFLP analysis. *New Phytologist*, **156**, 527–535.

Dobranic, J. K. & Zak, J. C. (1999). A microtiter plate procedure for evaluating fungal functional diversity. *Mycologia*, **91**, 756–765.

Donnison, L. M., Griffith, G. S., Hedger, J., Hobbs, P. J. & Bardgett, R. D. (2000). Management influences on soil microbial communities and their function in botanically diverse haymeadows of northern England and Wales. *Soil Biology and Biochemistry*, **32**, 253–263.

Durrall, D. M. & Parkinson, D. (1991). Initial fungal community development on decomposing timothy (*Phleum pratense*) litter from a reclaimed coal-mine spoil in Alberta, Canada. *Mycological Research*, **95**, 14–18.

Ellis, M. B. & Ellis, J. P. (1997). *Microfungi on Land Plants*, 2nd edition. Slough: Richmond.

Emmerton, K. S., Callaghan, T. V., Jones, H. E., et al. (2001). Assimilation and isotopic fractionation of nitrogen by mycorrhizal fungi. *New Phytologist*, **151**, 503–511.

Fenchel, T., Esteban, G. F. & Finlay, B. J. (1997). Local versus global diversity of microorganisms: cryptic diversity of ciliated protozoa. *Oikos*, **80**, 220–225.

Finlay, B. J., Maberly, S. C. & Cooper, J. I. (1997). Microbial diversity and ecosystem function. *Oikos*, **80**, 209–213.

Flanagan, P. W. & Scarborough, A. M. (1974). Physiological groups of decomposer fungi on tundra plant remains. *Soil Organisms and Decomposition in Tundra* (Ed. by A. J. Holding, O. W. Heal, S. F. MacLean Jr. & P. W. Flanagan), pp. 159–181. Stockholm: Tundra Biome Steering Committee.

Frankland, J. C. (1966). Succession of fungi on decaying bracken petioles. *Journal of Ecology*, **57**, 25–36.

Frankland, J. C. (1982). Biomass and nutrient cycling by decomposer basidiomycetes. *Decomposer Basidiomycetes: Their Biology and Ecology* (Ed. by J. C. Frankland, J. N. Hedger & M. J. Swift), pp. 241–261. Cambridge: Cambridge University Press.

Frankland, J. C. (1992). Mechanisms in fungal succession. *The Fungal Community*, 2nd edition. (Ed. by G. C. Carroll & D. T. Wicklow), pp. 383–401. New York: Marcel Dekker.

Frankland, J. C. (1998). Presidential Address. Fungal succession: unravelling the unpredictable. *Mycological Research*, **102**, 1–15.

Frankland, J. C., Dighton, J. & Boddy, L. (1990). Methods for studying fungi in soil and forest litter. *Methods in Microbiology*, Vol. 22 (Ed. by R. Grigorova & J. R. Norris), pp. 343–404. London: Academic Press.

Frankland, J. C., Poskitt, J. M. & Howard, D. M. (1995). Spatial development of populations of a decomposer fungus, *Mycena galopus*. *Canadian Journal of Botany*, **73**, S1399–S1406.

Frostegård, Å. & Bååth, E. (1996). The use of phospholipid fatty acid analysis to estimate bacterial and fungal biomass in soil. *Biology and Fertility of Soils*, **22**, 59–65.

Garland, J. L. & Mills, A. L. (1991). Classification and characterisation of heterotrophic microbial communities on the basis of patterns of community-level sole-carbon-source utilisation. *Applied and Environmental Microbiology*, **57**, 2351–2359.

Gaston, K. J. & Spicer, J. I. (1998). *Biodiversity: An Introduction*. Oxford: Blackwell Science.

Ghosh, A., Frankland, J. C., Thurston, C. F. & Robinson, C. H. (2003). Enzyme production by *Mycena galopus* mycelium in artificial media and in *Picea sitchensis* F_1 horizon needle litter. *Mycological Research*, **197**, 996–1008.

Gitay, H., Wilson, J. B. & Lee, W. G. (1996). Species redundancy: a redundant concept? *Journal of Ecology*, **84**, 121–124.

Gochenaur, S. E. (1984). Fungi of a Long Island oak–birch forest: II. Population dynamics and hydrolase patterns for the soil penicillia. *Mycologia*, **76**, 218–231.

Gray, S. N., Dighton, J., Olsson, S. & Jennings, D. H. (1995). Real-time measurement of uptake and translocation of ^{137}Cs within mycelium of *Schizophyllum commune* Fr. by autoradiography followed by quantitative image analysis. *New Phytologist*, **129**, 449–465.

Harley, J. L. & Waid, J. S. (1955). A method of studying active mycelia on living roots and other surfaces in the soil. *Transactions of the British Mycological Society*, **38**, 104–118.

Hawksworth, D. L. (1991). The fungal dimension of biodiversity: magnitude, significance, and conservation. *Mycological Research*, **95**, 641–655.

Hawksworth, D. L. (2001). The magnitude of fungal diversity: the 1.5 million species estimate revisited. *Mycological Research*, **105**, 1422–1432.

Hering, T. F. (1967). Fungal decomposition of oak leaf litter. *Transactions of the British Mycological Society*, **50**, 267–273.

Hobbie, E. A., Macko, S. A. & Shugart, H. H. (1999). Insights into nitrogen and carbon dynamics of ectomycorrhizal and saprotrophic fungi from isotopic evidence. *Oecologia*, **118**, 353–360.

Hobbie, E. A., Weber, N. S. & Trappe, J. M. (2001). Mycorrhizal vs. saprotrophic status of fungi: the isotopic evidence. *New Phytologist*, **150**, 601–610.

Hobbie, E. A., Weber, N. S., Trappe, J. M. & van Klinken, G. J. (2002). Using radiocarbon to determine the mycorrhizal status of fungi. *New Phytologist*, **156**, 129–136.

Hodge, A., Grayston, S. J., Campbell, C. D., Ord, B. G. & Killham, K. (1998). Characterisation and microbial utilisation of exudate material from the rhizosphere of *Lolium perenne* grown under CO_2 enrichment. *Soil Biology and Biochemistry*, **30**, 1033–1043.

Johnson, D., Leake, J. R. & Read, D. J. (2002a). Transfer of recent photosynthate into mycorrhizal mycelium of an upland grassland: short-term respiratory losses and accumulation of ^{14}C. *Soil Biology and Biochemistry*, **34**, 1521–1524.

Johnson, D., Leake, J. R., Ostle, N., Ineson, P. & Read, D. J. (2002b). In situ $^{13}CO_2$ pulse-labelling of upland grassland demonstrates a rapid pathway of carbon flux from arbuscular mycorrhizal mycelia to the soil. *New Phytologist*, **153**, 327–334.

Jones, T. H., Thompson, L. J., Lawton, J. H., *et al.* (1998). Impacts of rising atmospheric carbon dioxide on model terrestrial ecosystems. *Science*, **280**, 441–443.

Kaul, T. N. (2002). Conservation of mycodiversity in India: an appraisal. *Tropical Mycology*, Vol. 1, *Macromycetes* (Ed. by R. Watling, J. C. Frankland, A. M. Ainsworth, S. Isaac & C. H. Robinson) pp. 131–147. Wallingford: CAB International.

Kjøller, A. & Struwe, S. (1982). Microfungi in ecosystems: fungal occurrence and activity in litter and soil. *Oikos*, **39**, 391–422.

Klamer, M., Roberts, M. S., Levine, L. H., Drake, B. G. & Garland, J. L. (2002). Influence of elevated CO_2 on the fungal community in a coastal scrub oak forest soil investigated with terminal-restriction fragment length polymorphism analysis. *Applied and Environmental Microbiology*, **68**, 4370–4376.

Kowalchuk, G. A., Gerards, S. & Woldendorp, J. A. (1997). Detection and characterisation of fungal infections of *Ammophila arenaria* (marram grass) roots by Denaturing Gradient Gel Electrophoresis of specifically amplified 18S rDNA. *Applied and Environmental Microbiology*, **63**, 3858–3865.

Krsek, M. & Wellington, E. M. H. (1999). Comparison of different methods for the isolation and purification of total community

DNA from soil. *Journal of Microbiological Methods*, **39**, 1–16.

Lamar, R. T., Schoenike, B., vanden Wymelenberg, A., *et al.* (1995). Quantitation of fungal mRNAs in complex substrates by reverse transcription PCR and its application to *Phanerochaete chrysosporium*-colonised soil. *Applied and Environmental Microbiology*, **61**, 2122–2126.

Lindahl, B., Stenlid, J., Olsson, S. & Finlay, R. (1999). Translocation of ^{32}P between interacting mycelia of a wood-decomposing fungus and ectomycorrhizal fungi in microcosm systems. *New Phytologist*, **144**, 183–193.

Liu, W. T., Marsh, T. L., Cheng, H. & Forney, L. J. (1997). Characterization of microbial diversity by determining terminal restriction fragment length polymorphisms of genes encoding 16S rRNA. *Applied and Environmental Microbiology*, **63**, 4516–4522.

Lord, N. S., Kaplan, C. W., Shank, P., Kitts, C. L. & Elrod, S. L. (2002). Assessment of fungal diversity using terminal restriction (TRF) pattern analysis: comparison of 18S and ITS ribosomal regions. *FEMS Microbiology Ecology*, **42**, 327–337.

Marsh, T. L. (1999). Terminal Restriction Fragment Length Polymorphism (T-RFLP): an emerging method for characterizing diversity among homologous populations of amplification products. *Current Opinion in Microbiology*, **2**, 323–327.

May, R. M. (1991). A fondness for fungi. *Nature*, **352**, 475–476.

Miller, M., Palojärvi, A., Rangger, A., Reeslev, M. & Kjøller, A. (1998). The use of fluorogenic substrates to measure fungal presence and activity in soil. *Applied and Environmental Microbiology*, **64**, 613–617.

Miller, S. L. (1995). Functional diversity of fungi. *Canadian Journal of Botany*, **73**, S50–S57.

Muyzer, G. (1999). DGGE/TGGE a method for identifying genes from natural ecosystems. *Current Opinion in Microbiology*, **2**, 317–322.

Muyzer, G. & Smalla, K. (1998). Application of Denaturing Gradient Gel Electrophoresis (DGGE) and Temperature Gradient Gel Electrophoresis (TGGE) in microbial ecology. *Antonie van Leeuwenhoek International Journal of General and Molecular Microbiology*, **73**, 127–141.

Newsham, K. K., Frankland, J. C., Boddy, L. & Ineson, P. (1992a). Effects of dry-deposited sulphur dioxide on fungal decomposition of angiosperm tree leaf litter: I. Changes in communities of fungal saprotrophs. *New Phytologist*, **122**, 97–110.

Newsham, K. K., Ineson, P., Boddy, L. & Frankland, J. C. (1992b). Effects of dry-deposited sulphur dioxide on fungal decomposition of angiosperm tree leaf litter: II. Chemical content of litters. *New Phytologist*, **122**, 111–126.

Newsham, K. K., Boddy, L., Frankland, J. C. & Ineson, P. (1992c). Effects of dry-deposited sulphur dioxide on fungal decomposition of angiosperm tree leaf litter: III. Decomposition rates and fungal respiration. *New Phytologist*, **122**, 127–140.

Newton, A. C. & Haigh, J. M. (1998). Diversity of ectomycorrhizal fungi in Britain: a test of the species–area relationship, and the role of host specificity. *New Phytologist*, **138**, 619–627.

Paterson, R. R. M. & Bridge, P. D. (1994). *Biochemical Techniques for Filamentous Fungi*, International Mycological Institute Technical Handbook No. 1. Wallingford: CAB International.

Peter, M., Ayer, F. & Egli, S. (2001). Nitrogen addition in a Norway spruce stand altered macromycete sporocarp production and below-ground ectomycorrhizal species composition. *New Phytologist*, **149**, 311–325.

Pielou, E. C. (1975). *Ecological Diversity*. New York: Wiley-Interscience.

Preston-Mafham, J., Boddy, L. & Randerson, P. F. (2002). Analysis of microbial community functional diversity using sole-carbon-source utilisation profiles: a critique. *FEMS Microbiology Ecology*, **42**, 1–14.

Pryce Miller, E. J. (2002). Taxonomic biodiversity and community structure of saprotrophic fungi in an 'improved' and unimproved upland grassland: use of molecular methods. Unpublished Ph.D. thesis, King's College, University of London.

Radajewski, S., Ineson, P., Parekh, N. R. & Murrell, J. C. (2000). Stable-isotope probing as a tool in microbial ecology. *Nature*, **403**, 646–649.

Robinson, C. H., Dighton, J. & Frankland, J. C. (1993a). Resource capture by interacting fungal colonisers of straw. *Mycological Research*, **97**, 547–558.

Robinson, C. H., Dighton, J., Frankland, J. C. & Coward, P. A. (1993b). Nutrient and carbon dioxide release from straw by interacting species of fungi. *Plant and Soil*, **151**, 139–142.

Robinson, C. H., Dighton, J., Frankland, J. C. & Roberts J. D. (1994). Fungal communities on decaying wheat straw of different resource qualities. *Soil Biology and Biochemistry*, **26**, 1053–1058.

Robinson, C. H., Fisher, P. J. & Sutton, B. C. (1998). Fungal biodiversity in dead leaves of fertilised plants of *Dryas octopetala* from a high Arctic site. *Mycological Research*, **102**, 573–576.

Sigler, W. V. & Turco, R. F. (2002). The impact of chlorothalonil application on soil bacterial and fungal populations as assessed by denaturing gradient gel electrophoresis. *Applied Soil Ecology*, **21**, 107–118.

Smit, E., Leeflang, P., Glandorf, B., van Elsas, J. D. & Wernars, K. (1999). Analysis of fungal diversity in the wheat rhizosphere by sequencing of cloned PCR-amplified genes encoding 18S rRNA and temperature gradient gel electrophoresis. *Applied and Environmental Microbiology*, **65**, 2614–2621.

Swift, M. J. (1976). Species diversity and the structure of microbial communities in terrestrial habitats. *The Role of Aquatic and Terrestrial Organisms in Decomposition Processes* (Ed. by J. M. Anderson & A. MacFadyen), pp. 185–222. Oxford: Blackwell Scientific.

Swift, M. J. (1985). Microbial diversity and decomposer niches. *Current Perspectives in Microbial Ecology* (Ed. by M. J. King & C. A. Reddy), pp. 8–16. Washington, DC: American Society for Microbiology.

Swift, M. J., Heal, O. W. & Anderson, J. M. (1979). *Decomposition in Terrestrial Ecosystems*. Oxford: Blackwell Scientific.

Vainio, E. J. & Hantula, J. (2000). Direct analysis of wood-inhabiting fungi using denaturing gradient gel electrophoresis of amplified ribosomal DNA. *Mycological Research*, **104**, 927–936.

Valinsky, L., Della Vedova, G., Jiang, T. & Borneman, J. (2002). Oligonucleotide fingerprinting of rRNA genes for analysis of fungal community composition. *Applied and Environmental Microbiology*, **68**, 5999–6004.

Vandenkoornhuyse, P., Husband, R., Daniell, T. J., et al. (2002). Arbuscular mycorrhizal community composition associated with two plant species in a grassland ecosystem. *Molecular Ecology*, **11**, 1555–1564.

Visser, S. & Parkinson, D. (1975). Fungal succession on aspen poplar leaf litter. *Canadian Journal of Botany*, **53**, 1640–1651.

Warcup, J. H. (1957). Studies on the occurrence and activity of fungi in a wheat-field soil. *Transactions of the British Mycological Society*, **40**, 237–262.

Wardle, D. A. & Giller, K. E. (1996). The quest for a contemporary ecological dimension to soil biology. *Soil Biology and Biochemistry*, **28**, 1549–1554.

Watling, R. (1995). Assessment of fungal diversity: macromycetes, the problems. *Canadian Journal of Botany*, **73**, S15–S24.

White, D. C. (1995). Chemical ecology: possible linkage between macro- and microbial ecology. *Oikos*, **74**, 177–184.

Widden, P. & Parkinson, D. (1979). Populations of fungi in a high arctic ecosystem. *Canadian Journal of Botany*, **57**, 2408–2417.

Wood, D. A. (1998). Extracellular enzymes of *Agaricus bisporus*. *Proceedings of Sixth International Symposium of the Mycological Society of Japan*. UK/Japan Joint Symposium, Chibo, November 1998.

Zak, J. C. & Visser, S. (1996). An appraisal of soil fungal biodiversity: the crossroads between taxonomic and functional biodiversity. *Biodiversity and Conservation*, **5**, 169–183.

Zhou, D. & Hyde, K. D. (2001). Host-specificity, host-exclusivity, and host-recurrence in saprobic fungi. *Mycological Research*, **105**, 1449–1457.

CHAPTER TWELVE

Is diversity of mycorrhizal fungi important for ecosystem functioning?

J. R. LEAKE
University of Sheffield

D. JOHNSON
University of Aberdeen

D. P. DONNELLY
Cardiff University

L. BODDY
Cardiff University

D. J. READ
University of Sheffield

SUMMARY

1. Globally accelerating rates of species loss make it imperative that relationships between biodiversity and ecosystem function are analysed, yet resolution of these interactions has presented one of the most intractable challenges in ecological research.

2. Because biodiversity in soil is considerably greater than that above ground, and the identities and functions of many soil microorganisms are uncharacterised, the difficulties involved in establishing diversity–function relationships in the below-ground environment are compounded.

3. For some keystone groups of soil microorganisms, prominent among which are the mycorrhizal fungi, diversity–function relationships are starting to be elucidated. Mycorrhizal symbionts are present in virtually all terrestrial ecosystems where they are major components of the soil microbial biomass.

4. There is increasing evidence that mycorrhizal diversity is of central importance in agro-ecosystem functioning, and that intensification of agriculture and forestry, combined with air and soil pollution, is reducing their diversity and compromising their functioning. Two lines of evidence support the case that mycorrhizal diversity is of major functional significance, namely (1) that mycorrhizal associations are multi-functional and exhibit complementarity, assisting plants in nutrient acquisition, mediating carbon transfer between plants and protecting their roots from

Biological Diversity and Function in Soils, eds. Richard D. Bardgett, Michael B. Usher and David W. Hopkins. Published by Cambridge University Press. © British Ecological Society 2005.

pathogens; and (2) that on the basis of emerging evidence of a combination of high specificity and dependency in many mycorrhizal associations, especially those involving myco-heterotrophic plants, it is hypothesised that the extent of functional 'redundancy' is low.

Introduction

It is now recognised that the accelerating rates of species extinctions, which are arising from progressive intensification of land use, pose a global threat to biodiversity (Pimm *et al.* 1995). While the underlying drivers of biodiversity loss may be increasingly well understood, the consequences of such losses for ecosystem functioning are unclear. Indeed, the question of the extent to which biodiversity contributes to these functions has proven to be both contentious and intractable. Attempts to answer the question by experiment using simple assemblages of plants, usually of between 1 and 32 species, have led to conflicts of interpretation particularly over the issues of cause and effect (Loreau *et al.* 2001). Thus, while some such studies have suggested positive relationships between productivity and plant species diversity (Hector *et al.* 1999; Tilman 1999) they have been criticised (Huston 1997; Wardle 1999; Grime 2001) on the basis that the construction of more diverse communities leads inevitably, by the so-called 'sampling effect', to an increase in the chance that more productive species will be added to the assemblage.

The problems facing plant ecologists in their attempts to resolve diversity–function relationships pale into insignificance when compared with those confronting soil microbiologists. In addition to sampling effects, difficulties of interpretation are compounded by the much greater inherent diversity of soil microbial communities, which can contain, in the bacterial population alone, over 5000 genomes per gramme of soil (Chatzinotas *et al.* 1998). By comparison, the most diverse plant communities normally support only 40 plant species per square metre and the entire native higher plant flora of the British Isles comprises fewer than 1700 species in an area of 314 000 km^2 (Preston *et al.* 2002). While such quantitative issues are daunting enough, the diversity–function question is further complicated in microbial ecology by a fundamental lack of knowledge of the physiological properties of the organisms concerned, with less than 5% of microscopically observable bacterial cells being culturable (Sait *et al.* 2002). To the many uncharacterised bacteria present in any sample must be added the complex array of other microbes (archaea, fungi, protozoa, nematodes, insects, etc.) that make up the soil food web, each trophic group of which itself consists of a diverse assemblage of species.

Confronted by such complexity, microbial ecologists are forced to contemplate reductionist approaches as they begin to grapple with questions concerning the relationships between diversity and function in the soil ecosystem. One approach is to focus on specific groups of organisms such as mycorrhizas that are recognised to have especially important 'keystone' functions in ecosystems and that are sufficiently well characterised to enable relationships between diversity and function to be explored. Clearly this approach has inherent limitations, not the least being that it will not necessarily be appropriate to extrapolate biodiversity–function relationships detected in one keystone group to another such group. However, if loss of diversity in only one major group of microorganisms can be shown to have consequences for ecosystem functioning it should serve to provide impetus for a broader assault on these relationships in the soil. In this paper, mycorrhizal fungi are selected as a representative keystone group of soil microorganisms, and relationships between their biodiversity and functions in soil ecosystems are explored.

The mycorrhizal perspective

While there are many groups of soil organisms that may be considered to provide 'keystone' ecosystem functions, mycorrhizal fungi are arguably among the most important because of their direct access to the plant-derived carbon that fuels below-ground microbial communities.

Mycorrhizas are ubiquitous in virtually all ecosystems and since mycorrhizal associations can be traced back over 400 million years to the very origins of land plants (Remy *et al.* 1994), or even earlier (Redecker 2002), they have co-evolved with land plants and are integral to the functioning of terrestrial ecosystems. Through their intimate symbiotic associations with plants, they influence directly major ecosystem processes, as they provide the main pathways for plant–soil and soil–plant fluxes of labile carbon and nutrients respectively (Leake *et al.* 2002). In the case of ectomycorrhizas, where the fungal mycelium ensheaths the majority of fine root tips, virtually all nutrients passing into the host plants do so via the mycorrhiza and almost all the labile carbon released from roots to the soil passes through the fungi. Mycorrhizas are major contributors to soil microbial biomass, as they receive between 4 and 30% of net carbon fixation from the 80% of land plants that have these associations (Smith & Read 1997). For example, it has been estimated that the external mycelium of ectomycorrhizas can account for more than 30% of the microbial biomass in a pine forest soil and at least 50% of the soil respiration was attributed to mycorrhizal mycelium plus roots (Högberg & Högberg 2002). In addition to their roles in carbon and nutrient cycling, mycorrhizas have a major influence on plant establishment, and upon the species composition and diversity of plant communities (Grime *et al.* 1987; Francis & Read 1995).

Diversity of mycorrhizal fungi

In comparison to bacteria, protozoa, plants and insects, the global diversity of mycorrhizal fungi is very modest, and reasonably well characterised (Table 12.1). Of the two major types of mycorrhizas, ectomycorrhizas (EM), which are formed by basidiomycetes and a few ascomycetes, are estimated to number about

Table 12.1. *The global diversity of mycorrhizal fungi compared to that of plants, insects and soil bacteria.*

Group	Global no. of species	Reference
Arbuscular mycorrhizal fungi	~150	Walker and Trappe (1993)
Ectomycorrhizal fungi	~5500	Molina *et al.* (1992)
Plants	~250000	Wilson (1992)
Insects	~750000	Wilson (1992)
Soil bacteria	? >4000000	Curtis *et al.* (2002)

5000–6000 species world wide (Molina *et al.* 1992), whereas named arbuscular mycorrhizas (AM) currently number only 150 or so species of Glomeralean fungi (Walker & Trappe 1993). However, as molecular identification methods are more widely used and applied to natural ecosystems, new species of the latter are still regularly being found and the total number of AM species may run to several hundreds (e.g. see Husband *et al.* 2002).

At the ecosystem scale, mycorrhizal diversity is typically of the same order of magnitude as plant species richness as indicated by variations in the length of the internal transcribed spacer (ITS) region of ribosomal RNA after cutting into fragments with restriction enzymes (ITS-RFLP) and morphotype of root tips, and by sporocarp surveys. In well-established forest stands, up to 200 morphologically distinct mycorrhizal types have been detected in an area of just over 2 ha (Luoma *et al.* 1997) but the detected mycorrhizal diversity on roots is normally in the range between 10 and 50 species (Horton & Bruns 2001; Erland & Taylor 2002; Taylor 2002). Sporocarps indicate additional species that are not detected on roots. For example, in a Swedish pine stand 37 mycorrhizal types were found on over 5500 root tips, and 56 species of EM sporocarp were found in the same plots (Taylor 2002). This implies that most root-sampling studies need to include a larger sample size to detect the full diversity of mycorrhiza present because there are typically 2–4 million EM root tips per square metre (Dahlberg *et al.* 1997), a very large number of root tips must be sampled to detect all the rarer mycorrhizal species, and this problem is compounded by the very patchy spatial distribution of EM fungi (Horton & Bruns 2001).

There have now been enough detailed studies of EM fungal communities to enable generalisations to be made about their properties (Horton & Bruns 2001; Taylor 2002). One important feature is that in most ecosystems with EM plants, a small number (typically between 5 and 10 species) of ectomycorrhizal fungi accounts for by far the majority of mycorrhizas, whereas the remaining EM species, which comprise most of the mycorrhizal species richness, occur with varying rarity (Taylor 2002). However, the very high density of EM root tips in forest soils means that even species represented by less than 0.1% of root tips can be present on several thousand tips per square metre (Erland & Taylor 2002).

The AM fungi have also been found to comprise communities with a small number of dominant species and most species being rare (e.g. see Allen *et al.* 1995). AM fungi show similar, but generally slightly lower, diversity than EM at the scale of individual plant communities. In old-field sites, permanent grassland, temperate deciduous woodland and tropical rainforest 10–37 AM fungal species have been recorded in individual communities (Helgason *et al.* 1998; Bever *et al.* 2001; Hart & Klironomos 2002; Husband *et al.* 2002; Vandenkoornhuyse *et al.* 2002).

Anthropogenic impacts on mycorrhizal diversity

The evidence of loss of biodiversity due to intensification of agriculture and forestry, combined with air and soil pollution, is more compelling for mycorrhizal fungi than virtually any other major group of soil microorganisms. Pollutant nitrogen deposition and nitrogen fertilisation often reduce EM diversity both below ground on root tips (Erland & Taylor 2002) and in sporocarp production (Wallenda & Kottke 1998). Similar decreases in diversity have also been recorded in AM species (as indicated by spore extraction from soil) in response both to contemporary nitrogen deposition gradients (Egerton-Warburton & Allen 2000) and to historical increases in anthropogenic nitrogen deposition (Egerton-Warburton *et al.* 2001).

Diversity is also lost as a result of forestry practices and other disturbances of ectomycorrhizal communities (Taylor 2002; Jones *et al.* 2003). There is evidence of loss of species diversity, and of particular functional types, in clear-cut forests and in seedlings transplanted from forest to clear-cut sites (Hagerman *et al.* 2001; Kranabetter & Friesen 2002). The diversity of EM on individual trees is low at the seedling stage in comparison to the diversity seen on older trees (Kranabetter & Wylie 1998) and, furthermore, very old forests are not only more diverse, but contain species that are consistently absent from earlier successional stages following natural or anthropogenic disturbance (Horton & Bruns 2001).

Agricultural practices, such as ploughing and the addition of phosphate fertiliser, have for long been recognised as major suppressors of mycorrhizal infection in crop plants (Smith & Read 1997), through disruption of mycelial systems in the case of the former, and reduced mycorrhizal dependency in the case of the latter. Only recently has the impact of agricultural practices on AM *diversity* been recognised (Hendrix *et al.* 1995). The development of molecular-identification techniques has now allowed cultivation effects on diversity in roots, as distinct from spore populations, to be examined. For example, an analysis of arable fields on three farms supporting a range of crops infected by mycorrhizal fungi revealed only six species present, and all but one of these at frequencies of less than 4% (Helgason *et al.* 1998). The one common fungus (*Glomus mosseae*) accounted for more than 92% of AM fungal records. This fungus was absent from nearby woodland sites in which not only were there almost double the numbers of AM species (11) but there was also much higher evenness and species diversity. While some spore-based studies have shown less convincing evidence of cultivation effects on AM diversity, a dramatic loss of the genera *Acaulospora* and *Scutellospora* has been shown in studies by both Helgason *et al.* (1998) and Jansa *et al.* (2002) and may reflect particular sensitivity to cultivation in these genera. The depletion of species richness in land under arable cultivation is further reinforced by comparison of the number of AM species found in the former compared with semi-natural communities represented by a low-input, minimally managed long-term upland grassland (24; Vandenkoornhuyse *et al.* 2002) and tropical rainforest (30; Husband *et al.* 2002).

A wide range of other kinds of anthropogenic disturbance have been shown to affect ectomycorrhizal communities including elevated CO_2, ozone, heavy metals, acidification and liming (Erland & Taylor 2002). While the effects of CO_2, ozone and liming can be stimulatory to EM diversity, heavy metals and acidification appear to deplete it (Erland & Taylor 2002). Further work is required to understand more fully the impact of these variables on mycorrhizal communities.

Does mycorrhizal diversity matter?

In communities dominated by a few species but with a large number of rare species, a considerable loss of diversity might occur without major ecosystem consequences. This appears to be the case in a number of studies in agricultural soil in which the diversity of bacteria and saprotrophic fungi has been considerably depleted without catastrophic effects on ecosystem functioning (e.g. Griffiths *et al.* 2000), indicating that there is considerable 'redundancy' in soil microbial species. At a first glance this may appear to be the case with both EM and AM communities having a few dominant taxa and many rare ones (Allen *et al.* 1995; Taylor 2002). The idea of 'functional redundancy' among AM fungi has been reinforced by widely held misconceptions: (a) that the fungi involved show no host specificity (Sanders 2002); (b) that the anatomical simplicity of their mycelial systems implies little difference in their functioning (Hart & Klironomos 2002); and (c) that they have only one main function, namely to enhance the uptake of inorganic phosphorus from soil by plants (Bolan 1991).

Recent work on EM, AM and other mycorrhizas, however, has challenged these perceptions directly and refutes the notion that most of the diversity in mycorrhizal fungi must be functionally redundant. Significant progress in our understanding of these issues has recently arisen on two fronts:

1. Mycorrhizal fungi are now known to provide multiple functions within ecosystems. While there are functional guilds of mycorrhizal species, among which some redundancy may exist, there are instances, for example of extreme host specificity, where some species provide unique and distinct functions. Diverse communities of mycorrhizal fungi provide multifunctional benefits that cannot be provided by a single species on its own,

and this is supported by experimental studies that suggest that mycorrhizal biodiversity affects plant community functioning through complementarity effects of the symbionts.

2. Host-plant mycorrhiza specificity occurs much more frequently than thought previously, and some floristic components of ecosystems are controlled by specific mycorrhizal fungi.

Both of these lines of evidence (detailed below) provide support for the case that mycorrhizal biodiversity (and species composition) is of major significance for ecosystem functioning.

Multi-functionality and complementarity of mycorrhizal fungi

While early studies of mycorrhizal functioning emphasised the role of these symbioses in nutrient acquisition, with an almost exclusive emphasis on phosphorus nutrition of host plants (see Harley & Smith 1983), it is now apparent that all types of mycorrhiza are multi-functional, and include both nutritional and non-nutritional functions. Among the nutritional functions now recognised in both AM and EM are uptake and transfer to plants of the major nutrients nitrogen, phosphorus and potassium, together with trace elements such as iron, zinc and copper (Smith & Read 1997). Non-nutritional functions include reduced uptake of toxic heavy metals (Meharg & Cairney 2000), protection from soil-borne pathogens (Chakravarty *et al.* 1999), carbon subsidy to seedlings linked to established plants by a common mycorrhizal mycelium (Simard *et al.* 2002), the stabilisation of soil aggregates by hyphae and their hydrophobic proteins (Rillig *et al.* 2002), and increased plant drought resistance associated with mycorrhizal effects on plant nutrient status (Smith & Read 1997). All these activities contribute to higher-order effects of mycorrhizas on ecosystem functioning including control of the major pathways of nitrogen and phosphorus cycling (Leake *et al.* 2002) and control of plant community composition and plant productivity (Hart & Klironomos 2002).

As our understanding of the diverse range of functions provided by mycorrhizal fungi has rapidly increased in recent years, so has our awareness of major functional differences between different mycorrhizal fungi (Read 2002). The key functions provided by these fungi require specific morphological and/or physiological adaptations. EM fungi differ greatly in their anatomy, not only in their effects upon root tip structure and branching patterns (e.g. see Horton & Bruns 2001), but also in their external mycelial systems (Agerer 2001). These mycelia vary on the one hand from being almost entirely confined to the mycorrhizal sheath coating root tips, with virtually no mycelium in the soil, to highly complex mycelial systems with hyphal cords comprising differentiated 'vessel' and cortex hyphae that enable exploration and exploitation of very large soil volumes decimetres away from roots (Leake *et al.* 2001). While in AM fungi, variation in the structure of external mycelia in soil is much less than in EM, none-the-less, functionally important differences in the extent and activities of mycelia have been demonstrated in different species (Jakobsen *et al.* 1992; Dodd *et al.* 2000).

The structural diversity of ectomycorrhizas is accompanied by great variations in physiology and in their preferences and capacities to utilise different types of mineral and organic nutrient sources (Buscot *et al.* 2000). One especially import-ant development has been the recognition that the distinction between sapro-trophy and biotrophy in EM fungi is blurred (Read & Perez-Moreno 2003), since some EM fungi produce suites of extracellular enzymes that enable hydrolysis of the major nutrient-containing and structural polymers in plant and microbial litter (Leake & Read 1997). Furthermore, molecular phylogenetic studies of EM fungi indicate that while the majority of genera are evolved from saprotrophic fungi, in a few cases fungi have reverted to saprotrophy (Hibbert *et al.* 2000), showing remarkable instability in the evolution of symbiotic and saprotrophic traits. Some, but not all, EM fungi retain abilities to mobilise nitrogen from its major organic compounds that dominate nitrogen inputs and nitrogen capital in forest soils. As a result, these mycorrhizal fungi are able to short-circuit the mineralisation 'bottle-neck' in the nitrogen cycle (Fig. 12.1a, b). The total quan-tities of nutrients cycled by this route have not been precisely determined but, for example, 19–57% of the nitrogen and 53–91% of the phosphorus contained in dead nematodes and in pine pollen was extracted through the activities of a mycorrhizal fungus in microcosms containing non-sterile soil (Read & Perez-Moreno 2003). By taking into account the estimated quantities of nematodes and pollen produced in forest ecosystems, these components alone, as a result of the activities of mycorrhiza, may contribute 12% of the nitrogen and 15% of the total phosphorus uptake by trees.

In many forest ecosystems the rates of nitrogen mineralisation are much lower than the rates of nitrogen uptake by the trees (see, e.g., Schulze *et al.* 2000), and in unpolluted nitrogen-limited boreal forests there is sometimes no detectable nitrogen mineralisation, the nitrogen cycle being 'short-circuited' by organic nitrogen utilisation by ectomycorrhizas (Fig. 12.1a). However, in many parts of Europe, pollutant nitrogen deposition is leading to radical alterations to the nitrogen cycle so that it now closely resembles the conventional nitrogen cycle of intensive agriculture but 'fertiliser' is supplied by rain rather than by agro-chemicals (Fig. 12.1b). In the unpolluted, nitrogen-limited forests, the trees are dependent upon mycorrhizas that are specifically adapted for mobilisation of organic nitrogen and, by implication, their uptake of labile organic nitrogen (Nåsholm *et al.* 1998) is so effective that the normal mineralisation pathways are prevented (see Figs. 12.1a and 12.2; Northup *et al.* 1995; Schulze *et al.* 2000). In these situations a high proportion, but not all, of the EM fungi present on roots appear to have the ability to utilise polymeric organic nitrogen (Taylor *et al.* 2000). This implies that small amounts of mineral nitrogen such as that

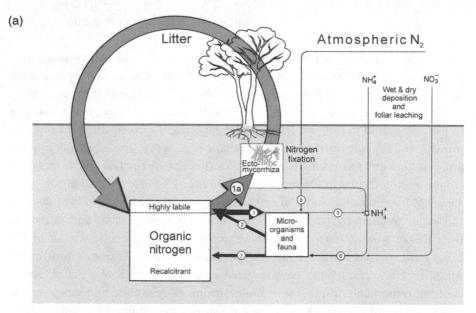

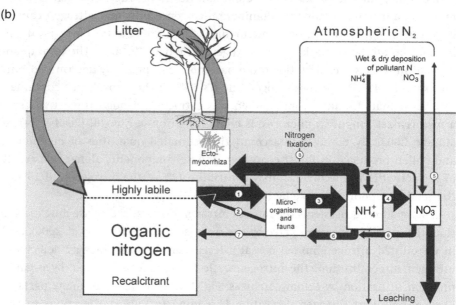

Figure 12.1. (a) The short-circuiting of the nitrogen cycle by use of organic nitrogen by diverse communities of ectomycorrhizas in nitrogen-limited boreal forests experiencing minimal inputs of reactive nitrogen deposition. In such soils, no nitrogen mineralisation is detected and there is no nitrification. This 'tight' nitrogen cycle is highly conservative with no opportunity for nitrogen loss by leaching or denitrification. There is intense competition between ectomycorrhizal fungi and saprotrophs for labile nitrogen (pathways 1a and 1).
(b) The nitrogen cycle in conifer forests that have sustained high inputs of reactive pollutant nitrogen, and suffered loss of ectomycorrhizal diversity, with elimination of the fungi that utilise polymeric organic nitrogen so that the cycle follows the conventional mineralisation pathways. This results in considerable loss of nitrogen through a combination of leaching and denitrification. The main microbially driven pathways are: 1, depolymerisation and microbial assimilation of organic nitrogen; 2, release of microbial-biomass nitrogen; 3, mineralisation (ammonification); 4, nitrification; 5, denitrification; 6, microbial immobilisation; 7, humification; 8, nitrogen fixation.

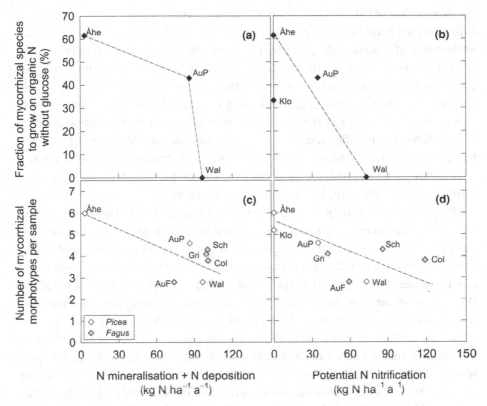

Figure 12.2. The effect of increasing nitrogen availability, resulting from reactive nitrogen deposition and associated mineralisation processes, on the proportion of mycorrhizal community that can use protein nitrogen and the diversity of mycorrhizas on root tips. The data are for sites ranging from unpolluted locations (Ahe, Aheden, Sweden; Klo, Klosterhede, Denmark) through moderately polluted sites (AuP, AuF, Aubure, France; Gri, Gribskov, Denmark) to highly nitrogen-enriched locations in central Germany (Wal, Walstein; Sch, Schacht) or Italy (Col, Collelongo). The decline in the proportion of species that are able to use protein without exogenous sugar decreases sharply with increasing availability of mineral nitrogen, and ectomycorrhizal diversity is halved in the sites with highest rates of mineralisation–deposition and nitrification. From Schulze *et al.* (2000). 'Interaction between the carbon and nitrogen cycles and the role of biodiversity: a synopsis of a study along a north–south transect through Europe', in Schulze, E. D. (Ed.) *Carbon and Nitrogen Cycling in European Forest Ecosystems*, Ecological Studies 142. Springer, Berlin Heidelberg New York, page 485.

provided in rainfall or by small-spatial-scale mineralisation processes that are not detected by conventional methods are being removed by those mycorrhizal fungi that cannot use polymeric nitrogen but are highly adapted to take up mineral nitrogen – and they provide functional complementarity. These communities of mycorrhizal fungi ensure extremely efficient nitrogen cycling, and through short-circuiting the mineralisation pathways prevent loss of nitrogen through leaching and denitrification (Fig. 12.1a, b).

Examination of EM diversity across a European-wide gradient of nitrogen deposition has shown that there is a systematic decline in EM diversity with increasing availability of mineral nitrogen (Fig. 12.2), and that the proportion of culturable EM fungi that are specifically adapted to utilise polymeric organic nitrogen falls dramatically with increasing mineral nitrogen availability (Schulze *et al.* 2000; Taylor *et al.* 2000). It appears that the shift in mycorrhizal community in these cases has resulted both in the increasing mineralisation of nitrogen and in the establishment of nitrification, both of which increase the 'leakiness' of nitrogen from the ecosystems. The consequences of loss of EM biodiversity, and of a major functional group of EM fungi in particular, can be tentatively linked to altered ecosystem functioning in this case (cf. Figs. 12.1a, b).

The functional diversity of EM communities reflects the enormous spatial and temporal heterogeneity of resources in forest soils. It is apparent from detailed studies of both the spatial distribution of EM communities in organic and underlying mineral horizons (Erland & Taylor 2002), and the temporal variations in EM communities associated with disturbance and succession (Horton & Bruns 2001), that EM communities are responsive to ecosystem heterogeneity. Furthermore, the application of molecular identification methods has enabled for the first time, the analysis of vertical stratification of populations of mycorrhizal hyphae in soil, and this has revealed clear evidence of niche differentiation by different species (Dickie *et al.* 2002). There is increasingly compelling evidence that no single EM species can combine all the structural or physiological adaptations that will provide optimal benefits to the host trees. There is considerable intraspecific variation in physiological functions in EM such as the production of extracellular enzymes with different thermal optima (Tibbett *et al.* 1999) and tolerance to heavy metals (Meharg & Cairney 2000), and this adds further complexity to consideration of biodiversity functioning. However, the functional diversity of EM fungi is much greater between species than within species so that species richness is likely to correlate strongly with functional diversity, and this in turn leads to functional complementarity.

Experimental tests of 'complementarity' of EM fungi have recently been attempted by inoculating pine or birch seedlings with single species and mixed-species combinations, but the results are somewhat equivocal (Baxter & Dighton 2001; Jonsson *et al.* 2001). However, when two host plants (*Pinus sylvestris* and *Betula pendula*) were combined with up to eight EM fungi at two levels of soil fertility, in two out of the four cases EM diversity positively increased plant productivity in excess of that achieved by monocultures of the same fungi, an effect most marked at low soil fertility (Jonsson *et al.* 2001).

In a similar kind of experimental manipulation, varying diversity and composition of AM fungi from 0 to 14 species in a plant species-rich 'calcareous grassland' had highly significant effects on plant diversity and biomass production, and these effects were explained by greater plant phosphorus uptake, and longer hyphal lengths in the more diverse AM communities (van der Heijden *et al.* 1998). Both shoot and root biomass increased progressively with increasing diversity of AM, these effects being more consistent than the biodiversity–productivity relationships seen in plant diversity studies (see Loreau *et al.* 2001; Pfisterer & Schmid 2002). However, the results of the AM manipulations are subject to the same intrinsic 'sampling effect' problem that has occurred in plant biodiversity–functioning studies (Wardle 1999), namely that the effects of each species on its own must be determined to establish for certain that the effects of species mixtures is not simply driven by the increasing chance, with increasing diversity, of including particular species that promote most growth. In this particular case (van der Heijden *et al.* 1998), the variability within means at each level of mycorrhizal diversity remains similar, whereas if the 'sampling' effect was the main factor driving the biodiversity–productivity response then we would expect the within-treatment variation to decrease with diversity.

The realised diversity of AM fungi was not determined by van der Heijden *et al.* (1998) but in the studies with EM, both Jonsson *et al.* (2001) and Baxter & Dighton (2001), found that realised EM diversity on the roots was much lower than intended. In the latter case, the establishment of individual EM species declined from 100% when inoculated singly, to 86% when inoculated in pairs, to 52% when inoculated in fours. As a result, the communities are not 'randomly assembled' so that it is difficult to distinguish species composition and biodiversity effects (Leake 2001).

In contrast to the aforementioned results, in a study in which individual plant species were inoculated with combinations of zero to three AM fungi, there was no evidence of increased biomass production in the three-species mixture compared to the best single AM species (van der Heijden *et al.* 2003). However, in this case all the fungi were species of *Glomus*, whereas in the earlier AM study not only was the overall diversity of AM fungi much higher, but they included representatives of five major genera: *Acaulospora*, *Entrophospora*, *Gigaspora*, *Glomus* and *Scutellospora*. As van der Heijden *et al.* (2003) suggest, depletion of diversity at the genus level may be of greater functional importance than loss of species in the same genera, due to the likelihood of greater functional differences at the genus level. This is supported by much greater coefficient of variation in plant growth (*Plantago lanceolata*) inoculated with different genera of AM fungi than with different species within the same genus or different isolates of the same species (Hart & Klironomos 2002). Consequently, the loss of diversity, seen in arable cropping systems, from which some genera such as *Scutellospora* and *Acaulospora* have been virtually eliminated (Helgason *et al.* 1998), is of grave concern.

Mycorrhiza specificity: linking function to particular species

While specificity is well established for many EM species, some of which show very narrow host ranges (Smith & Read 1997), the evidence for specificity in

AM has been more equivocal. Recently, however, molecular studies of mycor-rhizal communities have revealed that many plants have much higher fungal specificities than hitherto realised. For example, using AM-specific primers, the AM community composition infecting roots of two common grassland species – *Trifolium repens* and *Agrostis capillaris* – were distinct even though the plants were sampled from the same grassland community and grew in immediate proxim-ity to each other (Vandenkoornhuyse *et al.* 2002). Of the 24 phylogenetic species found, nine were exclusive to one or other, but not both species. Host specificity in AM fungi has also been revealed by determination of number and identity of spores in soil using plants and fungi that co-occurred at an old-field site, and monocultures of four plant species grown with a mixed inoculum of eight AM species, in five genera (Bever 2002). After one generation of plants, and after re-inoculation of a second generation of plant monocultures with the spore communities arising from the first generation, large and highly significant dif-ferences in AM spore communities were produced under the different plant species, demonstrating clear specificity in AM–plant interactions. These observa-tions add further weight to one of the most important findings of the studies by van der Heijden *et al.* (1998, 2003), that the effect of individual species of AM fungi on growth of monocultures of plants that co-exist in nature is host-plant specific.

In some plant–mycorrhiza associations there is absolute specificity to the extent that some plants will only occur if a specific mycorrhizal fungus is present. The best examples of this are provided by non-photosynthetic plants that obtain carbon from fungal partners and are therefore defined as myco-heterotrophs (Leake 1994). Over 400 species of higher plant are fully myco-heterotrophic, and many more, including the 18 000 species of orchids, are com-pletely dependent upon fungal partners for myco-heterotrophic nutrition for seedling establishment. Many myco-heterotrophic plants are epiparasitic upon green plants from which they obtain their carbon by 'cheating' a shared mycor-rhizal fungus (Bidartondo *et al.* 2002). In other words, the two plants have the same fungal partner, and while the fungus provides mineral nutrition to both, it only obtains carbon from the green plant and this is passed on to the myco-heterotrophic plant. This absolute dependence of epiparasitic myco-heterotrophs on their fungal partners is characterised by extreme specificity, many of these plants being exclusively associated with a single fungal species throughout their range (Taylor *et al.* 2002). It is among the myco-heterotrophs that the first exam-ple of absolute specificity of a plant for a single species or very narrow lineage of AM fungi has been demonstrated (Bidartondo *et al.* 2002). None of the fungal partners of myco-heterotrophs that have been identified to date are particularly common, and it appears that many of these fungi have patchy distributions, which are also reflected in the distributions of the plants that depend upon them (McKendrick *et al.* 2000, 2002). These findings demonstrate a pivotal role of mycorrhizas in the control of plant community composition, and complement the growing body of evidence that there are both positive and negative feed-backs between plant and mycorrhizal communities linked to specificity in these symbioses (Bever 2002).

Specificity in plant–mycorrhiza associations, coupled to the absolute dependence on particular fungi by some plants, has profound implications for the biodiversity–ecosystem functioning relationship. In such cases, the fungi have unique functions that cannot be replaced by other species. If the critical mycorrhizal fungus goes extinct at a site, the myco-heterotroph will go extinct too. If we place any intrinsic value upon conservation of dependent species such as myco-heterotrophs, then that value attaches also to the fungal partner, which cannot be considered 'functionally redundant'. As we gain increasing knowledge of these kinds of highly specific interdependencies between species, it is apparent that many of the mycorrhizal fungi that may be rare in the community cannot automatically be assumed to have no functional significance. Extreme host specificity, or highly evolved specific functions that may be restricted to very few taxa, may give disproportionate functional importance to rare mycorrhizal species.

It is clear that multi-trophic interactions between mycorrhizal fungi also can involve considerable specificity. For example, Chakravarty *et al.* (1999) have shown that the antagonistic effects of some EM fungi on some common root pathogens such as *Fusarium* sp. are species specific. We have found highly antagonistic effects of *Paxillus involutus* on mycelial growth of the wood-decay saprotroph *Phaerochaete velutina* (Leake *et al.* 2002), but this effect is not seen with *Suillus bovinus* (Leake *et al.* 2001). Mycorrhizas also interact with bacteria, and again some of these associations involve varying levels of specificity as, for example, in mycorrhiza–helper bacteria (Garbaye 1994) and in the case of the apparently endosymbiotic *Burkholderia*, which occurs inside some Glomeralean fungi (Bianciotto *et al.* 2000). In most cases, our knowledge of these kinds of interactions is poorly developed.

What is important for ecosystem functioning: species composition, species diversity or both?

The huge diversity in activities of mycorrhizal fungi highlights the awkward question: what exactly do we mean by 'ecosystem functioning'? With increasing evidence of management and pollutant impacts on mycorrhizal diversity, nutrient cycles and plant diversity in many agro-ecosystems, many functions may be lost without being detected if 'ecosystem functioning' is simply measured by plant productivity. The studies of mycorrhizal diversity and functioning, while placing considerable emphasis on plant productivity, have also revealed important effects on the plant species composition and structuring of their communities (Hart & Klironomos 2002). The increasing evidence of functional complementarity in mycorrhizal communities and the evidence that some species have very specific and unique functions suggests that both species composition and species diversity are important. Where species with keystone functions such as the mobilisation of polymeric organic nitrogen are lost, as in nitrogen-polluted European forests (Schulze *et al.* 2000), clearly the impact on ecosystem function will be large. Whereas in this case, and in the cases of loss of AM genera from cultivated land, the depletion of mycorrhizal diversity is associated with an even greater proportional loss of major functional groups – implying loss of significant functions. What is worrying is that many subtle functions are almost certainly lost as biodiversity is lost and where there are strong interdependencies between species then the loss of one species can lead to loss of another. Given the exceptional diversity of microbial communities in undisturbed soils, and our lack of knowledge of most of these organisms and their interdependencies, the ongoing losses of soil biodiversity in response to increasing disturbance and various pollutants is of major concern.

Future studies of the biodiversity–functioning issue need to look beyond biomass production to consider more fully issues such as ecosystem stability and 'recoverability'. Increasing attention needs to be directed to understanding specificity and interdependencies between species since it is clear that some rare species, such as the myco-heterotrophs, show absolute dependence upon specific and uncommon fungal partners. Failure to manage ecosystems to maintain the keystone fungi upon which these plants depend, for example, by replanting forests with trees that are incompatible hosts of the fungi, will lead to local extinction of these plants. There is increasing evidence that biodiversity begets biodiversity, and where there are complex multi-trophic interactions between species that have high interdependencies, loss of some species of mycorrhizas can result in the losses of other species. Until we understand more fully these complex multi-trophic interactions between mycorrhizas and other organisms, the functional significance of loss of particular mycorrhizal species may go unrecognised and undetected. As biodiversity losses continue, and rates of species extinctions increase, the 'recovery' of disturbed and polluted ecosystems back to their pristine condition is increasingly difficult to achieve. A major challenge lies ahead if we are effectively to manage and conserve soil biodiversity and understand the functional significance of mycorrhizal diversity.

Acknowledgements

We gratefully acknowledge funding from the Natural Environment Research Council UK grant code numbers NER/A/S/2000/00411 and NER/T/S/2001/177.

References

Agerer, R. (2001). Exploration types of ectomycorrhizae: a proposal to classify ectomycorrhizal mycelial systems according to their patterns of differentiation and putative ecological significance. *Mycorrhiza*, 11, 107–114.

Allen, E. B., Allen, M. F., Helm, D. J., *et al.* (1995). Patterns and regulation of mycorrhizal plant and fungal diversity. *Plant and Soil*, 170, 47–62.

Baxter, J. W. & Dighton, J. (2001). Ectomycorrhizal diversity alters growth and nutrient acquisition of gray birch (*Betula populifolia*) seedlings in host-symbiont culture conditions. *New Phytologist*, 152, 139–149.

Bever, J. D. (2002). Host-specificity of AM fungal population growth rates can generate feedback on plant growth. *Plant and Soil*, 244, 281–290.

Bever, J. D., Schulze, P. A., Pringle, A. & Morton, J. B. (2001). Arbuscular mycorrhizal fungi: more diverse than meets the eye, and the ecological tale of why. *BioScience*, 51, 923–931.

Bianciotto, V., Lumini, E., Lanfranco, L., *et al.* (2000). Detection and identification of bacterial endosymbionts in arbuscular mycorrhizal fungi belonging to the family Gigasporaceae. *Applied and Environmental Microbiology*, 66, 4503–4509.

Bidartondo, M. I., Redecker, D., Hijri, I., *et al.* (2002). Epiparasitic plants specialized on arbuscular mycorrhizal fungi. *Nature*, 419, 389–392.

Bolan, N. S. (1991). A critical review on the role of mycorrhizal fungi in the uptake of phosphorus by plants. *Plant and Soil*, 134, 189–207.

Buscot, F., Munch, J. C., Charcosset, J. Y., *et al.* (2000). Recent advances in exploring physiology and biodiversity of ectomycorrhizas highlight the functioning of these symbioses in ecosystems. *FEMS Microbiology Reviews*, 24, 601–614.

Chakravarty, P., Khasa, D., Dancik, B., *et al.* (1999). Integrated control of *Fusarium* damping-off in conifer seedlings. *Zeitschrift Für Pflanzenkrankheiten und Pflanzenschultz*, 106, 342–352.

Chatzinotas, A., Sandaa, R. A., Schonhuber, W., *et al.* (1998). Analysis of broad-scale differences in microbial community composition of two pristine forest soils. *Systematic and Applied Microbiology*, 21, 579–587.

Curtis, T. P., Sloan, W. T. & Scannell, J. W. (2002). Estimating prokaryotic diversity and its limits. *Proceedings of the National Academy of Sciences, USA*, 99, 10 494–10 499.

Dahlberg, A., Jonsson, L. & Nylund, J.-E. (1997). Species diversity and distribution of biomass above- and below-ground among ectomycorrhizal fungi in an old-growth Norway spruce forest in south Sweden. *Canadian Journal of Botany*, 75, 1323–1335.

Dickie, I. A., Xu, B. & Koide, R. T. (2002). Vertical niche differentiation of ectomycorrhizal hyphae in soil as shown by T-RFLP. *New Phytologist*, 156, 527–535.

Dodd, J. C., Boddington, C. L., Rodriguez, A., Gonzalez-Chavez, C. & Mansur, I. (2000). Mycelium of arbuscular mycorrhizal fungi (AMF) from different genera: form, function and detection. *Plant and Soil*, 226, 131–151.

Egerton-Warburton, L. M. & Allen, E. B. (2000). Shifts in arbuscular mycorrhizal communities along an anthropogenic nitrogen deposition gradient. *Ecological Applications*, 10, 484–496.

Egerton-Warburton, L. M., Graham, R. C., Allen, E. B. & Allen, M. F. (2001). Reconstruction of the historical changes in mycorrhizal fungal communities under anthropogenic nitrogen deposition. *Proceedings of The Royal Society of London, Series B, Biological Sciences*, 268, 2479–2484.

Erland, S. & Taylor, A. F. S. (2002). Diversity of ectomycorrhizal fungal communities in relation to the abiotic environment. *Mycorrhizal Ecology* (Ed. by. M. G. A. van der Heijden & I. Sanders), pp. 163–200. Berlin/Heidelberg: Springer-Verlag.

Francis, R. & Read, D. J. (1995). Mutualism and antagonism in the mycorrhizal symbiosis, with special reference to impacts on plant community structure. *Canadian Journal of Botany*, **73**, 1301–1309.

Garbaye, J. (1994). Helper bacteria: a new dimension to the mycorrhizal symbiosis. *New Phytologist*, **128**, 197–210.

Griffiths, B. S., Ritz, K., Bardgett, R. D., *et al.* (2000). Ecosystem response of pasture soil communities to fumigation-induced microbial diversity reductions: an examination of the biodiversity–ecosystem function relationship. *Oikos*, **90**, 279–294.

Grime, J. P. (2001). *Plant Strategies, Vegetation Processes and Ecosystem Properties*, 2nd edition. Chichester: Wiley.

Grime, J. P., Mackey, J. M. L., Hillier, S. H. & Read, D. J. (1987). Floristic diversity in a model system using experimental microcosms. *Nature*, **328**, 420–422.

Hagerman, S. H., Sakakibara, S. M. & Durall D. M. (2001). The potential for woody understory plants to provide refuge for ectomycorrhizal inoculum at an interior Douglas-fir forest after clear-cutting. *Canadian Journal of Forest Research*, **31**, 711–721.

Harley, J. L. & Smith, S. E. (1983). *Mycorrhizal Symbiosis*. London: Academic Press.

Hart, M. M. & Klironomos J. N. (2002). Diversity of arbuscular mycorrhizal fungi and ecosystem functioning. *Mycorrhizal Ecology* (Ed. by M. G. A. van der Heijden & I. Sanders), pp. 225–242. Berlin/Heidelberg: Springer-Verlag.

Hector, A., Schmid, B., Beierkuhnlein, C., *et al.* (1999). Plant diversity and productivity experiments in European grasslands. *Science*, **286**, 1123–1127.

Helgason, T., Daniell, T. J., Husband, R., Fitter, A. H. & Young, J. P. W. (1998). Ploughing up the wood-wide web? *Nature*, **394**, 431.

Hendrix, J. W., Guo, B. Z. & An, Z.-Q. (1995). Divergence of mycorrhizal fungal communities in crop production systems. *Plant and Soil*, **170**, 131–140.

Hibbert, D. S., Gilbert, L.-B. & Donoghie, M. J. (2000). Evolutionary instability of ectomycorrhizal symbioses in basidiomycetes. *Nature*, **407**, 506–508.

Högberg, M. N. & Högberg, P. (2002). Extramatrical ectomycorrhizal mycelium contributes one-third of microbial biomass and produces, together with associated roots, half the dissolved organic carbon in a forest soil. *New Phytologist*, **154**, 791–795.

Horton, T. R. & Bruns, T. D. (2001). The molecular revolution in ectomycorrhizal ecology: peeking into the black-box. *Molecular Ecology*, **10**, 1855–1871.

Husband, R., Herre, E. A., Turner, S. L., Gallery, R. & Young, J. P. Y. (2002). Molecular diversity of arbuscular mycorrhizal fungi and patterns of host association over time and space in a tropical forest. *Molecular Ecology*, **11**, 2669–2678.

Huston, M. A. (1997). Hidden treatments in ecological experiments: re-evaluating the ecosystem function of biodiversity. *Oecologia*, **110**, 449–460.

Jakobsen, I., Abbott, L. & Robson, A. D. (1992). External hyphae of vesicular arbuscular mycorrhizal fungi associated with *Trifolium subterraneum* L: I. Spread of hyphae and phosphorus inflow into roots. *New Phytologist*, **120**, 371–380.

Jansa, J., Mozafar, A., Anken, T., *et al.* (2002). Diversity and structure of AMF communities as affected by tillage in a temperate soil. *Mycorrhiza*, **12**, 225–234.

Jones, M. D., Durall, D. M. & Cairney, J. W. G. (2003). Ectomycorrhizal fungal communities in young forest stands regenerating after clearcut logging. *New Phytologist*, **157**, 399–422.

Jonsson, L. M., Nilsson, M. C., Wardle, D. A. & Zackrisson, O. (2001). Context dependent effects of ectomycorrhizal species richness on tree seedling productivity. *Oikos*, **93**, 353–364.

Kranabetter, J. M. & Friesen, J. (2002). Ectomycorrhizal community structure on western hemlock seedlings transplanted from forests into openings. *Canadian Journal of Botany*, **80**, 861–868.

Kranabetter, J. M. & Wylie, T. (1998). Ectomycorrhizal community structure across forest openings on naturally regenerated western hemlock seedlings. *Canadian Journal of Botany*, **76**, 189–196.

Leake, J. R. (1994). The biology of myco-heterotrophic ("saprophytic") plants. *New Phytologist*, **127**, 171–216.

Leake, J. R. (2001). Is diversity of ectomycorrhizal fungi important for ecosystem function? *New Phytologist*, **152**, 1–3.

Leake, J. R. & Read, D. J. (1997). Mycorrhizal fungi in terrestrial habitats. *The Mycota*, Vol. IV, *Environmental and Microbial Relationships* (Ed. by D. T. Wicklow & B. Söderström), pp. 281–301. Berlin: Springer-Verlag.

Leake, J. R., Donnelly, D. P., Saunders, E. M., Boddy, L. & Read, D. J. (2001). Rates and quantities of carbon flux to ectomycorrhizal mycelium following ^{14}C pulse labelling of *Pinus sylvestris* L. seedlings: effects of litter patches and interaction with a wood-decomposer fungus. *Tree Physiology*, **21**, 71–82.

Leake, J. R., Donnelly, D. P. & Boddy, L. (2002). Interactions between ectomycorrhizal and saprotrophic fungi. *Mycorrhizal Ecology* (Ed. by M. G. A. van der Heijden & I. Sanders), pp. 345–372. Berlin/Heidelberg: Springer-Verlag.

Loreau, M., Naeem, S., Inchausti, P., *et al.* (2001). Biodiversity and ecosystem functioning: current knowledge and future challenges. *Science*, **294**, 804–808.

Luoma, D. L., Eberhart, J. L. & Amaranthus, M. P. (1997). Biodiversity of ectomycorrhizal types from southwest Oregon. *Conservation and Management of Native Plants and Fungi* (Ed. by T. N. Kaye, A. Liston, R. M. Love, *et al.*), pp. 249–253. Corvallis, OR: Native Plant Society of Oregon.

McKendrick, S. L., Leake, J. R., Taylor, D. L. & Read, D. J. (2000). Symbiotic germination and development of myco-heterotrophic plants in nature: ontogeny of *Corallorhiza trifida* Châtel and characterisation of its mycorrhizal fungi. *New Phytologist*, **145**, 523–537.

McKendrick, S. L., Leake, J. R., Taylor, D. L. & Read, D. J. (2002). Symbiotic germination and development of the myco-heterotrophic orchid *Neottia nidis-avis* in nature and its requirement for locally distributed *Sebacina* spp. *New Phytologist*, **154**, 233–247.

Meharg, A. A. & Cairney, J. W. G. (2000). Co-evolution of mycorrhizal symbionts and their hosts to metal-contaminated environments. *Advances in Ecological Research*, **30**, 69–111.

Molina, R., Massicotte, H. & Trappe, J. M. (1992). Specificity phenomena in mycorrhizal symbioses: community–ecological consequences and practical implications. *Mycorrhizal Functioning* (Ed. by M. F. Allen), pp. 357–423. London: Chapman and Hall.

Näsholm, T., Ekblad, A., Nordin, A., *et al.* (1998). Boreal forest plants take up organic nitrogen. *Nature*, **392**, 914–916.

Northup, R. R., Zengshou, Y., Dahlgren, R. A. & Vogt, K. A. (1995). Polyphenol control of nitrogen release from pine litter. *Nature*, **377**, 227–229.

Pfisterer, A. B. & Schmid, B. (2002). Diversity-dependent production can decrease the stability of ecosystem functioning. *Nature*, **416**, 84–86.

Pimm, S. L., Russell, G. J., Gittleman, J. L. & Brooks, T. M. (1995). The future of biodiversity. *Science*, **269**, 347–350.

Preston, C. D., Pearman, D. A. & Dines, T. D. (2002). *New Atlas of the British and Irish Flora*. Oxford: Oxford University Press.

Read, D. J. (2002). Towards ecological relevance: progress and pitfalls in the path towards an understanding of mycorrhizal functions in nature. *Mycorrhizal Ecology* (Ed. by M. G. A. van der Heijden & I. Sanders), pp. 3–29. Berlin/Heidelberg: Springer-Verlag.

Read, D. J. & Perez-Moreno, J. (2003). Mycorrhizas and nutrient cycling in ecosystems: a journey towards relevance? *New Phytologist*, **157**, 475–492.

Redecker, D. (2002). Molecular identification and phylogeny of arbuscular mycorrhizal fungi. *Plant and Soil*, **244**, 67–73.

Remy, W., Taylor, T. N., Hass, H. & Kerp, H. (1994). Four hundred million year old vesicular arbuscular mycorrhizae. *Proceedings of the National Academy of Sciences, USA*, **91**, 11 841–11 843.

Rillig, M. C., Wright, S. F. & Eviner, V. T. (2002). The role of arbuscular mycorrhizal fungi and glomalin in soil aggregation: comparing effects of five plant species. *Plant and Soil*, **238**, 325–333.

Sait, M., Hugenholtz, P. & Janssen, P. H. (2002). Cultivation of globally distributed soil bacteria from phylogenetic lineages previously only detected in cultivation independent surveys. *Environmental Microbiology*, **4**, 654–666.

Sanders, I. (2002). Specificity in the arbuscular mycorrhizal symbiosis. *Mycorrhizal Ecology* (Ed. by M. G. A. van der Heijden & I. Sanders), pp. 415–437. Berlin/Heidelberg: Springer-Verlag.

Schulze, E.-D., Högberg, P., van Oene, H., *et al.* (2000). Interactions between the carbon and nitrogen cycles and the role of biodiversity: a synopsis of a study along a north–south transect through Europe. *Ecological Studies 142: Carbon and Nitrogen Cycling in European Forest Ecosystems* (Ed. by E.-D. Schulze), pp. 468–491. Berlin: Springer-Verlag.

Simard, S. W., Jones, M. D. & Durall, D. M. (2002). Carbon and nutrient fluxes within and between mycorrhizal plants. *Mycorrhizal Ecology* (Ed. by M. G. A. van der Heijden & I. Sanders), pp. 34–74. Berlin/Heidelberg: Springer-Verlag.

Smith, S. E. & Read, D. J. (1997). *Mycorrhizal Symbiosis*, 2nd edition. London: Academic Press.

Taylor, A. F. S., Martin, F. & Read, D. J. (2000). Fungal diversity in ectomycorrhizal communities of Norway Spruce (*Picea abies* (L.) Karst) and Beech (*Fagus sylvatica* L.) along north–south transects in Europe. *Ecological Studies 142: Carbon and Nitrogen Cycling in European Forest Ecosystems* (Ed. by E.-D. Schulze), pp. 343–365. Berlin: Springer-Verlag.

Taylor, A. F. S. (2002). Fungal diversity in ectomycorrhizal communities: sampling effort and species detection. *Plant and Soil*, **244**, 19–28.

Taylor, D. L., Bruns, T. D., Leake, J. R. & Read, D. J. (2002). Mycorrhizal specificity and function in myco-heterotrophic plants. *Mycorrhizal Ecology* (Ed. by M. G. A. van der Heijden & I. Sanders), pp. 374–413. Berlin/Heidelberg: Springer-Verlag.

Tibbett, M., Sanders, F. E., Cairney, J. W. G. & Leake, J. R. (1999). Temperature regulation of extracellular proteases in ectomycorrhizal fungi (*Hebeloma* spp.) grown in axenic culture. *Mycological Research*, **103**, 707–714.

Tilman, D. (1999). Ecological consequences of biodiversity: a search for general principles. *Ecology*, **80**, 1455–1474.

Vandenkoornhuyse, P., Husband, R., Daniell, T. J., *et al.* (2002). Arbuscular mycorrhizal community composition associated with two plant species in a grassland ecosystem. *Molecular Ecology*, **11**, 1555–1564.

Van der Heijden, M. G. A., Klironomos, J. N., Ursic, M., *et al.* (1998). Mycorrhizal fungal diversity determines plant biodiversity, ecosystem variability and productivity. *Nature*, **396**, 69–72.

Van der Heijden, M. G. A., Wiemken, A. & Sanders, I. R. (2003). Different arbuscular mycorrhizal fungi alter coexistence and resource distribution between co-occurring plant. *New Phytologist*, **157**, 569–578.

Walker, C. M. & Trappe, J. M. (1993). Names and epithets in the Glomales and Endogonales. *Mycological Research*, **97**, 339–344.

Wallenda, T. & Kottke, I. (1998). Nitrogen deposition and ectomycorrhiza. *New Phytologist*, **139**, 169–187,

Wardle, D. A. (1999). Is 'sampling effect' a problem for experiments investigating biodiversity–ecosystem function relationships? *Oikos*, **87**, 403–407.

Wilson, E. O. (1992). *The Diversity of Life*. London: Harvard University Press.

CHAPTER THIRTEEN

Trophic structure and functional redundancy in soil communities

HEIKKI SETÄLÄ

University of Helsinki

MATTY P. BERG

Vrije Universiteit

T. HEFIN JONES

Cardiff University

SUMMARY

1. Empirical and theoretical evidence suggests that the rate and magnitude of below-ground ecosystem processes depend on the architecture of the detrital food web. Although some species have an indisputable keystone role in determining soil processes, there is little evidence suggesting that species diversity per se has any major influence at a system level.
2. We review studies that shed light on the degree of functional redundancy in decomposer food webs – from microbes to soil fauna. As well as emphasising the need to define accurately functional redundancy (using both 'Hutchinsonian ecological niche' and 'functional niche' concepts), we also focus on features specific to soils and their communities that may affect the levels at which functional redundancy exists in detrital food webs.
3. We also explore the levels of ecological hierarchy (from species to trophic levels) at which diversity differences manifest themselves as altered ecosystem-level processes.
4. We conclude that the high degree of generalism – even omnivory – in resource-use among decomposer organisms, and the highly heterogeneous environment of soil organisms (reducing competition between species, thus allowing taxa with similar feeding preferences/environmental tolerances to co-exist), play major roles in explaining the high degree of functional complementarity found in decomposer communities.

Introduction

The accelerating loss of biodiversity in various global ecosystems (Lawton & May 1995; Lawton 2000; Schmid et al. 2002) and recent findings emphasising the close linkage between the above- and below-ground components of ecosystems

Biological Diversity and Function in Soils, eds. Richard D. Bardgett, Michael B. Usher and David W. Hopkins. Published by Cambridge University Press. © British Ecological Society 2005.

(Wardle 2002) have led many ecologists to direct their interest to soil ecology and processes. Increasing evidence suggests that both the rate and magnitude of important below-ground processes, such as decomposition of organic matter and the liberation of nutrients bound to it, depend not only on the biomass of the decomposer community but are also influenced by the architecture of the detrital food web (Berg *et al.* 2001; Scheu & Setälä 2002; Wardle 2002). For example, the addition and/or removal of trophic groups and manipulation of decomposer community feeding guilds affects decomposition rate and, ultimately, plant growth. Soil food webs, as all other food webs, include species that can have substantial effects both on other species and on system-level processes. Ecosystem engineers (Shachak *et al.* 1987; Jones & Lawton 1995), such as earthworms, and taxa capable of building large biomasses, such as enchytraeid worms (Didden 1993; Laakso & Setälä 1999a), provide the best examples of such organisms. Furthermore, even within functional groups or feeding guilds, species-specific properties, such as the physiological attributes of dominant species, can drive numerous soil processes (Faber 1991; Mebes & Filser 1998; Mikola & Setälä 1998; Wall & Virginia 1999; Cragg & Bardgett 2001; Wolters 2001; Mikola *et al.* 2002).

Although some species, such as earthworms, have been shown to be of fundamental importance in affecting particular processes, for example nutrient mobilisation, there is limited evidence suggesting that species diversity per se has a significant influence on soil system-level processes. Studies refer to either an idiosyncratic relationship between species diversity and soil processes (Mikola & Setälä 1998; Cragg & Bardgett 2001; Mikola *et al.* 2002; Heemsbergen *et al.* unpublished data) or show a high functional equivalence among decomposer species (Liiri *et al.* 2002; Setälä & McLean 2004; Heemsbergen *et al.* unpublished data). This contrasts markedly with experimental results from plant communities (Naeem *et al.* 1994; Tilman *et al.* 1996; Hector *et al.* 1999; Tilman 1999), and has led various authors to conclude that functional redundancy may be a common phenomenon among locally rich decomposer species assemblages (Andrén *et al.* 1995; Groffman & Bohlen 1999; Laakso & Setälä 1999a; Wardle 1999; Bradford *et al.* 2002; Liiri *et al.* 2002; Scheu & Setälä 2002).

It is therefore appropriate to ask whether soil communities really do behave differently from their above-ground counterparts. And if they do, why and what are the potential ecological consequences of this functional redundancy? In this chapter we explore whether functional redundancy in soil food webs is a common phenomenon. We first define the concept of functional redundancy and review the evidence for such redundancy in the soil. We then explore the concept of functional redundancy in the context of existing ecological theories and hypotheses (namely, the functional niche hypothesis) before considering a-priori reasons for expecting redundancy in the soil.

Functional redundancy

The concept of functional redundancy

The original idea of redundancy is based on the notion that most ecosystems have more species than the number of processes. Functional redundancy is said to occur when the omission of species from the community (by species extinction, etc.) does not result in changes in ecosystem process rates. Consequently, when more than one species performs a similar function within any one process, it is assumed that there is likely to be functional redundancy among the species or taxa within that functional group (Walker 1992; Cowling *et al.* 1994; Chapin & Körner 1996). Most processes are easily associated with functions carried out by trophic rather than taxonomic groups (Walker 1992; Lawton 2000).

Is there evidence for functional redundancy in soils?

Among the crucial functions carried out by soil decomposer communities are those related to the ability of the system to split the C–N bond in dead-organic material. It is the feeding interactions (or resource–consumer interactions in microbes) within the community that, to a large extent, regulate decomposition and nutrient mobilisation (e.g. Abrahamson 1990; Bardgett *et al.* 1998; Huhta *et al.* 1998; Edsberg 2000). These, in turn, control primary production upon which all the higher trophic levels are dependent (Wardle 2002).

Despite the fact that secondary decomposers (i.e. soil fauna) comprise a small percentage of the total decomposer biomass, their activity can significantly stimulate the decomposition activity of the primary decomposers (i.e. fungi and bacteria) and thereby affect plant growth (e.g. Ingham *et al.* 1985; Setälä & Huhta 1991). At this level of the ecological hierarchy the roles of primary and secondary decomposers are, by definition, functionally dissimilar, although the division may not always be clear cut (Eggers & Jones 2000; Scheu & Falca 2000).

As microbes represent a large majority of decomposer biomass and diversity, one might expect functional dissimilarities to be common among the primary decomposer trophic group. Indeed, the major groups of soil microbes, bacteria and fungi, are not only evolutionarily and morphologically dissimilar but they also have functionally divergent properties. Interestingly, they give rise to two different energy–nutrient channels (Coleman *et al.* 1983; Moore & Hunt 1988). Bacteria (and the bacterial-based energy channel relying on easily degradable substrates) dominate in the plant rhizosphere, whereas the fungal channel, capable of degrading recalcitrant matter, dominates in the soil bulk (note that mycorrhizae are not taken into account; Ingham *et al.* 1989; Whitford 1989). There is, therefore, little basis on which to assume that saprophytic fungi and bacteria are functionally equivalent, although collectively they comprise the same trophic group. Furthermore, the unique ability of some groups of bacteria to nitrify, denitrify and fix atmospheric nitrogen further emphasises the non-redundant nature of some soil microbes (Griffiths *et al.* 2000).

Bearing in mind the immense heterogeneity of resources in soils (Coleman & Crossley 1996; Wardle 2002) it is not surprising that secondary decomposers may also be split further. This is confirmed by various experiments that show how soil fauna trophic group manipulation can manifest itself as altered rates of nutrient mobilisation and net primary production (NPP; Alphei et al. 1996; Huhta et al. 1998; Laakso & Setälä 1999a; Setälä 2000; Bonkowski et al. 2001; Bradford et al. 2002). Similarly, manipulation of feeding guilds within trophic groups results in altered nitrogen mineralisation rates (Laakso & Setälä 1999b) and plant growth (Laakso & Setälä 1999a). Thus, there are few arguments to support treating the soil fauna as a uniform group having functionally equivalent roles.

The high species richness and obvious importance of feeding interactions among soil organisms in affecting nutrient mobilisation and even net primary production makes it very tempting to ask whether all this diversity and these interactions are needed for the 'proper' functioning of ecosystems. Several laboratory microcosm studies suggest that species within both soil microbe and faunal functional groups are not identical but perform differently (Faber & Verhoef 1991; Robinson et al. 1993; Mikola & Setälä 1998; Hedlund & Sjögren-Öhrn 2000; Cragg & Bardgett 2001). This suggests an idiosyncratic response of various functions to species richness (Bolger 2001; Wolters 2001; Mikola et al. 2002). Other workers have questioned the direct role of species diversity (Andrén et al. 1995; Ettema 1998; Groffman & Bohlen 1999; Laakso & Setälä 1999a; Wardle 1999; Ekschmitt et al. 2001; Griffiths et al. 2001) in maintaining soil function.

Surprisingly, few studies exist in which the relationship between species diversity and system level processes has been explicitly studied. In a microcosm study, Liiri et al. (2002) manipulated the species richness (1 to 51 species) of soil microarthropods (Acari and Collembola). Although two species were not sufficient to ensure full functioning (measured as the growth of silver birch, Betula pendula) of the system, tree growth in the presence of five or more species did not differ from systems with higher numbers of species. Setälä and McLean (2004) also manipulated the species richness of soil saprophytic fungi and showed that decomposition activity (measured as CO_2 production) was only weakly related to the number of fungal species. Importantly, significant differences were recorded at the species poor (1 to 5 species) end of the gradient whereas decomposition rate in systems with 5 to 43 fungal taxa showed no differences. These two studies imply a high degree of functional similarity among soil microarthropods and saprophytic fungi. Similar conclusions were drawn by Laakso and Setälä (1999a) when they manipulated the number of soil mesofaunal species. Both the Liiri et al. (2002) and Setälä and McLean's (2004) studies also tested the spare-wheel hypothesis of Andrén et al. (1995) and Naeem and Li (1997). Here, the role of functionally less important species increases after a system experiences a perturbation. However, the ability of the systems to resist and recover from a disturbance was unrelated to species diversity.

There is ample evidence suggesting the fundamental importance of structural complexity and diversity of detrital food webs in affecting below-ground processes and, ultimately, net primary production. However, evidence suggesting that species diversity per se is a significant factor controlling soil processes and plant growth is lacking. The few studies that do explore this, emphasise functional redundancy rather than functional dissimilarity or complementarity among species. This contradicts both the Hutchinsonian and functional niche theories (see below), both of which predict complementarity, rather than redundancy, prevailing within a functional group (or indeed at the species level).

Functional niche theory

The concepts of functional redundancy and functional groups relate to core ecological issues, in particular the Hutchinsonian niche concept and the limiting similarity of species. The Hutchinsonian niche (Hutchinson 1957) separates species based on habitats, resources and environmental tolerances, but does not incorporate a functional perspective. Rosenfeld (2002) proposed a new concept, that of the functional niche. Whereas the Hutchinsonian niche defines where and under which circumstances a particular species can exist, the functional niche defines the ecological effects that a species or taxon has within that habitat. These 'ecological effects' are, to a large extent, dependent upon the demographic attributes and environmental tolerances of a species.

The selection of axes used to define the functional niche is subjective and based on what are considered to be the key functions and processes important in ecosystem functioning (Bolger 2001). For example, although functional complementarity (or equivalence and flexibility) may be common in various communities (e.g. Frost *et al.* 1995; Swift *et al.* 1998), this does not necessarily imply that functional redundancy – a concept that takes into account a wide spectrum of species-specific attributes – will always be found within any given system. Thus, functional redundancy with respect to a particular process does not necessarily rule out other unique species-specific contributions to other unmeasured patterns (Morin 1995). Similarly, species may perform different functions under different conditions (Solbrig 1991). Thus, when considering functional niche and functional redundancy the specific ecological function in question, for example processes related to the rate of nitrogen mobilisation, must be explicitly defined (Bolger 2001; Rosenfeld 2002).

Consequences of functional niche theory for redundancy and stability

The functional niche concept does help us understand how the impact of diversity and functional redundancy on ecosystem processes are fundamentally related to the stability of the environment. Under changing and unpredictable conditions, the different environmental tolerances between species, as predicted by the Hutchinsonian niche theory, are likely to produce differential survival of the taxa. Whether a system is functionally stable or not may depend on the number of species and their varying tolerances to environmental changes. The more species that encompass the variety of functional response groups the system harbours, the higher the chance of there being taxa that are resistant or resilient to these changes (i.e. the insurance hypothesis; Yachi & Loreau 1999). Under stable conditions a more limited number of decomposer species within each functional group may suffice to ensure the proper functioning of a system. This implies high functional redundancy as a common feature in soil food webs (Laakso & Setälä 1999a; Laakso et al. 2000; Liiri et al. 2002; Setälä & McLean 2004).

To summarise, niche theory predicts that, to be truly functionally redundant, taxa must overlap not only in their population level functional effects but also in environmental tolerances. This leaves, at least under the theoretical framework, very little space for two species to be functionally redundant (Beare et al. 1995; Bolger 2001; Rosenfeld 2002). The limited data available suggest, however, that the species number–ecosystem function curve saturates at very low species numbers (Liiri et al. 2002; Setälä & McLean 2004; Heemsbergen et al. unpublished data). This does not rest easily with the functional niche approach that highlights the unique functional influence of each species. Ekschmitt et al. (2001) suggested that the effects of diversity on ecosystem functioning will only be observed with sessile organisms at large spatial scales in species-poor communities, and for processes with no feedback and on a short time scale and for organisms that have a decoupled activity and persistence. How do soil organisms fit within such a framework?

Are there a-priori reasons to expect functional redundancy among soil organisms?

Soils share an array of features that separate them from the above-ground component of ecosystems (e.g. Coleman 1996; Filser & Setälä 1999; Scheu & Setälä 2002). Do the specific attributes of the soil environment determine the dynamics of its inhabitants, and does this bring about the obvious lack of a positive effect of species diversity on ecosystem functioning?

Soils differ as a habitat from other ecosystems

Soils are opaque, porous and markedly heterogeneous both in space (horizontal and vertical) and time (Ghilarov 1977; Usher & Parr 1977; Anderson 1978; Giller 1996). This, often small-scale, heterogeneity (especially vertically) is believed to play a major role in determining the highly aggregated distribution of most organisms (Giller 1996). Furthermore, vertical heterogeneity leads to small-scale gradients that affect species composition and increases diversity but reduces competition. For example, in stratified habitats such as forest soils, assemblages of soil organisms show a clear microstratification (Faber 1991; Ponge 1991; Berg et al. 1998a, b; Didden & de Fluiter 1998). As most processes operate on a very small spatial scale, gaps emerging from species loss will quickly be occupied by other species (Ekschmitt et al. 2001). This should, at least in theory, lead to functional similarity between taxa and to potential redundancy in soil food webs.

Food web organisation and process control

A further feature of detrital food webs is the existence of two separate energy channels, bacterial and fungal based. This provides the system with alternative pathways (*sensu* MacArthur 1955) through which energy and nutrients flow from the base to the top of the detrital food web (Moore & Hunt 1988). This may be particularly important for omnivores such as some enchytraeids that feed on both channels (functional redundancy of energy channels; Didden 1993) or when the two basal biotic components react differently to disturbances, such as drought and agricultural practices (Moore & de Ruiter 1997). The presence of the two channels may provide systems with functional redundancy even at this high level of the ecological hierarchy.

Key soil processes (such as decomposition and nutrient mineralisation) are controlled by numerous interactions between organisms within and between various trophic positions (including the separate nutrient channels). Thus, the risk of losing important links, which influence soil processes, is spread among a diverse decomposer biota, and this is bound to increase the degree of functional redundancy in the soil food web.

Finally, the importance of abiotic factors (such as nutrients liberated by a consumer) in affecting food-web dynamics *via* indirect effects is a particularly typical feature of soils. As a consequence, the impacts of a consumer can propagate through the food web without changing the biomass of adjacent trophic levels. This contrasts to typical above-ground feeding interactions where the consumer directly affects the biomass of its victim (Wardle 2002), and where the trophic interactions often lead to adaptive, specialist feeding strategies.

Omnivory

In three-dimensional habitats such as soils, organisms, being enzymatic amateurs, cannot be overly 'choosy' as regards their feeding preference (e.g. Scheu & Setälä 2002; Wardle 2002). Instead, they must consume what is locally available. Some oribatid mites (e.g. Siepel & de Ruiter-Dijkman 1993) and Collembola (e.g. Newell 1984a, b; Verhoef *et al.* 1988; Hedlund *et al.* 1995; Klironomos & Kendrick 1996; Maraun *et al.* 2003), however, do feed selectively on, for example, fungal mycelia. There are also examples of predatory fauna that are specialised to feed on Collembola (Bauer 1982) or nematodes (Koehler 1997), and large detritivores, such as earthworms, are known to prefer particular litter types (Satchell & Lowe 1967). Although our knowledge of the detailed dietary preferences of soil organisms is scanty, these findings raise the possibility that idiosyncrasy may prevail in soil food webs and that functional redundancy should not be common (see Mikola *et al.* 2002).

That soil animal species can feed selectively (at least under laboratory conditions) does not necessarily imply that, in the field, they are specialist feeders *sensu stricto*. There is little, if any, evidence showing that any particular species of soil fauna has specialised to feed upon a particular fungal or bacterial 'species', which is in strong contrast to the above-ground situation in, for example, plant–herbivore interactions (Maraun *et al.* 2003). In a field study using relative abundance of ^{15}N, Scheu and Falca (2000) reported that arthropods can feed across more than one trophic level (see also Eggers & Jones 2000; Berg & Stoffer in press). Indeed, polyphagy and omnivory in soil appear to be the rule rather than the exception. This implication blurs the concept and definition of trophic levels in below-ground habitats (Swift *et al.* 1979; Laakso & Setälä 1999a; Scheu & Setälä 2002; Maraun *et al.* 2003) and suggests that, at least in soil, competitive exclusion is probably rare and is likely to promote co-existence. Functional compensation, at least as regards to processes related to feeding activities, appears common in soil food webs, thus promoting the potential for functional redundancy.

Spatial ecology, diversity and redundancy

Having considered more traditional soil biological perspectives, we now attempt to place our ideas in the context of more recent ecological theory explaining community assembly processes and how they relate to ecosystem functioning. Attempting to build on macroecological processes, Lawton (2000) and Srivastava (2002) predicted that structural stability, taxonomic diversity and functional redundancy are all interrelated. In practice, Srivastava (2002) suggested that the degree of functional redundancy within a community depends on whether those communities are 'saturated' or 'unsaturated'. If a community is saturated, lost species are rapidly replaced with new species because the species pool contains numerous species able to fill any particular niche. Density compensation takes place leading to functional compensation and thus functional redundancy. In unsaturated communities, however, there are few, if any, competitors for a given niche and lost species are not easily replaced. In other words, the defining question is how important are regional processes in affecting local processes? Weakly interactive communities are likely to show strong effects of local diversity on ecosystem functioning whereas in strongly interactive, saturated systems the relationship between species diversity and system functioning should be reduced.

The majority of soil systems appear to be saturated – the poor man's rainforest (Usher *et al.* 1979) – with luxurious diversity, and particularly high species packing (Giller 1996). This tight species packing within a multi-dimensional soil habitat is thought to arise from the immensely heterogeneous distribution of habitat types and food resources within the soil (e.g. Begon *et al.* 1990; Giller 1996; Scheu & Setälä 2002). Further evidence for tight species packing is the occurrence of a high α-diversity, with rapidly saturating species–area curve, implying that species are packed within a relatively small area. We believe soils represent typical saturated systems that are highly interactive (large species : niche ratio) and as such should promote 'functional stability' and a high degree of functional redundancy.

Synopsis

It is the relatively small spatial scale in environmental gradients and the lack of feeding specialists that are perhaps the most typical feature of soils and soil organisms. This is likely to stem from four factors. First, the food web is largely donor controlled and therefore direct co-evolution between 'dead' detritus and its consumers (microbes and detritivorous fauna) cannot be intense. Second, the heterogeneous nature of soils reduces resource competition between taxa, thus allowing co-existence of species with similar dietary requirements. Third, the porous structure of soils limits the movement of soil biota, and this in turn obliges them to feed upon what they encounter. It may also be that the high diversity of soil organisms within a habitat reflects their adaptation to microhabitats rather than their having different functions (Giller *et al.* 1997). In that sense, loss of diversity could result in expansion of the niches occupied by the remaining organisms, thus compensating for the functional activities of the lost species. Finally, in the light of current theory, soil communities appear saturated with high potential for functional redundancy.

Conclusions

There is both experimental evidence and a-priori reasons (stemming from observations and existing ecological theory) for suggesting that functional redundancy in soils is low – in many cases it may even be non-existent – at the coarser levels (such as trophic group) of the ecological hierarchy. Recent studies imply that this is less obvious at finer resolution (i.e. at the species level). Although many 'case studies' refer to idiosyncratic species number–ecosystem functioning responses in soils, emphasising the importance of species-specific attributes in affecting soil processes, the experiments have generally used artificially low diversity gradients with often less than three species.

We suggest that the degree of idiosyncrasy dilutes significantly as the systems become more natural, i.e. more species are present in a community. The main factor driving this 'dilution effect' is the high degree of polyphagy and omnivory among soil biota. The high degree of functional similarity within detrital food webs would, at least in part, lead to a relatively high degree of functional redundancy and this appears to contrast markedly with findings from other ecosystems. Also, as predicted by existing theory of spatial arrangement and community assembly processes of organisms (Lawton 2000; Srivastava

2002), soil communities, having tightly packed, saturated species assemblages in a heterogeneous environment, represent a typical case that promotes functional redundancy at the species level.

References

Abrahamson, G. (1990). Influence of *Cognettia sphagnetorum* (Oligochaeta, Enchytraeidae) on nitrogen mineralization in homogenized mor humus. *Biology and Fertility of Soils*, **9**, 159–162.

Alphei, J., Bonkowski, M. & Scheu, S. (1996). Protozoa, Nematoda and Lumbricidae in the rhizosphere of *Hordelymus europaeus* (Poaceae): faunal interactions, response of microorganisms and effects on plant growth. *Oecologia*, **106**, 111–126.

Anderson, J. M. (1978). Inter- and intra-habitat relationships between woodland Cryptostigmata species diversity and the diversity of soil and litter microhabitats. *Oecologia*, **32**, 341–348.

Andrén, O., Bengtsson, J. & Clarholm, M. (1995). Biodiversity and species redundancy among litter decomposers. *The Significance and Regulation of Soil Biodiversity* (Ed. by H. P. Collins), pp. 141–151. Dordrecht: Kluwer Academic.

Bardgett, R. D., Wardle, D. A. & Yeates, G. W. (1998). Linking above-ground and below-ground interactions: how plant responses to foliar herbivory influence soil organisms. *Soil Biology and Biochemistry*, **30**, 1867–1878.

Bauer, T. (1982). Predation by a carabid beetle specialized for catching Collembola. *Pedobiologia*, **24**, 169–179.

Beare, M. H., Coleman, D.C., Crossley, D. A. Jr., Hendrix, P. F. & Odum, E. P. (1995). A hierarchical approach to evaluating the significance of soil biodiversity to biogeochemical cycling. *Plant and Soil*, **170**, 5–22.

Begon, M., Harper, J. L. & Townsend, C. R. (1990). *Ecology: Individuals, Populations, and Communities*, 2nd edition. London: Blackwell Scientific.

Berg, M. P., Kniese, J. P. & Verhoef, H. A. (1998a). Dynamics and stratification of bacteria and fungi in the organic layers of a Scots pine forest soil. *Biology and Fertility of Soils*, **26**, 313–322.

Berg, M. P., Kniese, J. P., Bedaux, J. J. M. & Verhoef, H. A. (1998b). Dynamics and stratification of functional groups of micro- and mesoarthropods in the organic layer of a Scots pine forest. *Biology and Fertility of Soils*, **26**, 268–284.

Berg, M. P., de Ruiter, P. C., Didden, W. A. M., et al. (2001). Community food web, decomposition and nitrogen mineralisation in a stratified Scots pine forest soil. *Oikos*, **94**, 130–142.

Berg, M. P. & Stoffer, M. (in press). Feeding guilds in Collembola based on digestive enzymes. *Pedobiologia*, submitted.

Bolger, T. (2001). The functional value of species biodiversity: a review. *Biology and Environment: Proceedings of the Royal Irish Academy*, **101B**, 199–224.

Bonkowski, M., Geoghegan, I. E., Birch, A. N. E. & Griffiths, B. S. (2001). Effects of soil decomposer invertebrates (protozoa and earthworms) on an above-ground phytophagous insect (cereal aphid) mediated through changes in the host plant. *Oikos*, **95**, 441–450.

Bradford, M. A., Jones, T. H., Bardgett, R. D., et al. (2002). Impacts of soil faunal community composition on model grassland ecosystems. *Science*, **298**, 615–618.

Chapin, F. S., III & Körner, C. (1996). Arctic and alpine biodiversity: its patterns, causes and ecosystem consequences. *Functional Roles of*

Biodiversity: A Global Perspective (Ed. by H. A. Mooney, J. H. Cushman, E. Medina, O. E. Sala & E.-D. Schulze), pp. 7–32. Chichester: SCOPE/Wiley

Coleman, D. C. (1996). Energetics of detritivory and microbivory in soil in theory and practice. *Food Webs: Integration of Patterns and Dynamics* (Ed. by G. A. Polis & K. O. Winemiller), pp. 39–50. New York: Chapman and Hall.

Coleman, D. C. & Crossley, D. A., Jr. (1996). *Fundamentals of Soil Ecology*. San Diego: Academic Press.

Coleman, D. C., Reid, C. P. P. & Cole, C. V. (1983). Biological strategies of nutrient cycling in soil systems. *Advances in Ecological Research*, 13, 1–55.

Cowling, R. M., Mustart, P. J., Laurie, H. & Richards, M. B. (1994). Species diversity: functional diversity and functional redundancy in fynbos communities. *South African Journal of Science*, 90, 333–337.

Cragg, R. G. & Bardgett, R. D. (2001). How changes in soil faunal diversity and composition within a trophic group influence decomposition processes. *Soil Biology and Biochemistry*, 33, 2073–2081.

Didden, W. A. M. (1993). Ecology of terrestrial Enchytraeidae. *Pedobiologia*, 37, 2–29.

Didden, W. A. M. & de Fluiter, R. (1998). Dynamics and stratification of Enchytraeidae in the organic layer of a Scots pine forest. *Biology and Fertility of Soils*, 26, 305–312.

Edsberg, E. (2000). The quantitative influence of Enchytraeids (Oligochaeta) and microarthropods on decomposition of coniferous raw humus in microcosms. *Pediobiologia*, 44, 132–147.

Eggers, T. & Jones, T. H. (2000). You are what you eat . . . or are you? *Trends in Ecology and Evolution*, 15, 265–266.

Ekschmitt, K., Klein, A., Pieper, B. & Wolters, V. (2001). Biodiversity and functioning of ecological communities: why is diversity important in some cases and unimportant in others? *Plant Nutrition and Soil Science*, 164, 239–246.

Ettema, C. H. (1998). Soil nematode diversity: species coexistence and ecosystem function. *Journal of Nematology*, 30, 159–170.

Faber, J. H. (1991). Functional classification of soil fauna: a new approach. *Oikos*, 62, 110–117.

Faber, J. H. & Verhoef, H. A. (1991). Functional differences between closely-related soil arthropods with respect to decomposition processes in the presence or absence of pine tree roots. *Soil Biology and Biochemistry*, 23, 15–23.

Filser, J. & Setälä, H. (1999). Recent advances in decomposer food web ecology. *Perspectives in Ecology. A Glance from the VII INTECOL Congress of Ecology*, Florence, Italy (Ed. by A. Farina), pp. 355–368. Leiden: Backhuys.

Frost, T. M., Carpenter, S. R., Ives, A. R. & Kratz, T. K. (1995). Species compensation and complementarity in ecosystem function. *Linking Species and Ecosystems* (Ed. by C. G. Jones & J. H. Lawton), pp. 224–239. San Diego, CA: Chapman and Hall.

Ghilarov, M. S. (1977). Why so many individuals can exist in the soil? *Ecological Bulletins*, 25, 593–597.

Giller, P. S. (1996). The diversity of soil communities, the poor man's tropical rainforest. *Biodiversity and Conservation*, 5, 135–168.

Giller, K. E., Beare, M. H., Lavelle, P., Izac, A.-M. N. & Swift, M. J. (1997). Agricultural intensification, soil biodiversity and agroecosystem function. *Applied Soil Ecology*, 6, 3–16.

Griffiths, B. S., Ritz, K., Bardgett, R. D., *et al.* (2000). Ecosystem response of pasture soil communities to fumigation-induced microbial diversity reductions: an examination of the biodiversity–ecosystem function relationship. *Oikos*, 90, 279–294.

Griffiths, B. S., Ritz, K., Wheatley, R., *et al.* (2001). An examination of the biodiversity–ecosystem function relationship in arable soil microbial communities. *Soil Biology and Biochemistry*, **33**, 1713–1722.

Groffman, P. M. & Bohlen, P. J. (1999). Soil and sediment biodiversity. *BioScience*, **49**, 139–148.

Hector, A., Schmid, B., Beierkuhnlein, C., et al. (1999). Plant diversity and productivity experiments in European grasslands. *Science*, **286**, 1123–1127.

Hedlund, K. & Sjögren-Öhrn M. (2000). Tritrophic interactions in a soil community enhance decomposition rates. *Oikos*, **88**, 585–591.

Hedlund, K., Bengtsson, G. & Rundgren, S. (1995). Fungal odour discrimination in two sympatric species of fungivorous collembolans. *Functional Ecology*, **9**, 869–875.

Huhta, V., Persson, T. & Setälä, H. (1998). Functional implications of soil fauna diversity in boreal forests. *Applied Soil Ecology*, **10**, 277–288.

Hutchinson, G. E. (1957). Concluding remarks. *Cold Spring Harbour Symposium on Quantitative Biology*, **22**, 415–427.

Ingham, E. R., Coleman, D. C. & Moore, J. C. (1989). An analysis of food web structure and function in a shortgrass prairie, a mountain meadow, and a lodgepole pine forest. *Biology and Fertility of Soils*, **8**, 29–37.

Ingham, R. E., Trofymow, J. A., Ingham, E. R. & Coleman, D. C. (1985). Interactions of bacteria, fungi, and their nematode grazers: effects on nutrient cycling and plant growth. *Ecological Monographs*, **55**, 119–140.

Jones, C. G. & Lawton J. H. (1995). *Linking Species and Ecosystems*. New York: Chapman and Hall.

Klironomos, J. N. & Kendrick, W. B. (1996). Palatability of microfungi to soil arthropods in relation to the functioning of arbuscular mycorrhizae. *Biology and Fertility of Soils*, **21**, 43–52.

Koehler, H. H. (1997). Mesostigmata (Gamasina, Uropodina), efficient predators in agroecosystems. *Agriculture, Ecosystems and Environment*, **62**, 105–117.

Laakso, J. & Setälä, H. (1999a). Sensitivity of primary production to changes in the architecture of belowground food webs. *Oikos*, **87**, 57–64.

Laakso, J. & Setälä, H. (1999b). Population- and ecosystem-level effects of predation on microbial-feeding nematodes. *Oecologia*, **120**, 279–286.

Laakso, J., Setälä, H. & Palojärvi, A. (2000). Control of primary production by decomposer food web structure in relation to nitrogen availability. *Plant and Soil*, **225**, 153–165.

Lawton, J. H. (2000). *Community Ecology in a Changing World: Excellence in Ecology*. Oldendorf: Ecology Institute.

Lawton, J. H. & May, R. M. (1995). *Extinction Rates*. Oxford: Oxford University Press.

Liiri, M., Setälä, H., Haimi, J., Pennanen, T. & Fritze, H. (2002). Relationship between soil microarthropod species diversity and plant growth does not change when the system is disturbed. *Oikos*, **96**, 137–149.

MacArthur, R. H. (1955). Fluctuations of animal populations and a measure of community stability. *Ecology*, **36**, 533–536.

Maraun, M., Martens, H., Migge, S., Theenhaus, A. & Scheu, S. (2003). Adding to 'the enigma of soil animal diversity': fungal feeders and saprophagous soil invertebrates prefer similar food substrates. *European Journal of Soil Science*, **39**, 85–95.

Mebes, K.-H. & Filser, J. (1998). A method for estimating the significance of surface dispersal for population fluctuations of Collembola in arable land. *Pedobiologia*, **41**, 115–122.

Mikola, J. & Setälä, H. (1998). Relating species diversity to ecosystem functioning: mechanistic backgrounds and experimental approach with a decomposer food web. *Oikos*, **83**, 180–194.

Mikola, J., Bardgett, R. D. & Hedlund, K. (2002). Biodiversity, ecosystem functioning and soil decomposer food webs. *Biodiversity and Ecosystem Functioning: A Current Synthesis* (Ed. by M. Loreau, S. Naeem & P. Inchausti), pp. 169–180. Oxford: Oxford University Press.

Moore, J. C. & de Ruiter, P. C. (1997). Compartmentalization of resource utilization within soil ecosystems. *Multitrophic Interactions in Terrestrial Systems* (Ed. by A. C. Gange & V. K. Brown), pp. 375–393. London: Blackwell Science.

Moore, J. C. & Hunt, H. W. (1988). Resource compartmentation and the stability of real ecosystems. *Nature*, **333**, 261–263.

Morin, P. J. (1995). Functional redundancy, non-additive interactions, and supply-side dynamics in experimental pond communities. *Ecology*, **76**, 133–149.

Naeem, S. & Li, S. (1997). Biodiversity enhances ecosystem reliability. *Nature*, **390**, 507–509.

Naeem, S., Thompson, L. J., Lawler, S. P., Lawton, J. H. & Woodfin, R. M. (1994). Declining biodiversity can alter the performance of ecosystems. *Nature*, **368**, 734–737.

Newell, K. (1984a). Interaction between two decomposer basidiomycetes and a collembolan under Sitka spruce: grazing and its potential effects on fungal distribution and litter decomposition. *Soil Biology and Biochemistry*, **16**, 235–239.

Newell, K. (1984b). Interaction between two decomposer basidiomycetes and a collembolan under Sitka spruce: distribution, abundance and selective grazing. *Soil Biology and Biochemistry*, **16**, 227–233.

Ponge, J. F. (1991). Succession of fungi and fauna during decomposition of needles in a small area of Scots pine litter. *Plant and Soil*, **138**, 99–113.

Robinson, C. H., Dighton, J., Frankland, J. C. & Coward, P. A. (1993). Nutrient and carbon dioxide release by interacting species of straw-decomposing fungi. *Plant and Soil*, **151**, 139–142.

Rosenfeld, J. (2002). Functional redundancy in ecology and conservation. *Oikos*, **98**, 156–162.

Satchell, J. E. & Lowe, D. G. (1967). Selection of leaf litter by *L. terrestris*. *Progress in Soil Biology* (Ed. by O. Graff & J. E. Satchell), pp. 102–128. Amsterdam: North-Holland.

Scheu, S. & Falca, M. (2000). The soil food web of two beech forests (*Fagus sylvatica*) of contrasting humus type: stable isotope analysis of a macro- and a mesofauna-dominated community. *Oecologia*, **123**, 285–296.

Scheu, S. & Setälä, H. (2002). Multitrophic interactions in decomposer food webs. *Multitrophic Interactions in Terrestrial Systems* (Ed. by T. Tscharntke & B. A. Hawkins), pp. 223–264. Cambridge: Cambridge University Press.

Schmid, B., Joshi, J. & Schläpfer, F. (2002). Empirical evidence for biodiversity–ecosystem functioning relationships. *Functional Consequences of Biodiversity: Experimental Progress and Theoretical Extensions* (Ed. by A. Kinzig, S. Pacala & D. Tilman), pp. 120–150. Princeton, NJ: Princeton University Press.

Setälä, H. (2000). Reciprocal interactions between Scots pine and soil food web structure in the presence and absence of ectomycorrhiza. *Oecologia*, **125**, 109–118.

Setälä, H. & Huhta, V. (1991). Soil fauna increase *Betula pendula* growth: laboratory experiments with coniferous forest floor. *Ecology*, **72**, 665–671.

Setälä, H. & McLean, M. A. (2004). Decomposition rate of organic substrates in relation to the species diversity of soil saprophytic fungi. *Oecologia*, **139**, 98–107.

Shachak, M., Jones, C. G. & Granot, Y. (1987). Herbivory in rocks and the weathering of a desert. *Science*, **236**, 1098–1099.

Siepel, H. & de Ruiter-Dijkman, E. M. (1993). Feeding guilds of oribatid mites based on their carbohydrase activities. *Soil Biology Biochemistry*, **25**, 1491–1497.

Solbrig, O. T. (1991). *From Genes to Ecosystems: A Research Agenda for Biodiversity.* Cambridge, MA: IUBS/SCOPE/UNESCO.

Srivastava, D. S. (2002). The role of conservation in expanding biodiversity research. *Oikos*, **98**, 351–360.

Swift, M. J., Heal, O. W. & Anderson J. M. (1979). *Decomposition in Terrestrial Ecosystems.* Oxford: Blackwell.

Swift, M. J., Andrén, O., Brussaard, L., *et al.* (1998). Global change, soil biodiversity, and nitrogen cycling in terrestrial ecosystems: three case studies. *Global Change Biology*, **4**, 729–743.

Tilman, D. (1999). The ecological consequences of changes in biodiversity: a search for general principles. *Ecology*, **80**, 1455–1474.

Tilman, D., Wedin, D. & Knops, J. (1996). Productivity and sustainability influenced by biodiversity in grassland ecosystems. *Nature*, **379**, 718–720.

Usher, M. B. & Parr, T. (1977). Are there successional changes in arthropod decomposer communities? *Journal of Environmental Management*, **5**, 151–160.

Usher, M. B., Davis, P., Harris, J. & Longstaff, B. (1979). A profusion of species? Approaches towards understanding the dynamics of the populations of microarthropods in decomposer communities. *Population Dynamics* (Ed. by R. M. Anderson, B. D. Turner & L. R.

Taylor), pp. 359–384. Oxford: Blackwell Scientific.

Verhoef, H. A., Prast, J. E. & Verweij, R. A. (1988). Relative importance of fungi and algae in the diet of *Orchesella cincta* (L.) and *Tomocerus minor* (Lubbock). *Functional Ecology*, **2**, 195–201.

Walker, B. H. (1992). Biodiversity and ecological redundancy. *Conservation Biology*, **6**, 18–23.

Wall, D. H. & Virginia, R. A. (1999). Controls on soil biodiversity: insights from extreme environments. *Applied Soil Ecology*, **13**, 137–150.

Wardle, D. A. (1999). Is 'sampling effect' a problem for experiments investigating biodiversity–ecosystem function relationships. *Oikos*, **87**, 403–407.

Wardle, D. A. (2002). *Communities and Ecosystems: Linking the Aboveground and Belowground Components.* Princeton, NJ: Princeton University Press.

Whitford, W. G. (1989). Abiotic controls on the functional structure of soil food webs. *Biology and Fertility of Soils*, **8**, 1–6.

Wolters, V. (2001). Biodiversity of soil animals and its function. *European Journal of Soil Biology*, **37**, 221–227.

Yachi, S. & Loreau, M. (1999). Biodiversity and ecosystem productivity in a fluctuating environment: the insurance hypothesis. *Proceedings of the National Academy of Sciences, USA*, **96**, 1463–1468.

CHAPTER FOURTEEN

Plant–soil feedback and soil biodiversity affect the composition of plant communities

WIM H. VAN DER PUTTEN
Netherlands Institute of Ecology

SUMMARY

1. Plants affect the composition of their soil community and in return the soil community affects the productivity and composition of the plant community. I discuss patterns of plant–soil feedback and explore if the species and functional group diversity of the soil community matters for the type and magnitude of the soil feedback to the plant community.
2. There are two major pathways of soil feedback. One is direct, via root herbivores, pathogens and symbionts, and the second pathway is more indirect, through the effect of the soil decomposer subsystem on the supply of nutrients.
3. There are very few examples of experiments that have tested how diversity at the species, trophic or functional group level affects the composition of plant communities and the first results available do not enable generalisation.
4. I argue that the contribution of the different subsystems in the soil to plant–soil feedback depends on ecosystem characteristics, such as productivity, which is an important determinant of plant community composition. Then, I show why the different components of the soil subsystem can have variable, and even opposite, effects on plant community composition depending on the productivity level considered.
5. Finally, I predict how biodiversity in the different subsets of the soil community affects plant community composition. Effects of soil biodiversity depend on the context, spatial and temporal scales, as well as on the consequences of the functional group considered.
6. I conclude that future studies should pay particular attention to context and spatio-temporal scales, and that they should acknowledge that the possible role of species in communities changes over space and time. These

Biological Diversity and Function in Soils, eds. Richard D. Bardgett, Michael B. Usher and David W. Hopkins. Published by Cambridge University Press. © British Ecological Society 2005.

Table 14.1. *Abiotic and biotic drivers of plant community composition and ecosystem compartment where the major activity takes place (indicated with +).*

Driver of change in plant communities	Above-ground	Below-ground
Climate and (changes in) atmospheric composition	+	
Soil type		+
Species pool	+	+
Resource competition	+	+
Herbivores and pathogens	+	+
Symbiotic mutualists	+	+
Bioturbators		+
Decomposers		+

prerequisites are essential in order to be able to develop a general theory on consequences of soil biodiversity for plant community composition.

Introduction

Elucidating the factors that drive variation and change in the composition of natural plant communities, in space and time, is one of the longest-standing issues in ecology. The composition of plant communities depends on a large number of biotic and abiotic factors (Table 14.1). Since shifts in the abiotic conditions correlate reasonably well with shifts in vegetation patterns, abiotic factors have been interpreted as the drivers of the patterns observed (van Andel *et al.* 1993). Ecological experiments, such as exclosure studies that eliminate herbivores (e.g. McNaughton *et al.* 1988), however, show that biotic interactions between plants and herbivores, or pathogens, have important consequences for the composition of plant communities, because of the selective nature of these interactions (Burdon 1987; Crawley 1997). The primary focus on interactions between plants and animals, or plants and microorganisms, traditionally has been directed to the above-ground subsystem. More recently, it has been acknowledged that biotic interactions in the soil are major drivers of the composition of plant communities (Grime *et al.* 1987; van der Putten *et al.* 1993; Bever 1994; Packer & Clay 2000; Klironomos 2002; de Deyn *et al.* 2003), and that interactions within the soil also affect above-ground biota (Hooper *et al.* 2000; van der Putten *et al.* 2001; Wardle 2002; Bardgett & Wardle 2003; Wardle *et al.* this volume). Here, it is shown how soil organisms affect the composition of plant communities, and the relationship between soil biodiversity and the composition of natural plant communities is discussed.

There are two major types of interactions between plants and soil biota. First, there are interactions between dead plant materials and the decomposer subsystem; second, there are biotrophic interactions between living plant roots, herbivores and pathogens, and symbiotic mutualists. The decomposer food web is often thought to be the major energy channel in the soil subsystem (e.g. Moore *et al.* 1991) and receives its resources from dead plant material (detritus). The majority of the interactions between plant roots and the decomposer organisms are indirect. The decomposer food web converts dead plant tissues into mineral nutrients (Anderson 2000) and the availability of nutrients determines the outcome of competition between and within plant species (Tilman 1982).

It is less well known how much energy flows into the soil system through root herbivores and soil pathogens. Estimates have been based on the use of soil biocides and show that root-feeding nematodes may reduce 12–28% of the net primary production in prairie grasslands in the USA (Stanton 1988). However, there have been few, if any, attempts made to check the conclusions of these biocide experiments by studies including the addition of root-feeding nematodes. Exclusion and addition studies have been carried out in coastal sand dunes, where effects of nematicides turned out not to be indicative of the effects obtained when root-feeding nematodes were added to sterilised soil (van der Putten *et al.* 1990; de Rooij-van der Goes 1995; van der Stoel *et al.* 2002). Other researchers have calculated the flow of resources into root-feeding nematode communities and concluded that local hot spots of nematodes indeed may account for considerable losses from living plant roots (Verschoor 2002).

Root herbivores and pathogens also have direct effects on plant community composition through selective growth reduction, or killing of specific plant species, and indirect effects by changing the outcome of interspecific plant competition (van der Putten 2003). Symbiotic mutualists (mycorrhizal fungi, N_2-fixing microorganisms) influence the availability and uptake of resources (Northup *et al.* 1995; Smith & Read 1997; Giller this volume; Leake *et al.* this volume) and they also have the potential to change the outcome of interspecific plant competition, mostly in favour of their host (Turkington & Klein 1991).

The decomposer food web and the herbivores, pathogens and symbionts of plants are not independent of each other. The effectiveness of symbioses depends on nutrient availability in the soil (Smith & Read 1997) and root herbivory has a stronger impact on plant competition when nutrients are limited (van der Putten & Peters 1997). Many of the higher trophic level organisms in soil food webs are omnivorous and a number of the omnivores feed on microorganisms, small invertebrates, as well as plant roots (de Ruiter *et al.* 1995). Plants, therefore, are exposed to a range of soil biota and the ecological consequences of these interactions is a net effect on plant community development. Plants also influence the composition of the soil community, and these reciprocal interactions are called 'plant–soil feedback' (Bever *et al.* 1997). Feedback effects can be negative, neutral or positive, but the recent examples in literature point at a predominance of negative effects (Bever 2003), indicating a profound role of root herbivores and pathogens in structuring natural plant communities.

Current awareness of the rapid decline of biodiversity in natural ecosystems has led to a recent increase in studies examining the consequences of biodiversity loss for ecosystem functioning (Loreau *et al.* 2002). The experimental designs applied have been debated (Huston 1997; Wardle 1999; Huston *et al.* 2000; Loreau *et al.* 2001) and the ignorance of the nature of species assemblages has been criticised (Grime 1998). Another important shortcoming of the previous biodiversity studies is that the majority of attention has been on plant communities from temperate grasslands, whereas other vegetation types and higher trophic interactions have received relatively little attention (Rafaelli *et al.* 2002).

Most examples of effects of soil biodiversity on ecosystem functioning concern the effects of diversity within and between trophic levels in the decomposer food web on soil nutrient and carbon cycling (Laakso & Setälä 1999a, b; Mikola *et al.* 2002; Setälä *et al.* this volume). Studies on relationships, and feedbacks, between plant and soil biodiversity are rare (Wardle & van der Putten 2002), but some recent studies suggest that the nature of plant species affects soil communities more than plant diversity per se (Wardle *et al.* 2003; de Deyn *et al.* 2004). A few studies have tested the effects of diversity of mycorrhizal fungi on plants and plant communities (van der Heijden *et al.* 1998a, b; Jonsson *et al.* 2001; Leake *et al.* this volume), but studies on consequences of species diversity of nitrogen-fixing microorganisms, root feeding insects, nematodes or pathogens, are rare (but see Brinkman *et al.* 2005).

In this chapter, existing data are used to predict the possible outcomes of variations in biodiversity within functional groups of soil organisms (decomposers, herbivores, pathogens and mycorrhizal fungi) for the composition of natural plant communities. First, the principle of feedback between plant and soil communities is explained. Then spatio-temporal scales and context dependence of plant–soil feedback will be discussed. The chapter concludes with predictions about how soil biodiversity may affect the feedback between soil and plant communities. Finally, in order to stimulate the development of experimental studies on soil biodiversity in relation to plant community composition, challenges and priorities for future research are discussed.

Feedback between plant and soil communities

Feedback between plants and their soil communities is due to effects of root herbivores, root pathogens, symbiotic mutualists (Bever *et al.* 1997) and of the soil decomposer community (van Breemen 1998). Plant–soil feedback can also be due to changes in the nutrient supply status of the soil (Berendse 1998; Binkley & Giardina 1998) or to the exudation of specific allelopathic plant chemicals (Bais *et al.* 2003), but in this chapter the focus is mainly on feedbacks between plants and the soil biota. Plant roots are exposed to thousands of species of soil microorganisms around the roots (Torsvik *et al.* 1990), and tens of species of Enchytraeids, nematodes, microarthropods and earthworms (Hedlund *et al.* 2003; Porazinska *et al.* 2003). Fatty acid profiles (Bardgett *et al.* 1999b) and molecular signatures of the microbial community (Kowalchuk *et al.* 2002) of the rhizosphere show that plant roots amplify only specific subsets of the total soil microbial community, whereas the other soil microbes remain in the bulk soil at a relatively lower level of abundance.

Individual wild plant species have none, or some, specific species of root-feeding nematodes, and most share a number of generalist species with other plant species in the community (Yeates *et al.* 1993; Bongers & Bongers 1998; van der Putten & van der Stoel 1998; Verschoor *et al.* 2001; Porazinska *et al.* 2003). Dominant microbial species on the root surface, however, do vary with plant species (Kowalchuk *et al.* 2002) and change with root age (Duineveld & van Veen 1999), whereas soil organisms at higher trophic levels in the food web show less response to the identity of host plants than soil microbes, microbial feeders and root-feeding soil fauna (Porazinska *et al.* 2003; de Deyn *et al.* 2004). When the host plant disappears, however, the host-specific root feeders have limited capacity to survive, so they often disappear from the soil community after the death of their host plant. This may be less of a problem for soil-dwelling herbivorous insects than for nematodes, especially when they have life-history stages that allow above-ground dispersal (Mortimer *et al.* 1999). However, in spite of better dispersal capacity, soil-dwelling insects are also known to have a patchy and clustered occurrence (Brown & Gange 1990). Some studies have observed that plant–soil feedback phenomena were due to single pathogen species or genera of soil fungi (e.g. Bever 1994; Holah *et al.* 1997; Packer & Clay 2000; Westover & Bever 2001), nematodes (Olff *et al.* 2000) and insects (Strong 1999). In other systems the observed soil feedbacks appeared to be due to multi-species combinations of soil pathogens and root-feeding nematodes (de Rooij-van der Goes 1995; van der Putten & van der Stoel 1998; van der Stoel *et al.* 2002). Plant–soil feedbacks, therefore, can depend on key species (Mills & Bever 1998; Packer & Clay 2000), but also on complex communities of pathogens and root feeders, in which a number of species are involved in doing part of the job (de Rooij-van der Goes 1995).

Feedback from soil organisms to plants has been tested through various experimental approaches, ranging from complete soil sterilisation (e.g. Oremus & Otten 1981; van der Putten *et al.* 1988; Olff *et al.* 2000), partial exclusion using soil biocides (e.g. van der Putten *et al.* 1990; Newsham *et al.* 1994), comparative growth of plants in soil from the same and other species (van der Putten *et al.* 1993; Bever 1994; van der Putten & Peters 1997) to repetitive growth of the same plant species in soil samples (Klironomos 2002). Unplanned side effects, for example soil sterilisation resulting into nutrient flushes (Troelstra *et al.* 2001), may explain part of the differences in growth responses of plants between sterilised and non-sterilised soils. Such effects can also be checked for through the inclusion of nutrient addition treatments (van der Putten & Peters 1997) or by sequential plant–soil feedback tests throughout the development of the feedback effects over time (van der Stoel *et al.* 2002), but experimental setups need to be critically discussed in order to avoid the numerous pitfalls. Other problems with experimentation include the disruption of mycorrhizal networks and the relatively slow infection response of these fungi compared to pathogens or herbivores, which could result in overlooking positive soil feedback effects (Little & Maun 1996). However, despite experimental constraints, the plant–soil feedback studies all show that plant roots alter the abundance, and sometimes species composition, of the soil community resulting in a net feedback effect on plant growth and, potentially, plant community composition.

The classic approach to determine which of the soil organisms may have contributed to the results observed in experiments is to apply Koch's postulates, which dictate that the effects should disappear when the responsible organisms have been removed from the soil and that the effects re-appear when the responsible organisms have been added. Koch's postulates are well applicable in the case of major pathogens of a single species, but it has been questioned if the composition of multi-species communities of plant pathogens, where synergisms or antagonisms influence the effectiveness of individual species, may be unravelled similarly (Sikora & Carter 1987). For example, when examining the contribution of root-feeding nematodes and soil pathogens to the degeneration of the natural dune grass *Ammophila arenaria*, different subsets of the soil fungal community were all found to reduce plant growth, whereas individual species did not affect plant growth significantly (de Rooij-van der Goes 1995).

There are a number of ways to deal with Koch's postulates when exploring the possible contribution of different subsets of organisms from more complex soil communities. One is to assemble subcommunities of soil organisms based on their co-occurrence in the field (de Rooij-van der Goes *et al.* 1995). When applied to the dune grass *A. arenaria*, various subcommunities of soil fungi and root-feeding nematodes turned out to be functionally equivalent, but their predominance in geographically separated fore dune areas along the coastline (de Rooij-van der Goes *et al.* 1995) suggests that none of these subcommunities was functionally redundant. Other examples of handling complex soil communities in experiments involve assembling groups of soil organisms based on body size (Bradford *et al.* 2002), faunal origin (de Deyn *et al* 2003) or sifting arbuscular mycorrhizal fungi (AMF) from the soil microbial component by wet-sieving techniques (Klironomos 2002). In all cases, there should be awareness of experimental constraints that may obscure interpretation of the results.

The selective removal of soil organisms, such as root pathogens, herbivores or symbionts, requires specific agents or specific sieving techniques. In spite of claims that biocides are highly specific, side effects are unavoidable. The fungicide benomyl has been used to selectively eliminate AMF (Newsham *et al.* 1994; O'Connor *et al.* 2002; Callaway *et al.* 2003), whereas this broad-spectrum fungicide may also eliminate other soil fungi, including pathogenic species, and endoparasitic root-feeding nematodes (van der Putten *et al.* 1990). Benomyl might also affect fungal components of the soil decomposer subsystem, thereby altering decomposition and nutrient supply to the plants. Results from such studies, therefore, need to be interpreted as 'biocide addition effects' not to be confused with clear-cut effects on the supposed target organisms. Besides effects of bio-cides, there are a wide variety of other selective exclusion treatments. One inter-esting approach has been to apply different stress factors in succession in order to test possible consequences of reduced soil biodiversity for the stability of soil ecosystem processes (Griffiths et al. 2000). These stresses may be drought, frost, heavy metals, etc., differing in effectiveness and persistence. However, selective removal of soil organisms is a much more complicated tool for examining the relationship between soil biodiversity and plant community development than selectively removing plant species (Wardle et al. 1999a; Diaz et al. 2003).

With respect to potential shortcomings and pitfalls, the concept of plant–soil feedback is a powerful approach to study the possible consequences of bio-diversity of soil organisms for the composition of plant communities, because it includes the effects of plants on the soil community and, subsequently, the effects of the whole soil community on plant growth (Bever et al. 1997). It may also include effects of the soil community on plant–plant interactions (Bever 1994; van der Putten & Peters 1997), which determine the relative abundance of species in plant communities. Shortcomings in current plant–soil feedback stud-ies are, for example, that the contribution of the individual species of soil organ-isms to the responses observed in the plant communities has been determined with varying degrees of success and that isolation procedures have yielded the dominant and culturable species of soil organisms rather than the subordinate species (Klironomos 2002).

Very few studies (e.g. de Rooij-van der Goes 1995; Brinkman et al. 2005) have assessed functional similarity or functional redundancy within and between soil pathogen species complexes. Conceptual plant–soil feedback models lack spatial or temporal components, acknowledging that the presence and abundance of individual plant species changes continuously (Herben et al. 1997; Olff et al. 2000). Effects of plant–soil feedback change between sites (van der Putten et al. 1993), as well as during the colonisation history of individual plant species (van der Stoel et al. 2002). Therefore, before analysing the effect of soil biodiversity on the composition of plant communities, spatio-temporal scales in plant–soil feedback and its dependence on environmental context are first discussed.

Spatio-temporal scales and context-dependence of plant–soil feedback

The influence of soil communities on the composition of plant communities depends on the life history stage of the plant species involved; germination, establishment, growth, reproduction and dispersal are all affected by differ-ent species or groups of species of soil organisms. Seeds that do not germi-nate, immediately become resources for granivorous soil invertebrates or, after dying, for saprotrophic fungi from the decomposer subsystem. Germinating seedlings are vulnerable to pathogens that have less impact on established plants (Augspurger & Kelly 1984) and the establishment of offspring seedlings can be negatively affected by the mycorrhizal networks of mature plants (Moora & Zobel 1996). Selective root herbivores or pathogens reduce the competitive ability of their host plants (Chen et al. 1995; van der Putten & Peters 1997), whereas the competitive ability is enhanced by mutualistic root symbionts (Turkington & Klein 1991). The production of easily degradable litter increases nutrient turnover by decomposer organisms, which favours fast-growing plant species that outcompete slow-growing species (Berendse 1990).

There is little known on effects of soil organisms on seed production of plants, but mycorrhizal fungi may affect plant reproduction (e.g. Koide et al. 1988). Plants escape from soil pathogens through seed dispersal (Augspurger & Kelly 1984; Packer & Clay 2000, 2002) and seed dispersal may have evolved as a way to escape from exposure to the soil pathogens that have accumulated around the parents (van der Putten et al. 2001). Soil organisms are also able to change the vegetative expansion pattern of clonal species. While mycorrhizal fungi intensify clonal branching patterns (Streitwolf-Engel et al. 1997), soil pathogens stimulate unidirectional growth leading to escape from pathogen-infested sites (D'Hertefeldt & van der Putten 1998). The foraging pattern of clonal plants for soil resources (Stuefer et al. 1996) is also affected by the activity of soil burrowing and mining organisms (Blomqvist et al. 2000) and, as well, will be affected by the activity of the soil decomposer subsystem.

The influence of different functional groups of soil organisms (decomposers, mutualists, pathogens and herbivores) depends on the type of plant community. Seed germination and seedling establishment is crucial for all plant communities, but in some cases, such as Mediterranean herbal communities in which annual species play a more important role than in other temperate grasslands, yearly seedling establishment is a prerequisite for a major proportion of the plant community. In contrast, in wetlands or in forests the majority of the species are perennial, so that germination and establishment are more periodical or event-wise (Welling et al. 1988). Nevertheless, soil pathogens and root herbivores also contribute to primary succession (van der Putten et al. 1993), secondary succession (Brown & Gange 1990; de Deyn et al. 2003) and cyclic patterns in communities of long-lived plant species (Holah et al. 1997).

The influence of various functional groups of soil organisms is strongly context dependent. The net effect of organisms belonging to the decomposer subsystem depends on the result (net nutrient mineralisation vs. nutrient immobilisation), as well as on the productivity of the ecosystem. In fertile sites, nutrient mineralisation will further enhance productivity and, at the same time, reduce plant species diversity, while immobilisation is conducive to greater plant species diversity (Fig. 14.1a). In contrast, in infertile sites enhanced nutrient

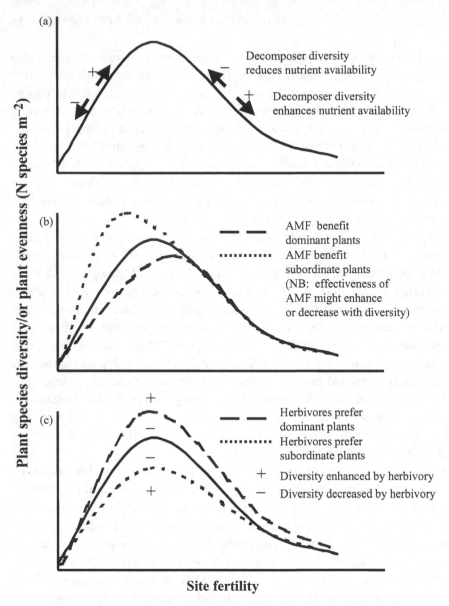

Figure 14.1. Relation between site fertility and plant species diversity (either the number of species or evenness of the plant community) according to Al-Mufti *et al.* 1977 and Marrs 1993 (solid line). (a) Consequences of decomposer diversity changes for site fertility and the diversity of the plant community (indicated by arrow). Effects of decomposer diversity on plant species diversity are expected to depend on the result of decomposer diversity (more or less nutrients become available for plant growth; see Mikola *et al.* (2002)) and site fertility. +, enhanced nutrient availability due to decomposer diversity; −, decreased nutrient availability. Setälä *et al.* (this volume) assumed that effects of enhancing soil decomposer diversity on nutrient availability will saturate at relatively low numbers of species. Heemsbergen *et al.* (2004) showed that diversity effects depend on taxonomic distance between the decomposer species added. (b) AMF enhances diversity when targeted at subordinate plant species (van der Heijden *et al.* 1998b), but reduces diversity when targeted at dominant plant species (Hartnett & Wilson 1999). A positive effectiveness of AMF diversity on dominant species diversity has not yet been demonstrated. (c) Root herbivores or soil pathogens increase plant species diversity when selectively attacking dominant plant species (de Deyn *et al.* 2003), but reduce diversity when specifically reducing growth of subordinate, or rare, plant species (Klironomos 2002); see also Wardle *et al.* (2004).

mineralisation enhances both productivity and species diversity, while immobilisation will reduce plant species diversity (Fig. 14.1a).

Mycorrhizal fungi and symbiotic nitrogen-fixing microorganisms play a crucial role in plant nutrition in nutrient-limited ecosystems, but they are of much less importance in highly productive ecosystems (Fig. 14.1b). They may specifically benefit subordinate plants (van der Heijden et al. 1998b) or dominant species (Hartnett & Wilson 1999). The effect of soil pathogens and root herbivores depends on the type of plant species (dominant, subordinate) that they attack (Fig. 14.1c). For example, in an old field, rare plant species rapidly accumulated soil pathogens, whereas dominant plants had a more neutral, or even positive, short-term soil feedback (Klironomos 2002), so that the capacity of plants to control soil pathogens and root herbivores directly or indirectly may influence plant dominance in natural communities (van der Putten 2003). Host selectivity reduces the competitiveness of the attacked plants (van der Putten & Peters 1997) leading to reduced dominance and, ultimately, to the replacement of species along succession gradients (de Deyn et al. 2003). These context-dependent responses complicate the capacity to draw general conclusions about current studies, so that there is a need for comparative experimental studies across ecosystems and other scales.

Soil communities are highly variable in space (both horizontally and vertically; Ettema & Wardle 2002) and in time (Bardgett et al. 1997; Bardgett et al. this volume). The observed patterns may be due to the dispersal and colonisation strategies of the soil organisms, their response to soil disturbance (Hedlund et al. 2003), successional stage of the ecosystem (Scheu & Schulz 1996) or to local variation in soil structure that may even occur in well-tilled agricultural fields (Robertson & Freckman 1995). Other structuring factors include the activity of the soil microbial community, which was shown to increase along a productivity gradient, while there was no corresponding relationship between plant productivity and the structure of the soil microbial community (Broughton & Gross 2000). The species composition of the nematode community might change fairly little during succession (Kaczmarek et al. 1995; Verschoor et al. 2001), but changes in abundances or plant species-specific responses to soil fauna have strong functional consequences for the diversity and succession of the plant community (de Deyn et al. 2003).

Spatio-temporal patterns of plant-associated organisms depend on the presence of specific host plant species, or on the composition of the plant community, so that net soil feedback effects vary in space and time (van der Putten et al. 2001). Spatial heterogeneity is also due to the specificity of the effects of plants on soil communities (van der Putten et al. 1993; Bever 1994), and to interactions between plants, soil organisms and above-ground vertebrates (Blomqvist et al. 2000) or invertebrates (Masters & Brown 1992; Bezemer et al. 2003). Plant species also differ in their capacity to stimulate sporulation of AMF (Bever et al.

1996), as well as in the diversity of the arbuscular mycorrhizal community in their root zone (Eom et al. 2000) and these effects feed back to plant community composition (Smith et al. 1999). Spatial patterns also exist within the root zone of certain plant species, depending on the ageing of roots and changing exudation patterns (Duineveld & van Veen 1999). Every plant amplifies a specific subset of the microbial species that are present in the bulk soil (Kourtev et al. 2002; Kowalchuk et al. 2002). Finally, similar treatments in geographically different sites may result in a variety of responses in the soil community. For example, a series of biodiversity experiments on ex-arable land across Europe with experimental plant communities of low and high diversity and early vs. mid-successional species resulted in some general, but many idiosyncratic, or site-specific, patterns in soil microbial biomass, nematodes, microarthropods and earthworms (Hedlund et al. 2003).

Temporal patterns can be observed by long-term time series or by analysing successional gradients. Throughout successional history, soil nutrient availability changes, resulting in shifts in plant productivity and diversity (Fig. 14.1; Al Mufti et al. 1977; Marrs 1993), resulting in shifts in the relative contribution of soil organisms over time. For example, in a study of secondary succession on limestone from a wheat field to a beech wood by Scheu and Schultz (1996), it was shown that there were changes in the composition and diversity of saprophagous invertebrates (Lumbricidae, Diplopoda, Isopoda). The species diversity of the Diplopoda and Isopoda first increased and then decreased, whereas species of earthworms were more similar along the chronosequence (Scheu & Schultz 1996). Moreover, these authors found that some groups of soil organisms, the oribatid mites, predominantly colonise the litter layer in the initial stages, whereas they inhabit the upper mineral soil in the later successional beech wood. These data suggest that different, but often functionally similar, groups of soil organisms respond very differently to successional changes (Scheu & Schultz 1996).

Some groups of soil organisms, for example arthropods (Wardle et al. 1999b) and mycorrhizal fungi (Helgason et al. 1998), are highly sensitive to management practices, while land abandonment does not seem to result in fast responses (for example, recovery) in the soil community (Malý et al. 2000; Korthals et al. 2001). In other cases, temporal changes are very fast, for example the colonisation of new roots of a clonal grass in coastal fore dunes and the consequent development of a negative plant–soil feedback (van der Stoel et al. 2002). This all indicates that time scales of plant–soil feedback can vary tremendously.

Life histories of AMF depend on the successional time scale (Hart et al. 2001). It is not well known if AMF are passengers, responding to the changes in the composition of the plant community, or whether they are drivers, causing the changes in the species composition of the plant community. Hart et al. (2001) proposed a classification of AMF in terms of their life-history strategies in order to test this hypothesis. Similarly, a driver–passenger hypothesis with a temporal component has been proposed for root pathogens and herbivores and above-ground natural enemies of plants (van der Putten *et al.* 2001). Since most root pathogens and herbivores are relatively immobile as compared to the above-ground plant enemies, the soil organisms could drive selection for dispersal, while the above-ground enemies could drive selection for plant defence. This 'below-ground dispersal, above-ground defence' hypothesis (van der Putten *et al.* 2001) needs further testing, and it may also be that drivers and passengers change seats every now and then.

Feedback between plant and soil communities: how does soil biodiversity fit in?

Decomposer community

Manipulations of different components of the soil community, for example microfauna (Alphei *et al.* 1996; Bardgett & Chan 1999), mesofauna (Setälä & Huhta 1991) and macrofauna (Brussaard *et al.* 1997), have been shown to result in significant effects on plant growth. Similarly, interactions between the various groups of soil organisms have been shown to impact on plant growth in model systems (Bardgett & Chan 1999; Bonkowski *et al.* 2000; Bradford *et al.* 2002). These effects act primarily through influencing the soil nutrient availability and, therefore, uptake of nutrients by plants. However, the influence of decomposer diversity on nutrient uptake has been mainly explored from the perspective of functional group diversity (Wardle & van der Putten 2002). Effects of species diversity within functional groups has received hardly any attention (but see Laakso & Setälä 1999b; Laakso *et al.* 2000; Liiri *et al.* 2002). Results on soil decomposer diversity, however, cannot be directly translated into consequences for plant growth due to a range of complicating factors. For example, soil micro-organisms compete with plants for nutrients as well as liberating them (Kaye & Hart 1997) and, moreover, the disappearance of one or another trophic group or species from the soil food web has been shown to result in highly unpredictable consequences for ecosystem processes (Mikola *et al.* 2002), thereby making generalisations difficult. Diversity effects of soil organisms on decomposition and nutrient mineralisation also depend on their taxonomic distance (Heemsbergen *et al.* 2004).

In a study by Alphei *et al.* (1996), protists and nematodes enhanced plant production, whereas earthworms reduced plant production. The effects of the protists depended on the presence of earthworms. Since the effects of plant production on plant community composition depends on the nutrient status of the site (Marrs 1993), enhanced soil nutrient availability will lead to reduced plant species richness in fertile sites, whereas it will have the opposite effect in unfertile sites (Fig. 14.1a). In other cases, when soil decomposer diversity reduces soil nutrient availability, soil diversity reduces plant species richness in unfertile sites, while it enhances plant species richness in fertile sites. In medium-fertile sites, effects of soil decomposer diversity will be relatively limited, since plant species diversity fluctuates around its maximum (Fig. 14.1a). Therefore, besides the unpredictability of the effects of soil decomposer diversity on nutrient release and plant production (Mikola *et al.* 2002), the eventual effects depend on site characteristics, the environmental context such as soil nutrient availability or fertility, and the relationship between soil decomposer diversity and soil nutrient availability to the plants. Moreover, the traits of the individual species of soil organisms that may or may not be present, as well as the nature of the species interactions, can strongly affect the effects of soil decomposer diversity on plant growth (Bardgett 2002). Therefore, we need better predictions of the consequences of losses of soil decomposer diversity in relation to the environmental context in order to improve our assessments of the consequences of soil decomposer diversity for the composition of plant communities.

Mutualists community

The effects of soil mutualists on the composition of a plant community have been focused mostly on the positive consequences of the presence of AMF (Grime *et al.* 1987) and the consequences of species diversity in this species group (van der Heijden *et al.* 1998b). Less is known on the consequences of plant species diversity for AMF (Bever *et al.* 1996). However, some negative effects of plant species diversity on AMF abundance have been reported (Hedlund & Gormsen 2002). There is quite a lot of evidence that suggests that mycorrhizal fungi may affect the composition of plant communities, but these effects are context dependent because AMF may act in a variety of ways. For example, AMF can improve the uptake of water or phosphorus by plants (Smith & Read 1997), protect plants against pathogen attack (Newsham *et al.* 1995) and above-ground herbivorous insects (Gange & West 1994) or, still debated (e.g. Fitter *et al.* 1999), translocate carbon between plants through the below-ground fungal network. Therefore, effects of mycorrhizal fungi on plant community composition will also strongly depend on the environmental conditions.

When the effects of AMF depend on improved phosphorus uptake and result in both enhanced biomass production and enhanced evenness in experimental plant communities (van der Heijden *et al.* 1998b), these effects may only be observed in relatively nutrient-poor environments (Fig. 14.1b). In medium-fertile and fertile soils, effects of AMF diversity will be neutral or even negative, when related to nutrient uptake only. This has been shown for ectomycorrhizal fungi by Jonsson *et al.* (2001). These authors found that in a low-fertile substrate, ectomycorrhizal diversity enhanced biomass production of *Betula pendula*, but no effects were found in the fertile substrate, while for *Pinus sylvestris* the effects of ectomycorrhizal species richness were negative in a highly fertile substrate. In a similar way, ericoid mycorrhizal fungi might even reduce plant species diversity in nutrient-poor soils due to monopolisation of the available nutrients to their host species (Northup *et al.* 1995). Effects of mycorrhizal diversity on protection of their host plants against pathogen attack have not yet been studied. There are no reports, as far as known, on consequences of diversity of N-fixing symbionts on the composition of plant communities.

Root herbivore and root pathogen community

In spite of the presence of a wide variety of root-feeding invertebrate herbivores on the roots of single plant species (Brown & Gange 1990; Mortimer *et al.* 1999), there has been little research on possible consequences of the diversity of root feeders on plant communities. Concepts that have been applied above-ground, such as facilitation or inhibition between herbivore species (Bakker & Olff 2003), have been tested for below-ground herbivores. For example, competition between root-feeding nematodes is more intense when they have similar feeding habits, whereas root-feeding nematodes with a more complex host relationship are usually more competitive than nematodes with a less complex interaction with their host (Eisenback 1993). Different species of nematodes and/or fungi have antagonistic, neutral or synergistic effects on each other, so that effects of biodiversity in root-feeding communities can be expected to range from negative to positive.

Bradford *et al.* (2002) suggested that large-sized soil organisms suppress the activity of smaller-sized soil organisms, whereas de Deyn *et al.* (2003) observed considerable effects on plant evenness and succession in early, mid- and late-successional communities of soil fauna. Individual fungal species from the roots of the dune grass A. *arenaria* did not cause growth reduction when added individually, but they reduced growth when added in combination with other species, whereas addition of ectoparasitic nematodes resulted in further growth reduction (de Rooij-van der Goes 1995). Similar additive effects of herbivore diversity have also been found for above-ground vertebrates (Duffy 2003, but see Bakker & Olff 2003), but root-feeding nematodes showed idiosyneratic diversity effects on plant growth (Brinkman *et al.* 2005).

Root herbivores and pathogens can affect plant communities by either reducing growth of dominant species (van der Putten *et al.* 1993; de Deyn *et al.* 2003) or subordinate (and rare) species (Klironomos 2002). When biodiversity changes in the root-feeding community, consequences depend on what plant types were affected most by the root feeders (Fig. 14.1c). There are at least four possible outcomes of root-feeder biodiversity changes, depending on whether this results in less or more root-feeding activity, and whether this activity concerns the dominant or subordinate plant species. In Fig. 14.1c, only effects on different plant types (subordinate vs. dominant) have been plotted, since this is what there is most information about. In order to study the ecological consequences of root herbivore or pathogen diversity on plant performance and plant community composition, such studies should start with describing the population dynamics of the root feeders, their spatial and temporal occurrence. This provides data on complementary distributions and timing of peak occurrence (Yeates *et al.*

1985; van der Stoel *et al.* 2002), to be used for the designing of realistic experiments that determine the consequences of the naturally observed patterns, for example by changing the order of root colonisation.

Questions for future studies

Research on the role of soil biodiversity in structuring the productivity and composition of natural plant communities is wide open for further study. At this moment, there are a handful of studies that have actually examined the consequence of soil biodiversity on the composition of plant communities. However, these studies should not be carried out in isolation, since mycorrhizal fungi, or root herbivores, do not act in isolation, and soil feedback effects depend on external factors, for example the successional stage of the ecosystem (de Deyn *et al.* 2003; Reynolds *et al.* 2003). Also, root herbivory can affect plant communities through other pathways such as the leakage of nutrients from roots into the rhizosphere (Bardgett *et al.* 1999a; Grayston *et al.* 2001), which has been shown to stimulate microbial activity and nutrient mineralisation, and also enhance nutrient transfer to neighbouring species.

Soils have considerable potential for top–down control of root herbivores and pathogens, and apparent competition between, for example, bacterivorous nematodes and root-feeding nematodes when they share the same predators has been suggested by Strong (1999), although this was not confirmed by subsequent studies (D. R. Strong personal communication). Therefore, interactions between the soil decomposer subsystem and the biotrophic subsystem need to be further explored. This may enable determination of the extent that top–down control of root feeders depends on specific natural enemies, or on more general predators. These and other studies in this area may also explain why in some cases soil feedback is due to major pathogens, or root herbivores, and why in other cases soil pathogens are part of more complex communities.

The soil decomposer subsystem also affects above-ground herbivores (Scheu *et al.* 1999; Bonkowski *et al.* 2001), and the role of soil biodiversity in the various types of above-ground-below-ground linkages (as shown by, for example, Masters & Brown 1992; Gange & West 1994; Masters *et al.* 2001; van der Putten *et al.* 2001; Wardle 2002; Bezemer *et al.* 2003) needs to be further explored. These results will elucidate when, how and which soil organisms contribute to vegetation processes, and how soil biodiversity may matter for the effectiveness of vegetation management and biodiversity conservation.

Acknowledgements

I thank Gerlinde de Deyn, Paul Kardol, Paul van Rijn, Martijn Bezemer, as well as Richard Bardgett and two anonymous referees, for helpful comments on previous versions. The development of this area contributes to the 'Energy, environment and sustainable development programme, key action global change, climate and biodiversity' of the European Commission, contracts EVK2-CT-2001-00123 (TLinks) and EVK2-CT-2001-00254 (CONSIDER).

References

Al-Mufti, M. M., Sydes, C. L., Furness, S. B., Grime, J. P. & Band, S. R. (1977). A quantitative analysis of shoot phenology and dominance in herbaceous vegetation. *Journal of Ecology*, **65**, 759–791.

Alphei, J., Bonkowski, M. & Scheu, S. (1996). Protozoa, Nematoda and Lumbricidae in the rhizosphere of *Hordelymus europaeus* (Poaceae): faunal interactions, response of microorganisms and effects on plant growth. *Oecologia*, **106**, 111–126.

Anderson, J. M. (2000). Food web functioning and ecosystem processes: problems and perceptions of scaling. *Invertebrates as Webmasters in Ecosystems* (Ed. by D. C. Coleman & P. F. Hendricks), pp. 3–24. Wallingford: CAB International.

Augspurger, C. K. & Kelly, C. K. (1984). Pathogen mortality of tropical tree seedlings: experimental studies of the effects of dispersal distance, seedling density, and light conditions. *Oecologia*, **61**, 211–217.

Bais, H. P., Vepachedu, R., Gilroy, S., Callaway, R. M. & Vivanco, J. M. (2003). Allelopathy and exotic plant invasion: from molecules and genes to species interactions. *Science*, **301**, 1377–1380.

Bakker, E. S. & Olff, H. (2003). The impact of different-sized herbivores on recruitment opportunities for subordinate herbs in grasslands. *Journal of Vegetation Science*, **14**, 465–474.

Bardgett, R. D. (2002). Causes and consequences of biological diversity in soil. *Zoology*, **105**, 367–374.

Bardgett, R. D. & Chan, K. F. (1999). Experimental evidence that soil fauna enhance nutrient mineralization and plant nutrient uptake in montane grassland ecosystems. *Soil Biology and Biochemistry*, **31**, 1007–1014.

Bardgett, R. D., Denton, C. S. & Cook, R. (1999a). Below-ground herbivory promotes soil nutrient transfer and root growth in grassland. *Ecology Letters*, **2**, 357–360.

Bardgett, R. D., Leemans, D. K., Cook, R. & Hobbs, P. J. (1997). Seasonality of the soil biota of grazed and ungrazed hill grasslands. *Soil Biology and Biochemistry*, **29**, 1285–1294.

Bardgett, R. D., Mawdsley, J. L., Edwards, S., et al. (1999b). Plant species and nitrogen effects on soil biological properties of temperate upland grasslands. *Functional Ecology*, **13**, 650–660.

Bardgett, R. D. & Wardle, D. A. (2003). Herbivore-mediated linkages between aboveground and belowground communities. *Ecology*, **84**, 2258–2268.

Berendse, F. (1990). Organic matter accumulation and nitrogen mineralization during secondary succession in heathland ecosystems. *Journal of Ecology*, **78**, 413–427.

Berendse, F. (1998). Effects of dominant plant species on soils during succession in nutrient-poor ecosystems. *Biogeochemistry*, **42**, 73–88.

Bever, J. D. (1994). Feedback between plants and their soil communities in an old field community. *Ecology*, **75**, 1965–1977.

Bever, J. D. (2003). Soil community feedback and the coexistence of competitors: conceptual frameworks and empirical tests. *New Phytologist*, **157**, 465–473.

Bever, J. D., Morton, J. B., Antonovics, J. & Schulz, P. A. (1996). Host-dependent sporulation and species diversity of arbuscular mycorrhizal fungi in mown grassland. *Journal of Ecology*, **84**, 71–82.

Bever, J. D., Westover, K. M. & Antonovics, J. (1997). Incorporation of the soil community into plant population dynamics: the utility of the feedback approach. *Journal of Ecology*, **85**, 561–573.

Bezemer, T. M., Wagenaar, R., van Dam, N. M. & Wäckers, F. L. (2003). Interactions between above- and belowground insect herbivores as mediated by the plant defence system. *Oikos*, **101**, 555–562.

Binkley, D. & Giardina, C. (1998). Why do tree species affect soils? The warp and woof of tree–soil interactions. *Biogeochemistry*, **42**, 89–106.

Blomqvist, M. M., Olff, H., Blaauw, M. B., Bongers, T. & van der Putten, W. H. (2000). Interactions between above- and belowground biota: importance for small-scale vegetation mosaics in a grassland ecosystem. *Oikos*, **90**, 582–598.

Bongers, T. & Bongers, M. (1998). Functional diversity of nematodes. *Applied Soil Ecology*, **10**, 239–251.

Bonkowski, M., Cheng, W., Griffiths, B. S., Alphei, J. & Scheu, S. (2000). Microbial–faunal interactions in the rhizosphere and effects on plant growth. *European Journal of Soil Biology*, **36**, 135–147.

Bonkowski, M., Geoghegan, I. E., Birch, A. N. E. & Griffiths, B. S. (2001). Effects of soil decomposer invertebrates (protozoa and earthworms) on an above-ground phytophagous insect (cereal aphid) mediated through changes in the host plant. *Oikos*, **101**, 441–450.

Bradford, M. A., Jones, T. H., Bardgett, R. D., *et al.* (2002). Impacts of soil faunal community composition on model grassland ecosystems. *Science*, **298**, 615–618.

Brinkman, E. P., Duyts, H. & van der Putten, W. H. (2005). Consequences of variation in species diversity in a community of root-feeding herbivores for nematode dynamics and host plant biomass. *Oikos*, **110**, 417–427.

Broughton, L. C. & Gross, K. L. (2000). Patterns of diversity in plant and soil microbial communities along a productivity gradient in a Michigan old-field. *Oecologia*, **125**, 420–427.

Brown, V. K. & Gange, A. C. (1990). Insect herbivory below-ground. *Advances in Ecological Research*, **20**, 1–58.

Brussaard, L., BehanPelletier, V. M., Bignell, D. E., *et al.* (1997). Biodiversity and ecosystem functioning in soil. *Ambio*, **26**, 563–570.

Burdon, J. J. (1987). *Diseases and Plant Population Biology.* Cambridge: Cambridge University Press.

Callaway, R. M., Mahall, B. E., Wicks, C., Pankey, J. & Zabinski, C. (2003). Soil fungi and the effects of an invasive forb on grasses: neighbor identity matters. *Ecology*, **84**, 129–135.

Chen, J. D., Bird, G. W. & Renner, K. A. (1995). Influence of *Heterodera glycines* on interspecific competition associated with *Glycine max* and *Chenopodium album. Journal of Nematology*, **27**, 63–69.

Crawley, M. J. (1997). Plant–herbivore dynamics. *Plant Ecology* (Ed. by M. J. Crawley), pp. 401–474. Oxford: Blackwell Science.

D'Hertefeldt, T. & van der Putten, W. H. (1998). Physiological integration of the clonal plant *Carex arenaria* and its response to soil-borne pathogens. *Oikos*, **81**, 229–237.

De Deyn, G. B., Raaijmakers, C. E., van Ruijven, J., Berendse, F. & van der Putten, W. H. (2004). Plant species identity and diversity effects on different trophic levels of nematodes in the soil food web. *Oikos*, **106**, 576–586.

De Deyn, G. B., Raaijmakers, C. E., Zoomer, H. R., *et al.* (2003). Soil invertebrate fauna enhances grassland succession and diversity. *Nature*, **422**, 711–713.

De Rooij-van der Goes, P. C. E. M. (1995). The role of plant-parasitic nematodes and soil-borne fungi in the decline of *Ammophila arenaria* (L.) Link. *New Phytologist*, **129**, 661–669.

De Rooij-van der Goes, P. C. E. M., van der Putten, W. H. & van Dijk, C. (1995). Analysis of nematodes and soil-borne fungi from *Ammophila arenaria* (marram grass) in Dutch coastal foredunes by multivariate techniques.

European Journal of Plant Pathology, **101**, 149–162.

De Ruiter, P. C., Neutel, A. M. & Moore, J. C. (1995). Energetics, patterns of interaction strengths, and stability in real ecosystems. *Science*, **269**, 1257–1260.

Diaz, S., Symstad, A. J., Chapin, F. S., Wardle, D. A. & Huenneke, L. (2003). Functional diversity revealed by removal experiments. *Trends in Ecology and Evolution*. **18**, 140–146.

Duffy, J. E. (2003). Biodiversity loss, trophic skew and ecosystem functioning. *Ecology Letters*, **6**, 680–687.

Duineveld, B. M. & van Veen, J. A. (1999). The number of bacteria in the rhizosphere during plant development: relating colony-forming units to different reference units. *Biology and Fertility of Soils*, **28**, 285–291.

Eisenback, J. D. (1993). Interactions between nematodes with root-rot fungi. *Nematode Interactions* (Ed. by M. W. Khan), pp. 134–174. London: Chapman and Hall.

Eom, A. H., Hartnett, D. C. & Wilson, G. W. T. (2000). Host plant species effects on arbuscular mycorrhizal fungal communities in tallgrass prairie. *Oecologia*, **122**, 435–444.

Ettema, C. H. & Wardle, D. A. (2002). Spatial soil ecology. *Trends in Ecology and Evolution*, **17**, 177–183.

Fitter, A. H., Hodge, A., Daniell, T. J. & Robinson, D. (1999). Resource sharing in plant–fungus communities: did the carbon move for you? *Trends in Ecology and Evolution*, **14**, 70.

Gange, A. C. & West, H. M. (1994). Interactions between arbuscular mycorrhizal fungi and foliar-feeding insects in *Plantago lanceolata* L. *New Phytologist*, **128**, 79–87.

Grayston, S. J., Dawson, L. A., Treonis, A. M., *et al.* (2001). Impact of root herbivory by insect larvae on soil microbial communities. *European Journal of Soil Zoology*, **37**, 277–280.

Griffiths, B. S., Ritz, K., Bardgett, R. D., *et al.* (2000). Ecosystem response of pasture soil communities to fumigation-induced microbial diversity reductions: an examination of the biodiversity–ecosystem function relationship. *Oikos*, **90**, 279–294.

Grime, J. P. (1998). Benefits of plant diversity: immediate, filter and founder effects. *Journal of Ecology*, **86**, 902–910.

Grime, J. P., Mackey, J. M., Hillier, S. H. & Read, D. J. (1987). Floristic diversity in a model system using experimental microcosms. *Nature*, **328**, 420–422.

Hart, M. M., Reader, R. J. & Klironomos, J. N. (2001). Life-history strategies of arbuscular mycorrhizal fungi in relation to their successional dynamics. *Mycologia*, **93**, 1186–1194.

Hartnett, D. C. & Wilson, G. W. T. (1999). Mycorrhizae influence plant community structure and diversity in tallgrass prairie. *Ecology* **80**, 1187–1195.

Hedlund, K. & Gormsen, D. (2002). Ectomycorrhizal colonization of Norway spruce (*Picea abies*) and beech (*Fagus sylvatica*) seedlings in a set aside agricultural soil. *Applied Soil Ecology*, **19**, 71–78.

Hedlund, K., Griffiths, B., Christensen, S., *et al.* (2003). Trophic interactions in a changing world: responses of soil food webs. *Oikos*, **103**, 45–58.

Hedlund, K., Santa Regina, I., van der Putten, W. H., *et al.* (2003). Plant species diversity, plant biomass and responses of the soil community on abandoned land across Europe: idiosyncracy or above-belowground time lags. *Oikos*, **103**, 45–58.

Heemsbergen, D. A., Berg, M. P., Loreau, M., *et al.* (2004). Biodiversity effects on soil processes explained by interspecific functional dissimilarity. *Science*, **306**, 1019–1020.

Helgason, T., Daniell, T. J., Husband, R., Fitter, A. H. & Young, J. P. Y. (1998). Ploughing up the wood wide web? *Nature*, **394**, 431.

Herben, T., Krahulec, F., Hadincova, V., Pechackova, S. & Kovarova, M. (1997). Fine-scale spatio-temporal patterns in a mountain grassland: do species replace each other in a regular fashion? *Journal of Vegetation Science*, **8**, 217–224.

Holah, J. C., Wilson, M. V. & Hansen, E. M. (1997). Impacts of a native root-rotting pathogen on successional development of old-growth Douglas fir forests. *Oecologia*, **111**, 429–433.

Hooper, D. U., Bignell, D. E., Brown, V. K., *et al.* (2000). Interactions between above- and belowground biodiversity in terrestrial ecosystems: patterns, mechanisms and feedbacks. *BioScience*, **50**, 1049–1061.

Huston, M. A. (1997). Hidden treatments in ecological experiments: re-evaluating the ecosystem function of biodiversity. *Oecologia*, **110**, 449–460.

Huston, M. A., Aarssen, L., Austin, M. P., *et al.* (2000). No consistent effect of plant diversity on productivity. *Science*, **289**, 1255a.

Jonsson, L., Nilsson, M.-C., Wardle, D. A. & Zackrisson, O. (2001). Context dependent effects of ectomycorrhizal species richness on tree seedling productivity. *Oikos*, **93**, 353–364.

Kaczmarek, M. A., Kajak, A. & Wasilewska, L. (1995). Interactions between diversity of grassland vegetation, soil fauna and decomposition process. *Acta Zoologica Fennica*, **196**, 236–238.

Kaye, J. P. & Hart, S. C. (1997). Competition for nitrogen between plants and soil microorganisms. *Trends in Ecology and Evolution*, **12**, 139–143.

Klironomos, J. N. (2002). Feedback to the soil community contributes to plant rarity and invasiveness in communities. *Nature*, **417**, 67–70.

Koide, R., Li, M., Lewis, J. & Irby, C. (1988). Role of mycorrhizal infection in the growth and reproduction of wild *vs* cultivated plants: 1. Wild *vs*. cultivated oates. *Oecologia*, **77**, 537–543.

Korthals, G. W., Smilauer, P., van Dijk, C. & van der Putten, W. H. (2001). Linking above- and belowground biodiversity: abundance and trophic complexity in soil as a response to experimental plant communities on abandoned arable land. *Functional Ecology*, **15**, 506–514.

Kourtev, P. S., Ehrenfeld, J. G. & Haggblom, M. (2002). Exotic plant species alter microbial community structure and function in the soil. *Ecology*, **83**, 3152–3166.

Kowalchuk, G. A., Buma, D. S., de Boer, W., Klinkhamer, P. G. L. & van Veen J. A. (2002). Effects of aboveground plant species composition and diversity on the diversity of soil-borne microorganisms. *Antonie van Leeuwenhoek, International Journal of General and Molecular Microbiology*, **81**, 509–520.

Laakso, J. & Setälä, H. (1999a). Population- and ecosystem-level effects of predation on microbial-feeding nematodes. *Oecologia*, **120**, 279–286.

Laakso, J. & Setälä, H. (1999b). Sensitivity of primary production to changes in the architecture of belowground food webs. *Oikos*, **87**, 57–64.

Laakso, J., Setälä, H. & Palojarvi, A. (2000). Influence of decomposer food web structure and nitrogen availability on plant growth. *Plant and Soil*, **225**, 153–165.

Liiri, M., Setälä, H., Haimi, J., Pennanen, T. & Fritze, H. (2002). Relationship between soil microarthropod species diversity and plant growth does not change when the system is disturbed. *Oikos*, **96**, 137–149.

Little, L. R. & Maun, M. A. (1996). The 'Ammophila problem' revisited: a role for mycorrhizal fungi. *Journal of Ecology*, **84**, 1–7.

Loreau, M., Naeem, S. & Inchausti, P. (2002). *Biodiversity and Ecosystem Functioning*. Oxford: Oxford University Press.

Loreau, M., Naeem, S., Inchausti, P., *et al.* (2001). Biodiversity and ecosystem functioning: current knowledge and future challenges. *Science*, **294**, 804–808.

Malý, S., Korthals, G. W., van Dijk, C., van der Putten, W. H. & de Boer, W. (2000). Effect of vegetation manipulation of abandoned land on soil microbial properties. *Biology and Fertility of Soils*, **31**, 121–127.

Marrs, R. H. (1993). Soil fertility and nature conservation in Europe: theoretical considerations and practical management solutions. *Advances in Ecological Research*, **24**, 241–300.

Masters, G. J. & Brown, V. K. (1992). Plant-mediated interactions between two spatially separated insects. *Functional Ecology*, **6**, 175–179.

Masters, G. J., Jones, T. H. & Rogers, M. (2001). Host-plant mediated effects of root herbivory on insect predators and their parasitoids. *Oecologia*, **127**, 246–250.

McNaughton, S. J., Ruess, R. W. & Seagle, S. W. (1988). Large mammals and process dynamics in African ecosystems. *BioScience*, **38**, 794–800.

Mikola, J., Bardgett, R. D. & Hedlund, K. (2002). Biodiversity, ecosystem functioning and soil decomposer food webs. *Biodiversity and Ecosystem Functioning* (Ed. by M. Loreau, S. Naeem & P. Inchausti), pp. 169–180. Oxford: Oxford University Press.

Mills, K. E. & Bever, J. D. (1998). Maintenance of diversity within plant communities: soil pathogens as agents of negative feedback. *Ecology*, **79**, 1595–1601.

Moora, M. & Zobel, M. (1996). Effect of arbuscular mycorrhiza on inter- and intraspecific competition of two grassland species. *Oecologia*, **108**, 79–84.

Moore, J., Hunt, H. W. & Elliott, E. T. (1991). Ecosystem perspectives, soil organisms and herbivores. *Microbial Mediation of Plant–Herbivore Interaction* (Ed. by P. Barbosa, V. A. Krischik & C. G. Jones), pp. 105–140. New York: Wiley.

Mortimer, S. R., van der Putten, W. H. & Brown, V. K. (1999). Insect and nematode herbivory below-ground: interactions and role in vegetation development. *Herbivores: Between Plants and Predators* (Ed. by H. Olff, V. K. Brown & R. H. Drent), pp. 205–238. Oxford: Blackwell Science.

Newsham, K. K., Fitter, A. H. & Watkinson, A. R. (1994). Root pathogenic and arbuscular mycorrhizal fungi determine fecundity of asymptomatic plants in the field. *Journal of Ecology*, **82**, 805–814.

Newsham, K. K., Fitter, A. H. & Watkinson, A. R. (1995). Multi-functionality and biodiversity in arbuscular mycorrhizas. *Trends in Ecology and Evolution*, **10**, 407–411.

Northup, R. R., Yu, Z. S., Dahlgren, R. A. & Vogt, K. A. (1995). Polyphenol control of nitrogen release from pine litter. *Nature*, **377**, 227–229.

O'Connor, P. J., Smith, S. E. & Smith, E. A. (2002). Arbuscular mycorrhizas influence plant species diversity and community structure in a semiarid herbland. *New Phytologist*, **154**, 209–218.

Olff, H., Hoorens, B., de Goede, R. G. M., van der Putten, W. H. & Gleichman, J. M. (2000). Small-scale shifting mosaics of two dominant grassland species: the possible role of soil-borne pathogens. *Oecologia*, **125**, 45–54.

Oremus, P. A. I. & Otten, H. (1981). Factors affecting growth and nodulation of *Hippophaë rhamnoides* L. spp. *rhamnoides* in soils from two successional stages of dune formation. *Plant and Soil*, **63**, 317–331.

Packer, A. & Clay, K. (2000). Soil pathogens and spatial patterns of seedling mortality in a temperate tree. *Nature*, **404**, 278–281.

Packer, A. & Clay, K. (2003). Soil pathogens and *Prunus serotina* seedling and sapling growth near conspecific trees. *Ecology*, **84**, 108–119.

Porazinska, D. L., Bardgett, R. D., Blaauw, M. B., et al. (2003). Relationships at the aboveground–belowground interface: plants, soil microflora and microfauna, and soil processes. *Ecological Monographs*, **73**, 377–395.

Rafaelli, D., van der Putten, W. H., Persson, L., et al. (2002). Multi-trophic dynamics and ecosystem processes. *Biodiversity and Ecosystem Functioning* (Ed. by M. Loreau, S. Naeem & P. Inchausti), pp. 147–154. Oxford: Oxford University Press.

Reynolds, H. L., Packer, A., Bever, J. D. & Clay, K. (2003). Grassroots ecology: plant–microbe–soil interactions as drivers of plant community structure and dynamics. *Ecology*, **84**, 2281–2291.

Robertson, G. P. & Freckman, D. W. (1995). The spatial distribution of nematode trophic groups across a cultivated ecosystem. *Ecology*, **76**, 1425–1433.

Scheu, S. & Schulz, E. (1996). Secondary succession, soil formation and development of a diverse community of oribatids and saprophagous soil macro-invertebrates. *Biodiversity and Conservation*, **5**, 235–250.

Scheu, S., Theenhaus, A. & Jones, T. H. (1999). Links between the detritivore and herbivore system: effects of earthworms and Collembola on plant growth and aphid development. *Oecologia*, **119**, 541–551.

Setälä, H. & Huhta, V. (1991). Soil fauna increases *Betula pendula* growth: laboratory experiments with coniferous forest floor. *Ecology*, **72**, 665–671.

Sikora, R. W. & Carter, W. W. (1987). Nematode interactions with fungal and bacterial plant pathogens: fact or fantasy. *Vistas on Nematology* (Ed. by J. A. Veech & P. W. Dickson), pp. 307–312. Hyattsville, MD: Society of Nematologists.

Smith, M. D., Hartnett, D. C. & Wilson, G. W. T. (1999). Interacting influence of mycorrhizal symbiosis and competition on plant diversity in tallgrass prairie. *Oecologia*, **121**, 574–582.

Smith, S. E. & Read, D. J. (1997). *Mycorrhizal Symbiosis*. London: Academic Press.

Stanton, N. L. (1988). The underground in grasslands. *Annual Review of Ecology and Systematics*, **19**, 573–589.

Streitwolf-Engel, R., Boller, T., Wiemken, A. & Sanders, I. R. (1997). Clonal growth traits of two *Prunella* species are determined by co-occurring arbuscular mycorrhizal fungi from a calcareous grassland. *Journal of Ecology*, **85**, 181–191.

Strong, D. R. (1999). Predator control in terrestrial ecosystems: the underground food chain of bush lupins. *Herbivores: Between Plants and Predators* (Ed. by H. Olff, V. K. Brown & R. H. Drent), pp. 577–602. Oxford: Blackwell Science.

Stuefer, J. F., de Kroon, H. & During, H. J. (1996). Exploitation of environmental heterogeneity by spatial division of labour in a clonal plant. *Functional Ecology*, **10**, 328–334.

Tilman, D. (1982). *Resource Competition and Community Structure*. Princeton, NJ: Princeton University Press.

Torsvik, V., Salte, K., Sorheim, R. & Goksoyr, J. (1990). Comparison of phenotypic diversity and DNA heterogeneity in a population of soil bacteria. *Applied and Environmental Microbiology*, **56**, 776–781.

Troelstra, S. R., Wagenaar, R., Smant, W. & Peters, B. A. M. (2001). Interpretation of bioassays in the study of interactions between soil organisms and plants: involvement of nutrient factors. *New Phytologist* **150**, 697–706.

Turkington, R. & Klein, E. (1991). Competitive outcome among four pasture species in sterilized and unsterilized soils. *Soil Biology and Biochemistry*, **23**, 837–843.

Van Andel, J., Bakker, J. P. & Grootjans, A. P. (1993). Mechanisms of vegetation succession: a review of concepts and perspectives. *Acta Botanica Neerlandica*, **42**, 413–433.

Van Breemen, N. (1998). Plant-induced soil changes: processes and feedbacks. Preface. *Biogeochemistry*, **42**, 1–2.

Van der Heijden, M. G. A., Boller, T., Wiemken, A. & Sanders, I. R. (1998a). Different arbuscular mycorrhizal fungal species are potential determinants of plant community structure. *Ecology*, **79**, 2082–2091.

Van der Heijden, M. G. A., Klironomos, J. N., Ursic, M., *et al.* (1998b). Mycorrhizal fungal diversity determines plant biodiversity, ecosystem variability and productivity. *Nature*, **396**, 69–72.

Van der Putten, W. H. (2003). Plant defense below ground and spatio-temporal processes in natural vegetation. *Ecology*, **84**, 2269–2280.

Van der Putten, W. H., Maas, P. W. Th., van Gulik, W. J. M. & Brinkman, H. (1990). Characterisation of soil organisms involved in the degeneration of *Ammophila arenaria*. *Soil Biology and Biochemistry*, **22**, 845–852.

Van der Putten, W. H. & Peters, B. A. M. (1997). How soil-borne pathogens may affect plant competition. *Ecology*, **78**, 1785–1795.

Van der Putten, W. H., van Dijk, C. & Peters, B. A. M. (1993). Plant-specific soil-borne diseases contribute to succession in foredune vegetation. *Nature*, **362**, 53–56.

Van der Putten, W. H., van Dijk, C. & Troelstra, S. R. (1988). Biotic soil factors affecting the growth and development of *Ammophila arenaria*. *Oecologia*, **76**, 313–320.

Van der Putten, W. H., Vet, L. E. M., Harvey, J. A. & Wäckers, F. L. (2001). Linking above- and belowground multitrophic interactions of plants, herbivores, pathogens and their antagonists. *Trends in Ecology and Evolution*, **16**, 547–554.

Van der Putten, W. H. & van der Stoel, C. D. (1998). Plant-parasitic nematodes and spatio-temporal variation in natural vegetation. *Applied Soil Ecology*, **10**, 253–262.

Van der Stoel, C. D., van der Putten, W. H. & Duyts, H. (2002). Development of a negative plant–soil feedback in the expansion zone of the clonal grass *Ammophila arenaria* following root formation and nematode colonisation. *Journal of Ecology*, **90**, 978–988.

Verschoor, B. C. (2002). Carbon and nitrogen budgets of plant-feeding nematodes in grasslands of different productivity. *Applied Soil Ecology*, **20**, 15–25.

Verschoor, B. C., de Goede, R. G. M., de Vries, F. W. & Brussaard, L. (2001). Changes in the composition of the plant-feeding nematode community in grasslands after cessation of fertiliser application. *Applied Soil Ecology*, **17**, 1–17.

Wardle, D. A. (1999). Is "sampling effect" a problem for experiments investigating biodiversity–ecosystem function relationships? *Oikos*, **87**, 403–407.

Wardle, D. A. (2002). *Communities and Ecosystems: Linking the Aboveground and Belowground Components*. Princeton, NJ: Princeton University Press.

Wardle, D. A., Bardgett, R. D., Klironomos, J. N., et al. (2004). Linkages between above-ground and below-ground biota. *Science*, **304**, 1629–1633.

Wardle, D. A., Bonner, K. I., Barker, G. M., et al. (1999a). Plant removals in perennial grassland: vegetation dynamics, decomposers, soil biodiversity and ecosystem properties. *Ecological Monographs*, **69**, 535–568.

Wardle, D. A., Nicholson, K. S., Bonner, K. I. & Yeates, G. W. (1999b). Effects of agricultural intensification on soil-associated arthropod population dynamics, community structure, diversity and temporal variability over a seven-year period. *Soil Biology and Biochemistry*, **31**, 1691–1706.

Wardle, D. A. & van der Putten, W. H. (2002). Biodiversity, ecosystem functioning and above-ground–below-ground linkages. *Biodiversity and Ecosystem Functioning* (Ed. by M. Loreau, S. Naeem & P. Inchausti), pp. 155–168. Oxford: Oxford University Press.

Wardle, D. A., Yeates, G. W., Williamson, W. & Bonner, K. I. (2003). The response of a three trophic level soil food web to the identity and diversity of plant species and functional groups. *Oikos*, **102**, 45–56.

Welling, C. H., Pederson, R. L. & van der Valk, A. G. (1988). Recruitment from the seed bank and the development of zonation of emergent vegetation during a drawdown in a prairie wetland. *Journal of Ecology*, **76**, 483–496.

Westover, K. M. & Bever, J. D. (2001). Mechanisms of plant species coexistence:

roles of rhizosphere bacteria and root fungal pathogens. *Ecology*, **82**, 3285–3294.

Yeates, G. W., Bongers, T., de Goede, R. G. M., Freckman, D. W. & Georgieva, S. S. (1993). Feeding habits in soil nematode families and genera: an outline for soil ecologists. *Journal of Nematology*, **25**, 315–331.

Yeates, G. W., Watson, R. N. & Steele, K. W. (1985). Complementary distribution of *Meloidogyne, Heterodera*, and *Pratylenchus* (Nematoda: Tylenchida) in roots of white clover. *Proceedings of the fourth Australian Conference on Grassland Invertebrate Ecology* (Ed. by R. B. Chapman), pp. 71–79. Christchurch: Caxton Press.

CHAPTER FIFTEEN

Response of the soil bacterial community to perturbation

ALAN J. McCARTHY

University of Liverpool

NEIL D. GRAY

University of Newcastle upon Tyne

THOMAS P. CURTIS

University of Newcastle upon Tyne

IAN M. HEAD

University of Newcastle upon Tyne

SUMMARY

1. Molecular biological methods have provided new insights into the true extent of bacterial diversity in soil, and here we focus on their application in tandem with soil process measurements.

2. Data from a field experiment are used to illustrate the impact of perturbation on the total bacterial community as well as those functional groups responsible for nitrification, denitrification, methanogenesis and methane oxidation. Increasing the organic matter by about 20% by sewage sludge addition had no statistically significant effect on soil respiration rates, and although methanogenesis and methane oxidation were both stimulated, the effect was short lived and variable. We argue that the methane transformations are particularly dependent on unevenly distributed microsite activity, unlike nitrification/denitrification rates, which were stimulated by liming and organic matter addition in a manner that was both reproducible and persistent. The genetic diversity of the ammonia-oxidising bacteria concomitantly decreased, implying classical selection or enrichment of competitive species upon perturbation.

3. Effects on the diversity of the soil bacterial flora quickly disappeared with time, and we argue that seasonal variation, and particularly its effect on plant growth, has a greater impact on the dynamics of bacterial populations in soil than single time-point perturbations aimed at stimulating general biological activity. However, time course experiments revealed that the bacterial diversity in untreated soils was more stable, as the 16S rRNA gene profiles were more stable than those that developed in disturbed soils.

Biological Diversity and Function in Soils, eds. Richard D. Bardgett, Michael B. Usher and David W. Hopkins.
Published by Cambridge University Press. © British Ecological Society 2005.

4. These attempts at correlating soil bacterial community function and structure (diversity) are hampered by our inability to quantify properly the extent of microbial diversity. We provide a rationale and series of derivations in which it is argued that reasonable assumptions on the relationship between the number and abundance of species enable diversity to be quantified by using two measurable parameters – the total number of bacterial cells and the abundance of the most abundant species. Existing experimental datasets are consistent with this proposal, and it is possible to apply this reasoning to different types of species distributions found in soil, depending upon the nature of the perturbation applied. Consequently, one approach that microbial ecologists might employ to explore the relationship between the diversity of bacterial communities and function in soil would be to assess the shape of the species abundance curve, rather than directly attempting to quantify the number of species present.

Introduction

Soil is one of the most intractable habitats from which to obtain information on microbial diversity and species abundance. Linking such information to function via process measurements either in the laboratory or field is an additional challenge, but one which must be met if we are to address properly the impact of environmental change and make valid predictions on the functions and stability of soil communities. Soil can be viewed as an immense heterogeneous collection of microsites throughout which microorganisms are unevenly distributed. This makes simple observations on selection as a response to environmental pressure practically redundant, especially with the added ingredient of dominance by uncultured species. In their recent perspective on the challenge of coupling function to structure in microbial communities, Paerl and Steppe (2003) prudently focused on the aquatic environment where habitats are compartmentalised and the distribution of microorganisms can be more readily correlated with physical and chemical parameters. Direct determinations of the occurrence and distribution of phenotypic species have been replaced by molecular biological analysis of communities defined by function and each with their own taxonomic composition. This has been relatively easy to apply to circumscribed communities where there is a good correlation between phylogeny and phenotype, for example methanogens or ammonia oxidisers, but more challenging for taxonomically disparate groups such as the denitrifying bacteria or cellulose degraders. Many of these studies have commenced with demonstrations of the poor representation of natural community structure by analysis of culturable species (for nitrifying bacteria see Prosser & Embley 2002). To overcome this problem, molecular biological tools that can directly fingerprint communities and provide information on diversity and detect shifts in species composition have been developed. Denaturing gradient gel electrophoresis (DGGE) of DNA amplified directly from soil samples is one example of a widely used fingerprinting method (Gray *et al.*

2003; Griffiths *et al.* 2003; Nicol *et al.* 2003; O'Donnell *et al.* this volume). We are currently entering a phase where quantitative data can be produced by the application of real-time PCR amplification of environmental DNA (Okano *et al.* 2004), gene expression in the environment can be addressed using stable isotope probing of nucleic acids (Manefield *et al.* 2002) and the soil metagenome can be analysed using DNA microarray technology (Voget *et al.* 2003).

The application of new molecular biological techniques provides a much improved description of the soil microbial community. In order to address the question of soil process stability, these same techniques are being applied to perturbed soils and coupled with measurements of process rates. There are three types of experimental perturbation commonly applied to soil that have both applied relevance and can be used to test hypotheses relating to the effect of microbial diversity on soil processes: xenobiotic contamination/pollution; management to enhance fertility or effect bioremediation; manipulation of temperature, moisture content and atmospheric CO_2 concentrations to simulate environmental change. Xenobiotic contamination is the simplest scenario and is well studied. Selection for members of the soil microbial community that resist and/or degrade added chemicals (typically aliphatic hydrocarbons, polyaromatic hydrocarbons and chlorinated compounds) has a predictable outcome that can be measured – mineralisation or transformation of the contaminant with, in many cases, a concomitant increase in the size of the specific biodegradative population whose growth can be encouraged by the addition of nutrients and agents that improve contaminant bioavailability (Rogers & McClure 2003). There is the opportunity to apply sensitive molecular biological methods to monitor the presence and expression of biodegradative genes *in situ* (e.g. Beaulieu *et al.* 2000) and there are many examples of applying this technology to examine the effects of contaminants on the overall bacterial community structure (e.g. MacNaughton *et al.* 1999; Zucchi *et al.* 2003; Evans *et al.* 2004). For perturbations that are deliberately applied for agronomic reasons, there is a wide range of functional microbial groups whose responses need to be measured and related to soil processes. These more general perturbations, such as liming or sewage sludge application, can provide data that indicate the stability of the bacterial community and therefore point to its ability to maintain soil processes under duress.

In this chapter, we attempt to address the effect of perturbation on soil bacterial communities and processes by discussing data generated from a field experiment at an upland grassland site. The perturbations applied were the addition of lime (6 t ha^{-1} annually in two successive years) to increase the pH and the addition of anaerobically digested domestic sewage sludge (100 000 or 200 000 dm^3 ha^{-1} annually in two successive years). The rationale was to impose perturbations that would have effects on as broad a range of soil bacterial species as possible in a context relevant to agricultural and environmental management practice. Determination of links between the specific soil processes of nitrification and denitrification, and of methanogenesis and methane oxidation and the structure of the bacterial community was the central objective, but we also examined overall bacterial diversity based on temperature gradient gel electrophoresis (TGGE) profiling of 16S rDNA amplified with universal eubacterial primers. Soil cores were taken at intervals from control, limed, sewage-treated and lime- plus sewage-treated plots (Gray *et al.* 2003). The results emphasised the importance of conducting assessments of both general and functional group diversity in tandem with overall effects on the rates of corresponding processes. Consideration of the data, together with the results of other studies, led to the conclusion that the responses of soil microorganisms are difficult to predict and site specific, or both, and different processes respond differently to the same perturbation. The persistence of any observed effects is also an important consideration. In contrast to the lack of uniformity emerging from experimental work in soil, we go on to propose that it is possible to make meaningful estimates of bacterial species diversity and abundance in soil, and provide methods that can now be tested experimentally and will hopefully prove to be universally applicable. In a sense then, the experimental data in the first part of the chapter illustrate the problem that may be at least partially solved by the theoretical framework proposed in the second part.

Soil processes

That soil microorganisms may maintain viable populations by growing slowly at low nutrient concentrations or by sustaining periods of starvation interspersed by rapid proliferation when nutrients become available is a classical view. It is often associated with the terms 'autochthonous versus zymogenous' coined by Winogradsky (1924), which have largely given way to 'copiotrophs versus oligotrophs' (Paul & Clark 1996) based on the substrate affinities of soil isolates. This is generally regarded as an oversimplification, as there are many examples of bacteria able to grow across a range of nutrient concentrations (Williams 1985; McCarthy & Williams 1992). Ecological theory provides a framework for considering microbial populations as '*r*' or '*K*' strategists, with distinct colonisation and survival characteristics (Graham & Curtis 2003) that are reminiscent of Winogradsky's 'zymogenous' and 'autochthonous', respectively. Nevertheless, within bacterial communities defined by function, nutrient concentration can exert a selective effect on the component species. For example, it is now well established that ammonia concentration has a profound effect on the species composition of ammonia-oxidising bacterial communities in soil (Prosser & Embley 2002). In a broader context, plant growth also has an important influence, with bacterial populations in the rhizosphere being much larger than in the surrounding soil (Paul & Clark 1996) and fast-growing species, such as pseudomonads, are often stimulated. The effect of plant growth quickly became apparent in our studies at an upland grassland site (Gray *et al.* 2003). In control plots that remained untreated, soil from below the zone of root penetration consistently exhibited very low rates of respiration, nitrification and denitrification compared with soil samples taken from the root zone. In the first year of sewage sludge application, respiration, nitrification and denitrification were consistently low in samples taken from below the root zone, indicating that the effects of both liming and sewage sludge application did not penetrate below the root zone in this upland grassland with a well-established dense sward. Soil moisture is another environmental factor that strongly influences soil micro-organisms (Hastings *et al.* 2000; Griffiths *et al.* 2003). The relatively high annual rainfall at the site contributed to moisture contents between 1.5 and 1.85 g water g^{-1} soil (dry wt.), so that effects due to periods of drought could be discounted.

The addition of sewage sludge to soil would be expected to increase basal respiration rates (CO_2 production), based on the assumption that the organisms were carbon limited and the observation that increased soil respiration rates often follow treatment with sewage sludge (e.g. Ortiz & Alcaniz 1993; Stamatiadis *et al.* 1999). We observed considerable variation in basal respiration rates (about 2–12 µmol CO_2 g^{-1} dry wt. soil h^{-1}), but ANOVA of these data for all sampling times demonstrated that treatment effects were never significant ($P > 0.05$). Addition of sewage sludge resulted in an approximate 20% increase in the organic carbon content of the soil and one would intuitively expect this to increase soil respiration rates (Oritz & Alcaniz 1993; Stamatiadis *et al.* 1999). The lack of any measurable effect in this study was therefore something of a surprise. We therefore conclude that the availability of readily decomposable organic matter was not limiting microbial activity in this soil, possibly because this is adequately supplied by plant roots. Methane is the other important end-product of organic carbon biodegradation in soil, and although this upland grassland is by no means waterlogged it is highly likely that there were anaerobic microsites where methanogenesis was supported. However, we only detected traces of methane being emitted from the control and limed soils, whereas soils that had received sewage sludge often produced methane upon incubation. The importance of microsites was, however, highlighted by the variability of these data. Although methanogenesis was only detected in sewage-sludge-treated soil samples (up to about 9 µmol CH_4 g^{-1} dry wt. of soil h^{-1} upon incubation), many sewage-sludge-treated soil samples did not generate any methane and even repli-cate samples from the same plots showed enormous variation. In all cases, the stimulation of methanogenesis was not persistent and had declined by two to three orders of magnitude 30 days post sludge addition. Data on methane oxidation potential of soil samples closely mirrored those for methanogenesis in variability, confinement to sludge-treated soils and disappearance of treatment effects after one month. This problem of heterogeneity or patchiness in methane cycling observed here is in line with the data of Nedwell *et al.* (2003) who high-lighted grassland as exhibiting particularly large variances in methane fluxes.

Methanogenic archaea are notoriously fastidious and dependent on syntrophic interactions (see Horn et al. 2003), and this coupled to a requirement for anaerobic microsites may explain the variation. Methane oxidation will inevitably follow the uneven distribution of methanogenesis, but there is an important interaction between the oxidation of methane and ammonia in soil that also needs to be borne in mind. Ammonia is a competitive substrate for the methane monooxygenase of methanotrophs, such that in upland soils where methane concentrations are low, this biochemical interaction can result in the inhibition of methane oxidation (Schimel 2000). In other ecosystems, such as rice-paddy soils, ammonia addition stimulates methane oxidation because it provides a nitrogen source for methanotrophs (Bodelier et al. 2000). This is a good illustration of the complexity of interactions between the bacterial communities responsible for soil processes, and provides some comfort for those who are concerned that the application of ammonium fertilisers could lead to increased methane emissions.

In contrast to the relatively indifferent response of soil respiration to perturbation and the highly variable data on methane cycling described above, nitrification and denitrification showed clear and sustained treatment effects. We determined the rates of these processes in incubated soil slurries, rather than from snapshot measurements of ammonium and nitrate concentrations in soil (that are uninformative and do not reflect the physiological state of the bacteria). Both processes were significantly stimulated by the addition of lime plus sewage sludge (Fig. 15.1) with a three- to four-fold increase in denitrification rate and an order of magnitude increase in nitrification potential. Addition of lime or sewage sludge alone resulted in small but measurable enhancement of both processes. The other feature of these data that set nitrification/denitrification apart from methane generation/oxidation is the persistence of the effect, which was still apparent up to 56 days after sludge application, and particularly so for denitrification (Fig. 15.1a). Nitrification is notoriously slow in acid soils and it is therefore predictable that liming would promote ammonia oxidation (Dancer et al. 1973; Nyborg & Hoyt 1978). The synergistic effect of also adding sewage sludge is likely to be multi-factorial but ammonia present in the sewage sludge may in part account for the positive interaction between lime and sewage sludge. However, all treatments produced an approximately two-fold increase in the ammonia concentration detected in soil samples (from 0.7 to 1.3–1.8 μmol NH_3 g^{-1} soil) even where sludge was not added. In contrast, total nitrogen concentrations for soil treated with sludge plus lime were only increased by about 12%. These are crude measures of nitrogen status and the interaction of liming and sludge addition may also relate to the nature of the organic matter released into soil from sludge at neutral pH. Liming effects on acid soils have previously been attributed to general effects on substrate availability resulting from the release of different forms of organic matter (Shah et al. 1989; Persson et al. 1990).

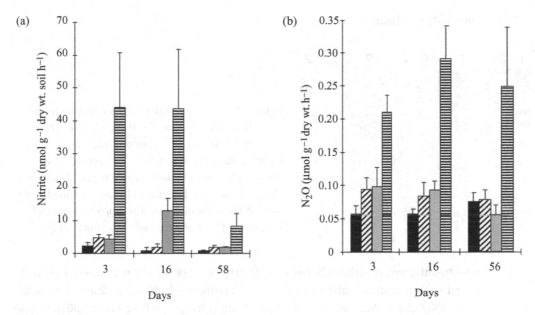

Figure 15.1. Ammonia oxidation (NO_2^{-2} accumulation in the presence of sodium chlorate) potentials (a) and denitrification (b) in treated and control soils incubated after removal from the site at intervals over the experimental sampling period. Control soils, black; sludge-treated soils, black with diagonal lines; lime-treated soils, grey; sludge + lime, horizontal lines. Error bars represent +2 s.e.

Bacterial community structure

Lime and sewage sludge application exhibit a clear and sustained effect on nitrification in upland grassland soil (Fig. 15.1), and comparison of the diversity within the ammonia-oxidising bacteria community should enable observations on species selection and redundancy in this functional group. Nitrification is driven by ammonia oxidation in soil and we can specifically relate this to autotrophic activity by using chlorate as an inhibitor of nitrite reduction in the assays to obviate the contribution of heterotrophic nitrification (Schimel *et al.* 1984). Autotrophic ammonia-oxidising bacteria (AOB) are phylogenetically coherent on the basis of 16S rRNA sequence analysis (Head *et al.* 1993) and the use of specific PCR amplification primers and oligonucleotide probes has confirmed the importance of β-proteobacterial AOB, principally members of the genera *Nitrosospira* and *Nitrosomonas*, in nitrification in a wide range of environments (Kowalchuk & Stephen 2001). Clusters can reproducibly be recognised within phylogenetic trees based on 16S rRNA gene sequences retrieved from the environment by PCR amplification of DNA extracts (Stephen *et al.* 1996; Purkhold *et al.* 2000) and this has enabled changes in community composition to be documented in a large number of ecological studies on terrestrial, freshwater and marine nitrification. Previous investigations of soils have demonstrated

Lime　Lime + sludge　Untreated　Sludge

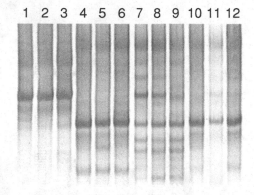

Figure 15.2. TGGE profiles of 16S rDNA fragments amplified from soil samples using PCR primers specific for autotrophic ammonia-oxidising bacteria. Samples were taken from plots one day after application of anaerobically digested sewage sludge. Lanes 1–3, limed soil; lanes 4–6, limed soil treated with sludge; lanes 7–9, untreated control soil; lanes 10–12, soil treated with sludge only.

community composition changes, including species selection following amendment with manures (Hastings *et al.* 1997), fertilisers (Phillips *et al.* 2000; Avrahami *et al.* 2002, 2003; Webster *et al.* 2002) or effluent irrigation (Oved *et al.* 2001). Imposition of drought stress affected AOB community structure in soil (Hastings *et al.* 2000) and changes have also been attributed to temperature effects (Avrahami *et al.* 2003) and microscale differences in oxygen availability (Briones *et al.* 2002). Although profiling AOB communities by applying temperature or denaturing gradient gel electrophoresis (TGGE or DGGE) underestimates diversity relative to sequencing environmental clone libraries (Whitby *et al.* 2003), it remains a useful fingerprinting technique for identifying community shifts, and we applied TGGE analysis to soil from the sewage sludge- and lime-amended plots (Gray *et al.* 2003). Webster *et al.* (2002), working on the same site and using DGGE analysis of PCR amplified AOB 16S rRNA gene fragments, found that diversity within this functional community was lower in soils that had received N fertiliser. This implies classical selection or enrichment of competitive species upon nutrient amendment. Our data demonstrated a similar response, with TGGE profiles of AOB 16S rRNA showing pronounced treatment-related differences in banding patterns and untreated soil yielding a more complex fingerprint indicative of greater diversity (Fig. 15.2). However, these clear differences were quickly dissipated as early as three days after the sewage sludge addition. All of the TGGE profiles were subjected to exhaustive statistical analysis by the application of Sorensen similarity coefficients, unweighted-pair-group method with arithmetic mean (UPGMA) cluster analysis and ANOVA. A central objective was to consider differences between AOB profiles from different treatments in relation to the variation between replicates within each treatment. Only when the latter is significantly lower than the former, is it possible to conclude a clear treatment effect on AOB community structure. Although the dramatically increased nitrification potentials in soils treated with lime and sewage sludge were maintained for up to 56 days post application (Fig. 15.1), there was no significant treatment effect on AOB diversity beyond three days. The similarity between replicates decreased progressively with time and this lack of association between elevated nitrification rates and a particular community structure suggests functional redundancy within a diverse AOB community.

Our studies on the methane-cycling bacteria were not so detailed, primarily because the process measurements did not provide any clear framework for interpretation of data on community structure. We recovered archaeal 16S rRNA gene fragments from the soil samples and then sequenced clones to confirm their identity as methanogen sequences and identify the taxa. More than 50% of the clones were *Crenarchaeota*, a lineage traditionally comprising thermophilic species. Crenarchaeote 16S rRNA gene sequences have now been recovered from a very diverse range of non-extreme environments and recently reported to be abundant in the upland grassland site studied here (Nicol *et al.* 2003). Analysis of the remaining sequences showed that members of the *Methanomicrobiales* predominated and there was no correlation of particular methanogen taxa with treatments. While only trace amounts of methane were emitted from soil from control plots, the methanogenic potential was revealed by sewage sludge addition and there was no evidence that an input of methanogens from the sludge itself was required since methanogenesis could be detected, albeit at low levels with heterogeneous distribution, even in soil which was not amended with sewage sludge. Similarly, all treatments at all sampling times harboured a diversity of methane oxidisers, on the basis of TGGE analysis of amplified 16S rRNA genes and frequent recovery of the soluble methane monooxygenase (sMMO) gene, irrespective of detectable methane oxidising activity in the soil.

Amplification of total DNA extracts from soils with primers targeting the majority of eubacterial 16S rRNA genes was used to generate TGGE profiles that provided an index of total bacterial diversity in the soil. The profiles generated in this way were predictably complex, and visual comparison of the banding patterns supported the data on soil respiration in the sense that there were no discernible treatment effects. This analysis did, however, suggest that the predominant bacterial species in the sewage sludge, identified as major bands in TGGE profiles of 16S rRNA gene fragments amplified from a sludge sample prior to application, did not become established as the predominant organism in the soil. More detailed statistical analysis of the banding patterns revealed small, but statistically significant effects, i.e. overall within treatment similarities were significantly higher ($P \leq 0.001$) than between treatment similarities.

Ecological studies of soil microorganisms are dominated by the problem of reproducibility, which is largely due to the fact that microbial processes occur at microsites and that soil contains hot spots of activity that may not be representatively sampled. We used a carefully designed sampling regime that included replication at plot, sample and subsample levels, and data were subjected to thorough statistical analysis prior to interpretation. The data on methanogenesis and methane oxidation discussed above demonstrate that despite these precautions, reproducibility can still be poor. In contrast, the effects of lime and sewage sludge on nitrification and denitrification were much less variable and conclusions were consequently unequivocal (Fig. 15.1). However, as the season progressed, the natural variation that occurred even in untreated control soils usually masked any treatment effects that were evident soon after lime and sewage sludge applications. While the gradual increase in ambient temperature throughout the summer and reduction in rainfall probably had an effect, we speculate that plant growth had a large influence on the microbial community. Above-ground biomass production approximately doubled over a two-month period in the early summer and limed plots were consistently and significantly more productive than the control plots. For both eubacterial diversity and AOB community structure, within-treatment similarity decreased progressively at each successive sampling time. The average similarity for eubacterial TGGE profiles drifted from 81 to 62% between 1 and 58 days, while for AOB profiles the decrease was from 71 to 28%. In both datasets, cluster analysis showed that any treatment effects were dissipated over time, with significant effects of treatments only obvious shortly after the perturbation was applied. Comparison of AOB gel profiles between successive sampling times gave extremely low average similarities, indicating that the AOB community structure was highly variable and was not simply exhibiting classical succession. That would have been suggested by profiles at adjacent time points having more in common than those that were from samples taken at time points that were further apart. This was not the case. Although the conclusion for the development of AOB communities is that their structure changes with time in a manner that is more significant than the initial liming and/or sewage-sludge-addition effect, perturbed soils do appear to contain a less stable population. The data in Fig. 15.3 show that untreated soils, with their generally more diverse AOB community (Fig. 15.2), maintain a much higher level of stability, expressed in terms of mean similarity between profiles obtained at different sampling times.

In general terms, the methods used to measure soil processes and determine microbial community structure have limited resolving power and introduce their own perturbation during sampling. In parallel studies at the same upland grassland site, membrane inlet mass spectrometry (MIMS) was used in soil monoliths to measure biogenic gas evolution *in situ* with probes (1.6 mm diameter) that could be very specifically located within the soil habitat. One of the most interesting aspects of the data emanating from these experiments was the demonstration that temporal variation in CO_2, CH_4 and O_2 evolution

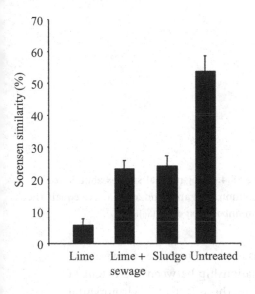

Figure 15.3. Average similarity between TGGE gel profiles of ammonia-oxidiser 16S rDNA amplified from soils across all sampling time points to indicate the degree of stability within the AOB population of each treated soil. Error bars represent ±2 s.e.

was considerable and, furthermore, exhibited a diurnal cycle (Sheppard & Lloyd 2002). It is against this background of very short-term shifts in microbial activity that we have been struggling to observe relationships between diversity and function on a much larger scale.

A fresh approach to estimating microbial diversity

The studies outlined above were primarily aimed at investigating any relationships between bacterial community structure and selected biogeochemical transformations. Questions of wider ecological significance, however, revolve around the relationship between microbial diversity per se and biogeochemical function. It is not known whether patterns observed in communities of larger organisms equally apply to microbial communities. For example, is there a relationship between community diversity and resistance of an ecosystem to perturbation (stability) as has been observed in some (Tilman *et al.* 1996), but not all (Grime 1997; Wardle 1999), communities of larger organisms, or is there a relationship between productivity and community diversity?

Addressing such questions is hampered to some extent by technical limitations when microbial communities are considered. Culture-independent approaches to assessing prokaryotic diversity (e.g. DGGE analysis of 16S rRNA gene fragments) often show clear qualitative changes in bacterial community structure in response to changes in many environmental variables, but quantitative metrics of diversity (e.g. Shannon indices) based on such data do not show significant changes in diversity per se. Such findings are, at least in part, a consequence of the massive diversity of prokaryotes (Torsvik *et al.* 2002), the true extent of which is widely held to be unknown and may even be unknowable for any habitat at

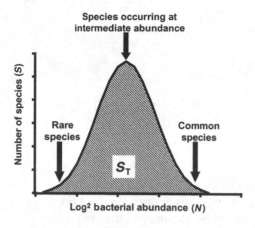

Figure 15.4. A log-normal species abundance distribution. The area under the curve equals the total number of species (S_T).

any time (Wilson 1994). As a consequence, it has not been possible to investigate even the simplest hypotheses regarding the relationship between the extent of prokaryotic diversity and ecosystem function, or the effect of environmental insults (e.g. pollution or climate change). An essential prerequisite for testing the relationship between diversity and ecosystem function is, therefore, development of methods or approaches to quantify reliably the extent of microbial diversity.

We have recently addressed this major gap in the microbial ecologist's toolbox by developing a new approach to estimate prokaryotic diversity quantitatively (Curtis *et al.* 2002). Microbial communities, like all other biological communities (May 1975), will be characterised by a particular species abundance distribution (e.g. Fig. 15.4). Thus, to estimate the microbial diversity of a community, it is not necessary to count every single species/taxon therein – it is only necessary to determine the area under the species abundance curve, which equates to the total number of species/taxa (S_T). This rationale raises two interlinked questions: What is the shape of the species abundance curve? How do we define the area under the curve?

To address the first question, we can examine empirical data and theoretical considerations that result in biological communities with different species abundance distributions. There are a number of possible species abundance curves commonly encountered. These include geometric, log-series, log-normal and 'broken-stick' distributions (May 1975). There are theoretical reasons to suppose that many prokaryotic communities may be characterised by a log-normal species abundance curve. This distribution is characteristic of communities with large numbers of species fulfilling diverse roles, and the log-normal distribution is a product of the effect of many random influences on large heterogeneous assemblages (MacArthur 1960; May 1975). Furthermore, it has also been shown that the log-normal distribution will hold for dynamic opportunistic

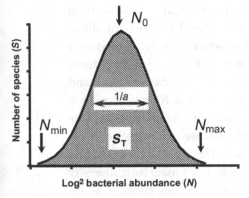

Figure 15.5. Key parameters required to determine the area under a log-normal distribution.

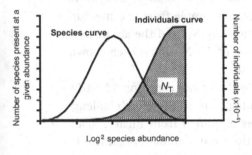

Figure 15.6. The species curve and the individuals curve derived from it. The area under the individuals curve is equal to the total number of individuals (N_T) or a direct cell count.

communities where the most abundant species will change in time and space (though the species abundance distribution will remain constant; MacArthur 1960) and also for more stable 'equilibrium' communities (Whittaker 1972). The broad applicability and importance of log-normal distributions to biological communities has been compellingly stated by Whittaker (1972): 'When a large number of samples is taken containing a good number of species, a log-normal distribution is usually observed . . .'.

Recent empirical analysis of the diversity in soils has supported the notion that a log-normal distribution best describes bacterial communities (Dunbar et al. 2002). Log-normal distributions are extremely well characterised mathematically and it is therefore possible, in principle, to calculate the area under a log-normal curve provided several parameters of the curve are known. These are the abundance of the most abundant organism (N_{max}), the abundance of the least abundant organism (N_{min}), the modal species abundance (N_0) and the spread of the distribution ($1/a$), (Fig. 15.5). All surveys of bacterial diversity conducted to date have only sampled the far right-hand side of the distribution and thus N_{min}, N_0 and $1/a$, cannot be measured directly. However, N_0 and $1/a$ are also a function of a second curve, known as the individuals curve (Fig. 15.6) and derived from the log-normal species abundance curve. This curve is a plot of the number of species present at a given abundance (the y-axis on the log-normal distribution) multiplied by the species abundance (the *x*-axis on the log-normal distribution). The area under this curve is equal to the total number of individuals in the community (N_T), which can be readily determined for prokaryotes using standard microscopic cell-counting procedures with a range of fluorescent stains. Since the individuals curve is a function of the log-normal curve, it, and hence N_T, is also a function of N_{max}, N_{min}, N_0 and $1/a$. Thus N_T can be related mathematically to N_0 and $1/a$, and N_{min} can be arrived at by application of logical reasoning (Curtis *et al.* 2002). Knowledge of what we can measure (N_{max}, N_T) therefore allows us to determine the area under the log-normal curve and hence the total number of species – one measure of the prokaryotic diversity of the system. Full details of the procedure that leads to this conclusion are presented by Curtis *et al.* (2002). The outcome is that it is possible to determine quantitatively the extent of prokaryotic diversity in a sample from two measurable parameters – the total number of prokaryote cells (N_T) and the abundance of the most abundant taxon (N_{max}) – since the ratio of (N_T/N_{max}) has been shown to be a function of the total number of taxa present.

Although the value of N_T can be determined by direct cell counting, identifying and counting the most abundant organism (N_{max}) in soil is more challenging. Cloning and sequencing of 16S rRNA genes amplified from nucleic acids extracted from soil potentially permits the putative most abundant organism to be identified. Some 16S rRNA-targeted fluorescent oligonucleotide probes may then be designed to target the organism represented by the most abundant sequence in a clone library and quantified using fluorescence *in-situ* hybridisation (FISH). An important assumption in this analysis is that the most frequently recovered sequence in a clone library represents the most abundant organism. While there is good evidence that this is often the case, organisms with a high copy number of the rRNA operon, for example, will be over represented in environmental 16S rRNA gene clone libraries. There is also empirical evidence that, in some cases, the most frequent clone in a library does not always represent the most abundant organisms when the same sample is further analysed by FISH (e.g. Eilers *et al.* 2000). Furthermore, the application of FISH in soil is not straightforward (e.g. Zarda *et al.* 1997) and advances in quantitative PCR techniques (Stults *et al.* 2001) may provide a more robust method. The practical determination of N_{max} in the soil environment may also be hampered by the extreme diversity that is apparent in soil bacterial communities (Torsvik *et al.* 1990; Dunbar *et al.* 2002). Often every clone sampled from a soil-derived 16S rRNA gene library is unique (i.e. a singleton; e.g. McCaig *et al.* 1999). This highlights the poor sampling of soil bacterial communities that is afforded by analysis of 16S rRNA gene clone libraries – one estimate has suggested that between 16 000 and 45 000 clones would need to be screened to detect 50% of the taxa present (Dunbar *et al.* 2002). Nevertheless initial, albeit somewhat crude, empirical tests of the N_T/N_{max} approach are consistent with the experimental data that are

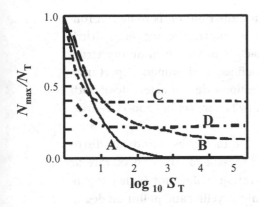

Figure 15.7. The effect of number of species (S_T) on N_{max}/N_T for a range of species abundance distributions. A, broken stick; B, log normal; C, geometric series; D, log series. Adapted from May (1975). Reproduced with permission.

currently available (see for example Torsvik *et al.* 1990; Dykhuizen 1998; Curtis *et al.* 2002).

Although the above arguments have centred on the log-normal distribution, different species abundance distributions are characteristic of prevailing ecological conditions, and changes in the distributions may be a useful measure of the effects of perturbation on a microbial community. Log-normal distributions are typical of situations where many factors influence a biological community, but where a single dominant resource or condition (e.g. a pollutant) affects the community, a geometric or log-series distribution would be expected. The concept of the N_T/N_{max} ratio can be adapted to any of these distributions and their innate properties mean that the relationship between the ratio of the most abundant type of organism to the total population, relative to the total number of taxa is different (Fig. 15.7). We can therefore describe the relationship between prokaryotic diversity and ecosystem properties in slightly abstract terms, i.e. the shape of the species abundance curve. Thus, to assess the relationship between diversity and properties that reflect ecosystem function, the question shifts from 'how many species?' to 'what is the shape of the species abundance curve?' In more practical terms, it is possible to consider whether the form of the species distribution as influenced by specific perturbations (pollution or environmental change) can be correlated with effects on productivity or specific biogeochemical processes. These arguments may appear somewhat theoretical and could perhaps be considered esoteric, with little practical impact. However, the ability to estimate quantitatively prokaryotic species diversity and also infer the likely shape of the species distribution curve has important implications for microbial ecology. First, due to technical limitations, it has not been hitherto possible to assess the effects of environmental perturbations on species composition and hence the consequences of stability or plasticity in prokaryotic diversity on ecosystem stability. The method that we outline above should enable this fundamental question to be addressed and presents an opportunity for obtaining sensible comparative analyses of the diversity/function relationship.

Furthermore, the shape of the species abundance distribution is rooted in fundamental ecological properties of the system; geometric and log-series distributions are indicative of a small number of factors driving community structure and might be expected following a well-defined environmental perturbation, whereas log-normal or broken-stick distributions are characteristic of both dynamic and 'equilibrium' systems in which many factors act in concert to drive community composition.

The relative abundance of species, as exemplified in species abundance distribution curves, has important implications for the issue of common vs. rare or exotic taxa. This is not only relevant to conservation biology for higher organisms, but has many implications for biogeochemical cycling and pollutant degradation. For example, organisms with the metabolic potential to degrade some pollutants may be sufficiently common as to be regarded as ubiquitous. Pollutants targeted by common organisms are less likely to pose a long-term threat to the environment. In contrast, organisms with the ability to attack certain contaminants may be sufficiently rare that such pollutants exhibit a degree of persistence that is directly related to the abundance of the requisite catabolically competent organisms. Thus, the development of methods to quantify the diversity of prokaryote communities and describe their species distribution patterns has important implications for determining the relationship between prokaryotic community structure and function, which is the most important challenge facing microbial ecologists.

Acknowledgements

We are grateful to Richard Hastings, Paul Loughnane and Paul Donohue for some of the data presented, Sarah Buckland and Rob Hunter for field site management, and the Natural Environment Research Council, UK, for supporting the research that underpins the contents of this chapter. The revision suggestions provided by the anonymous referees are also appreciated.

References

Avrahami, S., Conrad, R. & Braker, G. (2002). Effect of soil ammonium concentration on N_2O release and on the community structure of ammonia oxidizers and denitrifiers. *Applied and Environmental Microbiology*, **68**, 5685–5692.

Avrahami, S., Liesack, W. & Conrad, R. (2003). Effects of temperature and fertilizer on activity and community structure of soil ammonia oxidizers. *Environmental Microbiology*, **5**, 691–705.

Beaulieu, M., Becaert, V., Deschenes, L. & Villemur, R. (2000). Evolution of bacterial diversity during enrichment of PCP-degrading activated soils. *Microbial Ecology*, **40**, 345–355.

Bodelier, P. L. E., Roslev, P., Henckel, T. & Frenzel, P. (2000). Stimulation by ammonium-based fertilizers of methane oxidation in soil around rice roots. *Nature*, **403**, 421–424.

Briones, A. M., Okabe, S., Umemiya, Y., *et al.* (2002). Influence of different cultivars on populations of ammonia-oxidizing bacteria in the root environment of rice. *Applied and Environmental Microbiology*, **68**, 3067–3075.

Curtis, T. P., Sloan, W. & Scannell, J. W. (2002). Estimating prokaryotic diversity and its limits. *Proceedings of the National Academy of Sciences, USA*, **99**, 10 494–10 499.

Dancer, W. S., Peterson, L. A. & Chesters, G. (1973). Ammonification and nitrification of N as influenced by soil pH and previous N treatments. *Soil Science Society of America Proceedings*, **37**, 67–69.

Dunbar J., Barns S. M., Ticknor L. O. & Kuske C. R. (2002). Empirical and theoretical bacterial diversity in four Arizona soils. *Applied and Environmental Microbiology*, **68**, 3035–3045.

Dykhuizen, D. E. (1998). Santa Rosalia revisited: why are there so many species of bacteria? *Antonie van Leeuwenhoek International Journal of General and Molecular Microbiology*, **73**, 25–33.

Eilers H., Pernthaler J., Glöckner F. O. & Amann R. (2000). Culturability and *in situ* abundance of pelagic bacteria from the North Sea. *Applied and Environmental Microbiology*, **66**, 3044–3051.

Evans, F. F., Seldin, L., Sebastian, G. V., *et al.* (2004). Influence of petroleum contamination and biostimulation treatment on the diversity of *Pseudomonas* spp. in soil microcosms as evaluated by 16S rRNA based PCR and DGGE. *Letters in Applied Microbiology*, **38**, 93–98.

Graham, G. W. & Curtis, T. P. (2003). Ecological theory and bioremediation. *Bioremediation: a Critical Review* (Ed. by I. M. Head, I. Singleton & M. G. Milner), pp. 61–92. Norwich: Horizon Scientific Press.

Gray, N. D., Hastings, R. C., Sheppard, S. K., *et al.* (2003). Effects of soil improvement treatments on bacterial community structure and soil processes in an upland grassland soil. *FEMS Microbiology Ecology*, **46**, 11–22.

Griffiths, R. I., Whiteley, A. S., O'Donnell, A. G. & Bailey, M. J. (2003). Physiological and

community responses of established grassland bacterial populations to water stress. *Applied and Environmental Microbiology*, **69**, 6961–6968.

Grime, J. P. (1997). Biodiversity and ecosystem function: the debate deepens. *Science*, **277**, 1260–1261.

Hastings, R. C., Butler, C., Singleton, I., Saunders, J. R. & McCarthy, A. J. (2000). Analysis of ammonia-oxidizing bacteria populations in acid forest soil during conditions of moisture limitation. *Letters in Applied Microbiology*, **30**, 14–18.

Hastings, R. C., Ceccherini, M. T., Miclaus, N., *et al.* (1997). Direct molecular biological analysis of ammonia oxidising bacteria populations in cultivated soil plots treated with swine manure. *FEMS Microbiology Ecology*, **23**, 45–54.

Head, I. M., Hiorns, W. D., Embley, T. M., McCarthy, A. J. & Saunders, J. R. (1993). The phylogeny of autotrophic ammonia-oxidising bacteria as determined by analysis of 16S ribosomal RNA gene sequences. *Journal of General Microbiology*, **139**, 1147–1153.

Horn, M. A., Matthies, C., Kusel, K., Schramm, A. & Drake, H. L. (2003). Hydrogenotrophic methanogenesis by moderately acid-tolerant methanogens of a methane-emitting acidic peat. *Applied and Environmental Microbiology*, **69**, 74–83.

Kowalchuk, G. A. & Stephen, J. R. (2001). Ammonia-oxidizing bacteria: a model for molecular microbial ecology. *Annual Review of Microbiology*, **55**, 485–529.

MacArthur, R. H. (1960). On the relative abundance of species. *American Naturalist*, **94**, 25–36.

MacNaughton, S. J., Stephen, J. R., Venosa, A. D., *et al.* (1999). Microbial population changes during bioremediation of an experimental oil spill. *Applied and Environmental Microbiology*, **65**, 3566–3574.

Manefield, M., Whiteley, A. S., Griffiths, R. I. & Bailey, M. J. (2002). RNA stable isotope probing, a novel means of linking microbial community function to phylogeny. *Applied and Environmental Microbiology*, **68**, 5367–5373.

May, R. M. (1975). Patterns of species abundance and diversity. *Ecology of Species and Communities* (Ed. by M. Cody & J. M. Diamond), pp. 81–120. Cambridge, MA: Harvard University Press.

McCaig, A. E., Glover, L. A. & Prosser, J. I. (1999). Molecular analysis of bacterial community structure and diversity in unimproved and improved upland grass pastures. *Applied and Environmental Microbiology*, **65**, 1721–1730.

McCarthy, A. J. & Williams, S. T. (1992). Actinomycetes as agents of biodegradation in the environment: a review. *Gene*, **115**, 189–192.

Nedwell, D. B., Murrell, J. C., Ineson, P., *et al.* (2003). Microbiological basis of land use impact on the soil methane sink: molecular and functional analysis. *Genes in the Environment* (Ed. by R. S. Hails, J. E. Beringer & H. C. J. Godfray), pp. 150–166. Oxford: Blackwell Science.

Nicol, G. W., Glover, L. A. & Prosser, J. I. (2003). The impact of grassland management on archaeal community structure in upland pasture rhizosphere soil. *Environmental Microbiology*, **5**, 152–162.

Nyborg, M. & Hoyt, P. B. (1978). Effects of soil acidity and liming on mineralization of soil nitrogen. *Canadian Journal of Soil Science*, **58**, 331–338.

Okano, Y., Hristova, K. R., Leutenegger, C. M., *et al.* (2004). Application of real-time PCR to study effects of ammonium on population size of ammonia-oxidising bacteria in soil. *Applied and Environmental Microbiology*, **70**, 1008–1016.

Ortiz, O. & Alcaniz, J. M. (1993). Respiration potential of microbial biomass in a calcareous soil treated with sewage sludge. *Geomicrobiology Journal*, **11**, 333–340.

Oved, T., Shaviv, A., Goldrath, T., Mandelbaum, R. T. & Minz, D. (2001). Influence of effluent irrigation on community composition and function of ammonia-oxidizing bacteria in soil. *Applied and Environmental Microbiology*, **67**, 3426–3433.

Paerl, H. W. & Steppe, T. F. (2003). Scaling up: the next challenge in environmental microbiology. *Environmental Microbiology*, **5**, 1025–1038.

Paul, E. A. & Clark, F. E. (1996). *Soil Microbiology and Biochemistry*, 2nd edition. San Diego, CA: Academic Press.

Persson, T., Wiren, A. & Andersson, S. (1990). Effects of liming on carbon and nitrogen mineralization in coniferous forests. *Water Air and Soil Pollution*, **54**, 351–364.

Phillips, C. J., Harris, D., Dollhopf, S. L., *et al.* (2000). Effects of agronomic treatments on structure and function of ammonia-oxidizing communities. *Applied and Environmental Microbiology*, **66**, 5410–5418.

Prosser, J. I. & Embley, T. M. (2002). Cultivation-based and molecular approaches to characterisation of terrestrial and aquatic nitrifiers. *Antonie van Leeuwenhoek International Journal of General and Molecular Microbiology*, **81**, 165–179.

Purkhold, U., Pommerening-Roser, A., Juretschko, S., *et al.* (2000). Phylogeny of all recognised species of ammonia oxidizer based on comparative 16S rRNA and *amoA* sequence analysis: implications for molecular surveys. *Applied and Environmental Microbiology*, **66**, 5368–5382.

Rogers, S. L. & McClure, N. (2003). The role of microbiological studies in bioremediation process optimisation. *Bioremediation: A Critical Review* (Ed. by I. M. Head, I. Singleton & M. G. Milner), pp. 27–59. Norwich: Horizon Scientific Press.

Schimel, J. (2000). Rice, microbes and methane. *Nature*, **403**, 375–377.

Schimel, J. P., Firestone, M. K. & Killham, K. S. (1984). Identification of heterotrophic nitrification in a Sierran acid forest soil. *Applied and Environmental Microbiology*, **48**, 802–806.

Shah, Z., Adams, W. A. & Haven, C. D. V. (1989). Composition and activity of the microbial population in an acidic upland soil and effects of liming. *Soil Biology and Biochemistry*, **22**, 257–263.

Sheppard, S. K. & Lloyd, D. (2002). Direct mass spectrometric measurement of gases in soil monoliths. *Journal of Microbiological Methods*, **50**, 175–188.

Stamatiadis, S., Doran, J. W. & Kettler, T. (1999). Field and laboratory evaluation of soil quality changes resulting from injection of liquid sewage sludge. *Applied Soil Ecology*, **12**, 263–272.

Stephen, J. R., McCaig, A. E., Smith, Z., Prosser, J. I. & Embley, T. M. (1996). Molecular diversity of soil and marine 16s rRNA gene sequences related to β-subgroup ammonia-oxidizing bacteria. *Applied and Environmental Microbiology*, **62**, 4147–4154.

Stults, J. R., Snoeyenbos-West, O., Methe, B., Lovley, D. R. & Chandler, D. P. (2001). Application of the 5' fluorogenic exonuclease assay (TaqMan) for quantitative ribosomal DNA and rRNA analysis in sediments. *Applied and Environmental Microbiology*, **67**, 2781–2789.

Tilman D., Wedin, D. & Knops, J. (1996). Productivity and sustainability influenced by biodiversity in grassland ecosystems. *Nature*, **379**, 718–720.

Torsvik, V., Ovreas, L. & Thingstad, T. F. (2002). Prokaryotic diversity: magnitude, dynamics, and controlling factors. *Science*, **296**, 1064–1066.

Torsvik, V., Goksøyr, J. & Daae, F. L. (1990). High diversity in DNA of soil bacteria. *Applied and Environmental Microbiology*, **56**, 782–787.

Voget, S., Leggewie, C., Uesbeck, A., et al. (2003). Prospecting for novel biocatalysts in a soil metagenome. *Applied and Environmental Microbiology*, **69**, 6235–6242.

Wardle, D. A. (1999). Biodiversity, ecosystems and interactions that transcend the interface. *Trends in Ecology and Evolution*, **14**, 125–127.

Webster, G., Embley, T. M. & Prosser, J. I. (2002). Grassland management regimens reduce small-scale heterogeneity and species diversity of beta-protobacterial ammonia oxidizer populations. *Applied and Environmental Microbiology*, **68**, 4058–4067.

Whitby, C. B., Meade, R. A., Hall, G., et al. (2003). Molecular genetic analysis of the ammonia-oxidising bacterial community in a defined hypereutrophic freshwater lake. *Genes in the Environment* (Ed. by R. S. Hails, J. E. Beringer & H. C. J. Godfray), pp. 167–186. Oxford: Blackwell Science.

Whittaker, R. H. (1972). Evolution and measurement of species diversity. *Taxon*, **21**, 213–251.

Williams, S. T. (1985). Oligotrophy in soil: fact or fiction? *Bacteria in their Natural Environments* (Ed. by M. M. Fletcher & G. D. Floodgate), pp. 81–110. Orlando, FL: Academic Press.

Wilson, E. O. (1994). *The Diversity of Life*. New York: Penguin.

Winogradsky, S. (1924). Sur la microflora autochthone de la terre arable. *Compte Rendu Academie Science, Paris*, **178**, 1236–1239.

Zarda, B., Hahn, D., Chatzinotas, A., et al. (1997). Analysis of bacterial community structure in bulk soil by *in situ* hybridization. *Archives of Microbiology*, **168**, 185–192.

Zucchi, M., Angiolini, L., Borin, S., et al. (2003). Response of a bacterial community during bioremediation of an oil-polluted soil. *Journal of Applied Microbiology*, **94**, 248–257.

PART V

Applications of soil biodiversity

CHAPTER SIXTEEN

Soil biodiversity in rapidly changing tropical landscapes: scaling down and scaling up

KEN E. GILLER
Wageningen University

DAVID BIGNELL
University of London

PATRICK LAVELLE
Institut de Recherches pour le Développement Paris

MIKE SWIFT
CIAT Nairobi

EDMUNDO BARRIOS
CIAT Cali

FATIMA MOREIRA
Universidade Federal de Lavras

MEINE VAN NOORDWIJK
World Agroforestry Centre Bogor

ISABELLE BARIOS
Instituto de Ecología A.C.

NANCY KARANJA
University of Nairobi

JEROEN HUISING
CIAT Nairobi

SUMMARY

1. Habitat modification and fragmentation of remaining pristine areas in the tropics is occurring at a speed that threatens to compromise any serious attempt to assess their value in the biosphere, and catalogue their true biological diversity.
2. Knowledge about the functional significance of soil biodiversity has been strongly influenced by emphasis on temperate climates and by focusing on particular processes of significance to high-input, intensive agriculture. We do not know how robust our methodologies and our concepts are when applied to low-input systems.

Biological Diversity and Function in Soils, eds. Richard D. Bardgett, Michael B. Usher and David W. Hopkins.
Published by Cambridge University Press. © British Ecological Society 2005.

3. Links between diversity and function are clearer for functions that are relatively specific, such as the roles of ecosystem engineers, or specific nutrient transformations compared with generalist functions, such as decomposition, micrograzing, predation and antibiosis.
4. Substantial redundancy exists in relation to general functions that could be important for functional stability.
5. When considering the legume–rhizobium symbiosis as a specific case, rhizobial diversity based on molecular phylogeny is only weakly correlated with specific functions such as ability to form nodules (infectiveness), to fix N_2 (effectiveness) and to survive in the soil (adaptation).
6. Major challenges for the future include developing tools for managing soil biodiversity through manipulation of above-ground vegetation and soil amendments, and understanding the effects of scale to design land use systems for optimal future conservation of the biodiversity of tropical soils.

Introduction

If the soil is said to be the 'poor man's rainforest' in terms of the bewildering biodiversity it harbours (Usher 1985), then what status should the soil in the tropical rainforest be assigned? Our knowledge of the relationships between biodiversity in soil is increasing rapidly in ecosystems in temperate regions, and for the sake of simplicity research initiatives have often chosen to focus on ecosystems with restricted plant species diversity such as Antarctica (see Wall this volume) and the depauperate upland grassland of Sourhope, UK. The new research network described in this chapter focuses on critical transition environments in and around tropical forests that perhaps represent ecosystems that are currently under the greatest threat of biodiversity loss.

Over recent decades many pristine tropical forests have been replaced with agricultural and silvicultural systems (Noble & Dirzo 1997), while the amount of degraded land has increased rapidly (Dobson *et al.* 1997). The underlying causes of deforestation in the tropics are highly complex, but more often than not include commercial logging and large-scale plantations (Lambin *et al.* 2001; Geist & Lambin 2002). A further major driver is the pressure of increasing human population, strongly coupled with the problems of poverty and lack of other opportunities for earning income through agriculture or paid employment (Moran 1993). Poverty and population pressure lead to the expansion of formerly forested areas under slash-and-burn agriculture or pasture, often coupled with increased access for migrants due to building of roads for timber extraction. Further, when forests are cleared, there can also be agricultural intensification (e.g. simplification of cropping regimes and shortening of fallow periods on forest-derived soils, *sensu* Matson *et al.* 1997), leading to rapid loss of soil and soil fertility. Remaining areas of pristine forest tend to be in the remotest regions where human population densities are low and the local people are highly dependent on forest products and have little access to the broader market economy.

In this chapter we discuss recent and ongoing research conducted by the Below Ground Biodiversity (BGBD) Network of the Tropical Soil Biology and Fertility (TSBF) Institute of the International Centre for Tropical Agriculture (CIAT). Our broad goal is to explore the relationships between above- and below-ground biodiversity along gradients of agricultural intensification around undisturbed primary forests in the tropics. A particular emphasis of our ongoing research is on linkages to demonstrable ecosystem functions that can be communicated to local people living in these environments, and other local and national stakeholders, encouraging a community effort towards the long-term conservation of biodiversity. We review some of our existing knowledge of changes in soil biodiversity and ecosystem function along gradients of intensification across tropical areas of Latin America, Africa and Asia, and discuss some of the problems of scale associated with such analyses.

Agricultural intensification, soil biodiversity and ecosystem function

The impetus for our research on agricultural intensification dates back to a meeting held in 1996 in Hyderabad, India, the proceedings of which were published as a special volume of *Applied Soil Ecology* (Swift 1997). Two 'global hypotheses' were identified in this workshop, namely

H_1: agricultural intensification results in a reduction in soil biodiversity leading to a loss of function detrimental to resilience and sustained productivity, and

H_2: agricultural diversification enhances ecosystem resilience and sustained productivity by increasing soil biodiversity.

These were elaborated in a series of more specific hypotheses (Giller *et al.* 1997). H_1 and H_2 are not simply opposites in the sense that we expect that a strong hysteresis could occur between the destructive effects of agricultural intensification and the restoration of soil biodiversity through management. Direct causal links between intensification, soil biodiversity and functioning of ecosystems are starting to emerge, but do not seem to be universal for all groups of organisms, as we discuss below.

A comparative analysis between the different agroecological conditions in the different countries requires a means of quantifying the degree of intensity of agriculture. The term 'intensification' has different interpretations in relation to land use, economics and agronomic efficiency. An attempt was made earlier by the TSBF network, where intensification (I) was defined as a function of indices of land use intensity (L) as defined by Ruthenberg (1980), dependence on mineral fertilisers (N), pesticides (P), fossil energy (E) and on irrigation water (W)

$$I = L \times N \times P \times E \times W \quad \text{(Giller et al. 1997)}$$

This definition is currently being revised to encompass other aspects and dimensions of intensification (M. van Noordwijk unpublished results). Several aspects are being considered, among which the most important are (i) extending the concept of the Ruthenberg index for land use which simply reflects the relative area under cultivation to recognise that some parts of the landscape are permanently under forest, or left as refugia, corridors or filters; (ii) creating an 'offtake index' to reflect the amount of crop or tree litter or residue that is recycled within the land unit compared with harvesting or fire; and (iii) evaluating the balance between use of fossil energy compared with intensity of labour use for activities such as soil tillage, weeding and harvesting of crop or forest products.

A functional approach to biodiversity

It is not practical to work on all aspects of below-ground biodiversity in detail across the full range of tropical environments. Therefore we should focus on aspects of below-ground biodiversity that have a strong linkage to tangible functions which underpin 'soil-based' ecosystem services, for example production of useful products (food, fruits, fuelwood and other products such as housing materials), effects on atmospheric composition (emission of gases involved in global warming in relation to sequestration in vegetation and soil), hydrological flows (provision of clean water and control of soil erosion) and conservation of biodiversity. This is a similar approach to that used elsewhere (Bignell *et al.* 2003). The three main 'soil-based' functions underpinning these ecosystem services will be studied. First, there are the effects of soil organisms, largely the ecosystem engineers among the macrofauna, on soil structure and effects on hydrological flows through and over the soil, strongly linked to erosion control. Second, there is the suppression of pests and diseases through maintenance of diverse soil communities and stable soil food webs. Finally, decomposition and nutrient cycling includes the breakdown and stabilisation of organic matter and net release of nutrients for plant growth, as well as free-living and symbiotic nutrient-transforming bacteria responsible for increasing nutrient availability and net emissions of gases with global warming potential.

The potential for management of soil organisms to alter functioning of tropical ecosystems is summarised in Fig. 16.1. The possible intervention points operate at different levels in a hierarchy, that is at the vegetation or 'cropping system' level through the diversity and arrangement of plant species in space and time; at soil level through management of inputs of organic resources (in terms of both quantity and quality) of mineral fertilisers and other amendments, or through tillage or irrigation; and at organism level through introduction of biological control agents or inoculation of microsymbionts. These can be characterised as top–down and bottom–up approaches. The top–down approach is the analysis of how the design of the farming systems and the management of the crops, the organic inputs and the major soil ecosystem engineers can move the system towards physical stability and sustainability. The bottom–up approach is how

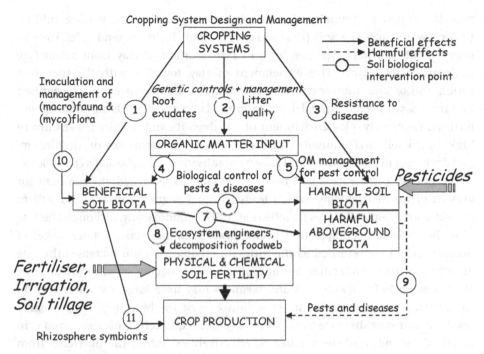

Figure 16.1. Potential for managing soil fertility related functions and properties through manipulation of soil processes and populations. The numbers in circles refer to the potentials for intervention and are explained in the diagram. Modified by Susilo et al. (2004), after Swift (1998). Reproduced with permission of CAB International.

we can protect and encourage the specific small-scale organisms that defend against pests and mediate the key specific nutrient transformations. The extent to which these potential interventions are understood or have been realised is addressed in the following sections.

Effects mediated through soil structure: the ecosystem engineers

The role of ecosystem engineers

Earthworms, ants and termites are the soil macrofauna that have the greatest effects on soil structure and therefore mediate soil physical properties influencing infiltration, drainage, water-holding capacity and aeration (Lavelle 1997; Lavelle et al. 1997). They modify soil structure through selective ingestion of organic and mineral particles that are deposited, often at large distances (>10–30 m) from where they feed, as faecal pellets and organomineral aggregates. At a larger scale, building and tunnelling by earthworms and termites contribute directly to the matrix of macropores, and generate a significant vertical turn-over of mineral material (Wood 1996). In addition, they create a diverse range of biogenic structures, mixing organic and mineral material (Decaëns et al. 2001a). The subterranean galleries of both earthworms and termites can be extensive and, together with the increased surface roughness and porosity caused by faunal activities, play a major role in increasing infiltration of water into the soil (Holt & Lepage 2000). Termites tend selectively to ingest small soil particles, resulting in translocation of clay from subsurface horizons to termitaria. This enrichment of clay, together with the increased cation and organic matter content of termitaria of some species, is recognised by farmers who often spread termite mounds on their fields to enhance soil fertility. Conversely, clay enrichment of ant deposits and termite sheathings in highly acid soil environments can lead to high concentrations of aluminium, and decreases in base cations and slow mineralisation rates (Decaens *et al.* 2001a).

The guts of earthworms and termites provide a specialised environment for growth of microorganisms, which leads to rapid degradation of up to 90% of cellulose and hemicelluloses (Lavelle *et al.* 1997). Paradoxically, although the rate of early stages of decomposition is enhanced by faunal activity, later stages of decomposition are retarded as the intimate mixing with soil increases the rate at which organic molecules become physically protected by clays (Heal *et al.* 1997; Lavelle 1997). Some ants and termites may have large spatial effects on the distribution of organic matter within ecosystems by transporting organic residues over long distances, and in many cases sequestering nutrients and C in localised mounds and deep galleries, effectively excluding the nutrients from cycling over time scales of tens to hundreds of years (but for a contrary view see Coventry *et al.* 1988).

Links between biodiversity of ecosystem engineers and ecosystem function
Along an intensification gradient in Cameroon, the diversity of soil-feeding termites was strongly reduced due to disturbance associated with increasing intensity of cultivation, whereas the diversity of wood-feeding termites was little affected and, indeed, slightly increased (Eggleton *et al.* 2002). Among the macrofaunal groups monitored across seven land uses from forest to *Imperata* savanna or cassava fields in eastern Sumatra, termites showed the greatest sensitivity to increasing intensification, their diversity diminishing with the ratio of (aboveground) plant species richness to plant functional types (PFT, a measure of the morphological complexity of the plant community; Fig. 16.2a), and with woody basal area of trees (Fig. 16.2b). There was a strong switch from dominance of endogeic to epigeic macrofauna across this disturbance gradient and the bulk density of soil increased indicating decreases in soil porosity and permeability (Bignell *et al.* 2005).

Loss of soil macrofauna biodiversity on conversion of rainforest to pasture (from more than 160 species to less than 40 species) coupled with the invasion of a compacting earthworm species (*Pontoscolex corethrurus*) resulted in extreme problems of degradation (Chauvel *et al.* 1999). *Pontoscolex corethrurus* produces compact casts that accumulate in the upper 5 cm of soil and form continuous crusts covering the soil surface, preventing water infiltration and plant growth.

(a)

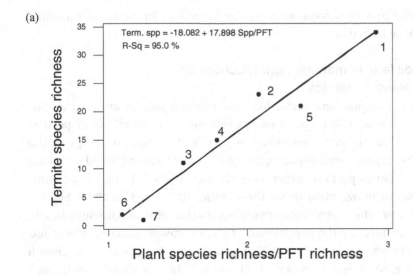

(b)

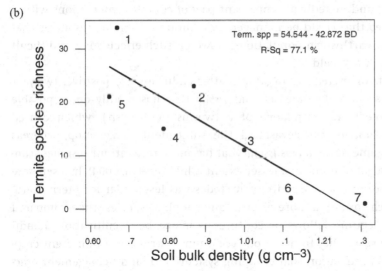

Figure 16.2. Relationships between termite species richness across a forest disturbance gradient in lowland Sumatra and (a) the ratio plant species richness/plant functional types and (b) soil bulk density. The numbers next to data points indicate the land use systems: 1, primary rainforest; 2, logged over rainforest; 3, industrial soft-wood plantation; 4, eight-year-old rubber plantation; 5, old growth jungle rubber tree mosaic; 6, *Imperata* grassland; 7, ten-year-old cassava plantation. After Gillison *et al.* (2003) and Jones *et al.* (2003).

Barros *et al.* (2001) showed that the situation is reversible in less than one year through an elegant experiment in which compacted soil monoliths were translocated from the pasture to the forest and vice versa. Under normal conditions, decompacting species of ants and termites dig holes in the casts to maintain their own galleries and accesses to surface, or re-ingest them to take advantage of the post-digestion enzymatic activities by microflora in casts, thus limiting their abundance in soil.

Effects of soil fertility management practices on pest and disease dynamics

Maintenance of complex and stable food webs in soil (see de Ruiter *et al.* this volume) may be essential in preventing explosion of populations of particular organisms that can cause problems as pests and diseases on crops (Susilo *et al.* 2004). Tree legume–maize rotational systems in nitrogen-limited soils have shown a significant impact on reduction of the parasitic weed *Striga*, responsible for large losses in maize yield in southern Africa (Barrios *et al.* 1998). Specific interactions occur where beneficial organisms control pests and diseases in soil. Earthworms can decrease the populations of plant parasitic nematodes by reducing viability of eggs in ingested cysts (Yeates 1981) and this can improve growth of rice in pot experiments (J. Boyer, M. Blouin & P. Lavelle unpublished). Such interactions are undoubtedly an important part of complex interactions within the soil food web that could be managed by addition of organic resources that will favour large earthworm populations. However, such effects may be difficult to demonstrate in the field.

Not all effects of increasing organic matter additions are positive: two-year fallows of the shrubby legume *Sesbania sesban* that has readily decomposable residues promoted the abundance of cutworms (*Agrotis* sp.), which caused extreme root damage and devastated the subsequent maize crop, whereas another tree legume *Acacia angustissima* that has more recalcitrant residues stimulated cutworm populations to a lesser extent (Chikowo *et al.* 2004). In a separate study, abundance of plant parasitic nematodes was less under long-term natural fallows which contain a more diverse range of plant species when compared with improved legume fallows or continuous maize–bean cultivation (Kandji *et al.* 2001). The variability in nematode community structure in different cropping phases is an indication that cropping systems and land management practices have a significant impact on nematodes. These examples suggest that the diversity of the organic inputs could be a key in maintaining complex food webs. Current research within the TSBF–BGBD network is targeted at a range of soil-borne nematodes, insect pests (including some termites) and fungal pathogens (e.g. root rots and wilts).

Decomposition and nutrient cycling

Decomposition can be defined as the progressive breakdown of organic resources into their constituent molecules, but this is a simplistic view as simple constituents (sugars, amino acids and minerals) are continually assimilated into complex molecules through microbial growth. Therefore the initial organic resources, which are the starting point for decomposition, are continually modified, forming new molecules and resources which in turn enter the decomposition cycle in what has been termed the 'resource cascade' (Swift *et al.* 1979). Release of reactive polyphenolic molecules from fresh residues and their formation by microorganisms, plus generation of free radicals leads to condensation and complexation reactions with carbohydrates and proteins that can give rise to chemical stabilisation of large polymeric molecules characteristic of 'humus' (see Hopkins *et al.* this volume). Rates of decomposition are governed by three interacting factors, the chemical and physical properties of organic resources or 'quality' (Q), the prevailing conditions of the physico-chemical environment (P) and the soil organisms (O).

The importance of quality in determining rates of decomposition and nitrogen mineralisation from organic residues has received comprehensive treatment (e.g. Cadisch & Giller 1997; Palm *et al.* 2001). Rates of decomposition and net nitrogen mineralisation can be fairly accurately predicted on the basis of parameters such as carbon : nitrogen ratio, lignin and reactive polyphenol contents of the organic resource. The main factors of the physico-chemical environment that regulate rates of decomposition are temperature and moisture. Decomposition can be retarded in strongly acid soils rich in aluminium, resulting in podzolisation and the development of mor humus, although such effects are best known from acid forest soils of boreal regions. The soil texture and structure influence rates of decomposition as well as rates of physical stabilisation of soil organic matter (Feller & Beare 1997). The activities of soil macro- and mesofauna, including many smaller invertebrates such as collembola and enchytraeid worms, and of soil fungi, have significant effects on aggregation in soils that can contribute to the physical protection of organic matter from decomposition.

Managed rates of decomposition and nutrient release can be achieved through the manipulation of the quality of organic resources, both through attention to the composition of the vegetation, and thus the litter layer, and through deliberate addition of organic residues of differing quality. The diversity of functional litter types has been shown to be more important in regulating nitrogen mineralisation than species richness of litter per se (Wardle *et al.* 1997). Decision trees have been developed that can be used by farmers to recognise functional types of plant residues and to demonstrate and manage organic residues (Giller 2000; Palm *et al.* 2001). The hierarchical control of nitrogen content (leaf colour) and then lignin content (crushability when dry) and reactive polyphenol content (astringent taste) is recognised in determining potential uses of the organic resources for nutrient provision or mulch. Poor-quality residues with slow decomposition rates are suitable for mulching and have been shown to stimulate activity of soil fauna more strongly than residues that decompose rapidly and are good for nutrient supply (Tian *et al.* 1993). The diversity of functional litter types is thus important in creating diverse microhabitats in soil, and coupled with the role of soil macrofauna and mesofauna on aggregation creating a heterogeneous soil will cause localisation of hot spots of microbial activity (see Schimel this volume). Heterogeneity of microhabitats is critical for determining relative rates of emission of gaseous forms of carbon (CO_2 and CH_4) and nitrogen (NO_x, N_2O) that are involved in global warming.

The decomposition function is therefore amenable to productive manipulation for improved soil fertility through management of the organic input diversity. In such management the biotic diversity is treated as a 'black box'. Yet it is well established that in natural environments the decomposition of any unit of organic matter is brought about by a community of organisms that is disparate at the level of phyla and frequently diverse down to the species level. For example, decomposing leaves often show a complex succession of fungi, interrupted and diverted by a variety of invertebrates at various points and in association with a diverse range of bacteria (Swift 1976, 1987). None-the-less, it has proved easy to demonstrate that this degree of diversity is inessential to the decomposition process. Single species of organisms such as basidiomycete fungi can fully decompose leaves or twigs in the laboratory. In the field, experimental exclusion of whole functional groups, by the classic litter bag technique, which prevents access and activity of the soil macro- and mesofauna and thus inhibits the comminution function, rarely leads to strong retardation of decomposition (Swift *et al.* 1979). Further, the majority of soil microorganisms can be eliminated through heavy metal contamination without measurable effects on overall decomposition of organic matter (Giller *et al.* 1998).

From a management viewpoint these results mean that decomposition is less amenable to manipulation through the organisms (factor *O*) than through modification of factors *P* and *Q*. At a more basic level it is as yet not at all well established what significance, if any, such diversions, which effectively remove layers of the 'resource cascade', have on soil fertility. The definitive experiments have yet to be done to demonstrate the relationship between changes in diversity, the rate of decomposition (in terms of its constituent processes of catabolism, comminution and leaching) and, most importantly, the chemical changes that result. This unresolved issue has resulted in the hypothesis of very substantial redundancy among the decomposer organisms (Swift 1976, 1987; Giller *et al.* 1997; Setälä *et al.* this volume). This may be expected to hold for the large number of organisms with rather 'generalist' functions in decomposition. It can be hypothesised, however, that the extent of redundancy is significantly lower in some specialist (e.g. microbes adapted to high polyphenol levels) or 'keystone' (e.g. earthworms and termites) organisms which strongly determine the subsequent pattern of both process and biotic succession, and that this may influence the outcome of decomposition in a significant way. The lack of certainty in these concepts has been a product of the historical existence of two approaches in soil biology – the organismal and the functional. Only by linking the two approaches can these issues be resolved. A major factor inhibiting this has undoubtedly been the lack of methods to link identity and function. The advent of molecular methods for typing soil organisms promises potential to resolve this, but still remains to be demonstrated.

Specific nutrient cycling functions: the example of N₂-fixing bacteria

Molecular phylogeny of bacteria provides a poor basis for functional ecology: rhizobial diversity in the tropics

An effective taxonomy is often said to be the basis of understanding the ecology of organisms. The advent of molecular phylogeny of rhizobia using 16S rDNA sequences (Martínez-Romero & Caballero-Mellado 1996) thus promised opportunities for unravelling some of the complex relationships between rhizobial diversity and ecological adaptation. The number of rhizobial species has risen dramatically in the past few years, with many new species and genera being described from strains isolated from the nodules of tropical legumes. Examination of the phylogenetic tree in relation to host range for nodulation shows poor correspondence with the phylogenetic position. Indeed an alternative phylogeny based on the genes required for nodulation gives better correspondence with the taxonomy of host legumes that the strains can nodulate (Young & Haukka 1996). The burgeoning of research on molecular phylogenies has led to discoveries of widely diverse bacteria that are able to form nodules on legumes, including bacterial species that are opportunistic human pathogens (Chen et al. 2001) and other soil bacteria of the genus *Burkholderia* (Moulin et al. 2001) that belong to the β-proteobacteria, whereas all rhizobia previously described belonged to the α-proteobacteria.

Rhizobia are highly competent, heterotrophic free-living bacteria that can survive as large populations for decades in the absence of host legumes (Giller 2001); some of the bacteria are facultative denitrifiers (e.g. *Bradyrhizobium* and *Sinorhizobium*; Daniel et al. 1982) or methylotrophs (Sy et al. 2001a). Some discoveries have arisen specifically from a greater focus on rhizobia from tropical legumes. For example, species within the legume genus *Crotalaria* (and perhaps *Lotononis*; Moulin et al. 2001) appear to be specifically nodulated by strains of *Methylobacterium* (Sy et al. 2001b). Studies in soils from the Amazon region of Brazil reveal a large diversity of *Bradyrhizobium* types and presumably other rhizobial species that await full description (Moreira et al. 1993, 1998). However, analysis of the distribution of the locations where new species of N₂-fixing bacteria have been isolated tells us more of the ecology of scientists than of the ecology of the bacteria (Giller 2001). Given that the family Leguminosae contains 16 000–19 000 species, of which the vast majority are found in the tropics, it is likely that we have only just started to reveal the full diversity of bacteria with the potential for symbiotic N₂ fixation.

Despite the seminal work of Wilson (1944) who exposed the pitfalls of assuming a close host-strain specificity between legumes and rhizobia, this has remained a driving theme for research. Contrary to the received wisdom developed by intensive study of a limited range of (mainly temperate) crop legumes, promiscuity in host range appears to be the norm for tropical legumes and rhizobia (Giller 2001; Sprent 2001). As a consequence, simply sampling of rhizobia using a range of host legumes is insufficient to differentiate rhizobial types in soil and assess the genetic diversity, although this can distinguish functional groups of rhizobia in terms of ability to form nodules and fix N_2 with particular legume species of importance.

Bacterial biogeography?

It is reasonable to suggest that legumes and their compatible rhizobia have co-evolved, and thus that centres of diversity of host legumes should provide the greatest diversity of their compatible rhizobia (Lie 1981). This theory was examined in relation to the biogeography of rhizobia nodulating some fast-growing tree legumes across continents of the tropics (Bala *et al.* 2003a). Several economically important agroforestry legumes (e.g. *Leucaena leucocephala*, *Calliandra calothyrsus* and *Gliricidia sepium*), with centres of diversity in Central America, were widely introduced throughout south-east and south Asia in the sixteenth century on the Spanish galleons and later spread throughout Africa. The success of these legumes undoubtedly results from their ability to fix N_2 actively, thus raising two possible explanations. First, the success of these agroforestry legumes depended on either their promiscuous ability to nodulate with a diverse range of indigenous rhizobia such that compatible strains are encountered in virtually all soils when the legumes are introduced. Alternatively, the legumes are nodulated by a narrow range of rhizobia that are widely distributed across the tropics. Comparison of nodulation ability of these three legume trees in soils from Africa, Asia and Latin America where the legumes had not previously been grown demonstrated successful nodulation and N_2 fixation in most of the soils (Bala *et al.* 2003b). However, there was no supporting evidence for greater diversity of rhizobia in soils from the centres of diversity of these legumes, although the least diverse populations were found in a recently formed volcanic soil from a crater rim in eastern Java, Indonesia (Bala *et al.* 2003a).

Complex relationships existed between the rhizobia in terms of cross-nodulation and N_2-fixation efficiency ability (Bala & Giller 2001). Comparison with another important agroforestry tree, *Sesbania sesban*, that is indigenous to Africa, showed this species to be much more specific in its symbiosis (Bala *et al.* 2002), both in relation to the three tree legumes from Central America and a wide range of other herbaceous and tree legumes (Bala & Giller 2001). An examination of the phylogenetic spread of the strains isolated from all four legume trees showed that all of them were nodulated with rhizobia broadly distributed across the three main fast-growing rhizobial genera: *Rhizobium*, *Mesorhizobium* and *Sinorhizobium* (Bala *et al.* 2003a). Overall the strongest relationship observed was between rhizobial diversity and the exchangeable acidity (and aluminium

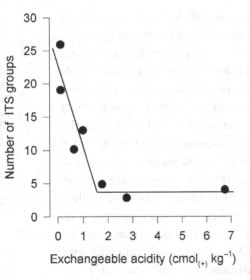

Figure 16.3. The relationship between biodiversity of indigenous populations of rhizobia (number of internally transcribed spacer (ITS) groups) from tropical legume trees and soil exchangeable acidity. From Bala *et al.* (2003a).

toxicity) in the soils (Fig. 16.3); this was not obviously related to rhizobial function in terms of either nodulation or N_2 fixation (Bala *et al.* 2003b).

These results support the suggestion of Finlay (2002) and Finlay and Clarke (1999) based on eukaryotic microorganisms, that biogeographical patterns do not occur for microorganisms as their huge abundance directly results in global dispersal. This intriguing theory indicates that the global diversity of soil microorganisms may be much smaller than that supposed by extrapolation from studies of individual soils, and suggests that local diversity will always be sufficient to drive ecosystem functions.

Potential for rhizobial autecology?
Several early results from molecular characterisation of rhizobia suggested real breakthroughs in ecological understanding. Of particular interest is the grain legume *Phaseolus vulgaris* that originated in the New World but is now widely distributed throughout the globe and is an important protein source for the poor. This legume is renowned in the scientific literature as a poor fixer (although it fixes atmospheric N_2 as well as any legume under controlled conditions). Part of the reasons for poor N_2 fixation in the field appeared to be due to nodulation by relatively ineffective strains. Strains from *Phaseolus* nodules (originally collectively termed *Rhizobium phaseoli*) were also notoriously unstable and tended to change in nodulation and N_2-fixation efficiency after continuous subculture, as well as differing enormously in host range and in acid tolerance.

The discovery that nodulation ability could be transferred among strains of *Phaseolus* rhizobia (*Rhizobium phaseoli*), pea rhizobia (*R. leguminosarum*) and clover rhizobia (*R. trifolii*) led to the description of one species *R. leguminosarum* with three 'biovars' depending on which symbiotic plasmid they carried (Jordan 1984).

Detailed investigation of strains from Latin American soils led to the description of *Rhizobium tropici* and later *Rhizobium etli* (Martínez-Romero *et al.* 1991; Segovia *et al.* 1994). *Rhizobium tropici* differed in its phenotypic properties, having a broad host range, stable in laboratory culture, as well as demonstrating remarkable acid tolerance (Graham *et al.* 1994). By contrast, the specific strains of *Phaseolus* rhizobia were sensitive to acid soil conditions, and multiple re-arrangements of the symbiotic plasmid in strains of these species provided a clue to the instability of strains (Soberon-Chavez *et al.* 1986; Romero *et al.* 1988). At last a vision of predictive molecular ecology appeared to be emerging! Studies of rhizobial populations from Kenyan soils supported this conclusion, with an acid aluminium-saturated tea soil dominated by strains of *Rhizobium tropici*, while a soil with near-neutral pH was dominated by *R. etli* type strains (Anyango *et al.* 1995). *Rhizobium tropici* was also better able to compete for nodulation (assessed using β-glucuronidase (*GUS*)-gene fusion marked strains) in acid soil conditions whereas *R. leguminosarum* competed better in alkaline soil (Anyango *et al.* 1998).

Unfortunately, these promising trends that indicated that adaptive or functional traits were coupled with particular species of rhizobia have not been supported by further research. First, further studies showed that the host range of *R. etli* strains was not so specific, as some strains could nodulate legumes such as *Leucaena* thought to be useful in differentiating *R. etli* from *R. tropici* (Hernandez-Lucas *et al.* 1995). Further, a study of rhizobial populations in Brazil indicated that *Rhizobium leguminosarum* strains dominated in the most acid soils with highest aluminium saturation, whereas the abundance of *R. tropici* strains increased with decreasing aluminium stress – the opposite from what was expected from the earlier studies (Andrade *et al.* 2002a). Indeed it has since been shown that *Phaseolus vulgaris* is a relatively permissive host that is nodulated by at least 16 species of which only 6 have been fully described (Giller 2001).

Relationships between bacterial biodiversity and function

Although phylogenetic classification is an excellent tool for determining evolutionary relationships among bacteria, there is little correspondence between phylogeny and function in terms of both ability to nodulate (infectiveness) and ability to fix N_2 (effectiveness) with legume hosts. Nor are there clear relationships between phylogenetic groupings and adaptation to the environment, both in the free-living state and in terms of adaptation to the soil environment influencing ability to form nodules (as adaptation is a major factor in competitiveness). This is not to say that the bacterial biomass is simply comprised of a 'gene soup', as barriers to gene transfer between bacterial groups do exist (Young 1994). But the available evidence clearly demonstrates that phylogeny is likely to be a weak predictor of adaptation and function. Biodiversity per se is unlikely to have clear relationships with function but could well be of major importance in functional *stability* in terms of resilience and resistance to change in the face of further perturbations.

Discussion

Effects of human intervention: the potential for management of below-ground biodiversity

A major effect of human activity is the increased predisposition for invasive species to spread, resulting in strongly negative effects on soil functioning. A survey of the distribution of the earthworm communities in eight different land uses on the Caribbean coast of Costa Rica revealed that the invasive pantropical species *Pontoscolex corethrurus* was dominant in all land uses except in the large primary forest (Lapied & Lavelle 2003). The density of *P. corethrurus* was greatest in banana plantations (361 individuals m^{-2}), but when it reached a density above 70 individuals m^{-2} a threshold for the earthworm community was reached whereby the rest of the earthworm species disappeared almost completely. This invasive species is causing a great loss of the earthworm biodiversity in the tropics as a result of human intervention (Fragoso *et al.* 1999) with associated problems due to the formation of crusts mentioned above.

The clearest opportunities that exist for management of below-ground biodiversity are through the manipulation of organic inputs, or through manipulation of the soil physico-chemical environment as indicated above. Studies of macroinvertebrate communities of the well-drained savannas (Decaëns *et al.* 2001b) and Andean hillsides (Feijoo *et al.* 2001) of Colombia were shown to be very sensitive to environmental changes associated with agricultural intensification. In extensively grazed native pastures, earthworms are favoured by grazing, but traditional management by burning has the opposite effect on termites. This suggests that the earthworm/termite ratio may be a sensitive indicator of soil health. Introduced forage grasses and legumes and increasing animal production have a very important impact on soil macrofauna, especially in the earthworm population, which showed a ten-fold increase compared to the earthworm population of native pastures. Annual crops showed a dramatic impact on earthworms and arthropod populations, with marked decreases in biomass, population density and taxonomic richness. Mulching can restore termites to degraded, semi-arid landscapes, with a resulting restoration of good water infiltration and increased soil stability (Mando 1997; Holt & Lepage 2000).

A study of rhizobial numbers in soil along an acidity gradient clearly demonstrates the interactive effects of soil conditions and plant species (Fig. 16.4). Numbers of rhizobia compatible with *Phaseolus vulgaris* decreased drastically with increasing soil acidity unless the compatible host was present (Fig. 16.4a), although rhizobial numbers decreased in the most acidic treatment if the soils were incubated in the absence of the host plant for 12 months (Fig. 16.4b). The diversity of rhizobia also varied along this gradient, with changes in abundance

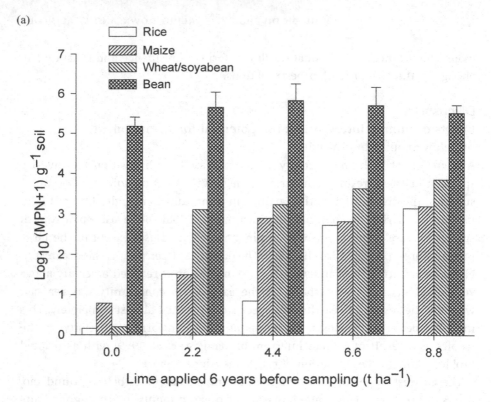

(a)

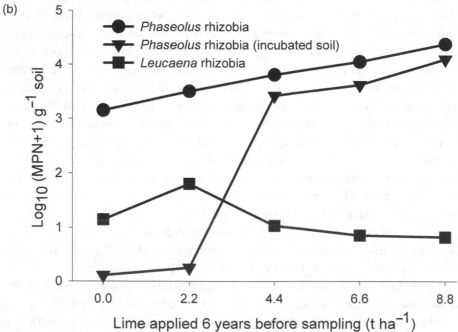

(b)

Figure 16.4. (a) Populations of *Phaseolus* rhizobia in Brazilian soils limed six years previously in replicated experiments cropped continuously with *Phaseolus vulgaris*, soyabean/wheat intercrop, rice or maize. (b) Populations of *Phaseolus* rhizobia after incubation of the soil in a moist condition for 12 months. Redrawn from Andrade *et al.* (2002b).

of different species of rhizobia (Andrade *et al.* 2002a). This example demonstrates the potential for intervention to manage the abundance and diversity of rhizobia through the interacting effects of cropping system design and amendments such as liming.

The effects of short-term perturbations or 'disturbance' caused by management such as change in above-ground vegetation composition or tillage are distinct from effects of long-term chronic toxicities or 'stress' such as soil acidity, aluminium or heavy metal toxicity (see Giller *et al.* 1998; Wardle 2002). This is of particular importance in relation to management of biodiversity as effects of disturbances are much more likely to be reversible than effects of stress. The distinction between such effects can sometimes become blurred as repeated or persistent disturbance can have similar effects to those of chronic stress.

Issues of scale

Ecosystem processes and functions vary with scale, the most obvious examples being those influenced by water movement such as runoff and erosion. Measurements of erosion on small plots that are strongly influenced by litter layers and associated macrofaunal activity can indicate large losses of soils and nutrients. At larger scales, contour hedgerows and field boundaries can act as filters capturing nutrients and causing deposition so that losses at the landscape scale may be negligible.

Scales relevant for soil functioning start from micrometres (less than 100 μm) where consortia of microorganisms are formed, for example in a microaggregate of mineral and organic material, or in a micropore or as a biofilm on a particle of substratum. Such consortia may mediate multiple processes such as carbon and nitrogen transformations. At the next scale upwards, the micropredators move between the consortia of microorganisms regulating their numbers and growth, with varying degrees of feeding specialisation (thus creating soil food webs). Above this, a group of larger organisms, mainly plants (through their roots) and invertebrate animals, are responsible either for maintaining the physical heterogeneity of the substratum, especially mediating the mixing and re-mixing of organic and mineral material in the role of ecosystem engineers, or for inputs of organic detritus in the role of litter transformers. The biogenic structures created by the ecosystem engineers can vary in size from microaggregates to large hills (greater than 30 m radius) in the case of termitaria. This means that spatial heterogeneity can be complex even when considering single vegetation units, land management units or ecosystems, which are normally studied at the 'plot level'. At higher scales, mosaics of such units are found within the landscape that contribute to diversity at higher scales, and results in emergent properties as small areas of particular vegetation types can act as refugia, filters or corridors.

Soils have low openness (also called connectivity) when compared with aquatic systems. This concept relates to the potential of an event in one unit of the ecosystem to influence other units in the ecosystem, or other ecosystems in the same domain, rapidly. The concept also includes rates of exchange of propagules and materials. Rapidly, means days, weeks and months, rather than years, decades or longer periods. Thus, the freshwater and marine domains, which are constituted in a fluid medium and where there are continuous net displacements of materials and medium from one place to another, are open in the sense of the passive or active mobility of their component organisms, whereas soil is not open in the same way. Despite this, soil organisms do have a finite mobility (i.e. local connectivity) on perhaps the metre scale for the larger members of the biota. Relative mobility of organisms has large implications for invasion and for replacement. The relative mobility of materials has large implications for the propagation of perturbations and stresses away from their sites of occurrence. In the sense that the marine system has less local connectivity, it has greater inherent recoverability. By contrast, most connections in the soil system are local (although a degree of connection at greater spatial scales is possible via the atmosphere, vagile animals and human interventions), and there is therefore less inherent recoverability. The system as a whole should have the ability to survive the temporary loss or suppression of any one of its component organisms, but the destruction of biogenic structure, for example by compaction or the removal of topsoil, can be ameliorated only in the medium to long term by replacing the hierarchy of organisms and processes. The challenge of scale is to understand the dimensions of the stage on which each member of the biota makes its unique contribution – without this understanding they cannot be managed to any useful purpose.

Simple linear decreases of diversity with increasing intensification should not be expected, as already indicated above for the ecosystem engineers. Nor is extrapolation from biodiversity of soil organisms in small plots within vegetation types a simple matter. Although α-diversity (within patch) of oribatid mites 1 m^{-2} sample was greatest in a central African rainforest, overall γ-diversity was greatest in savanna, due to heterogeneity of habitats within the savanna leading to very high β-diversity (turn-over between patches; Noti *et al.* 2003). A working hypothesis is that the microorganisms, particularly the smaller fungi and bacteria, are largely scale independent as dispersal and colonisation can be very rapid, often mediated through the air or through water flows. By contrast, some groups of soil fauna are strongly affected by disturbance and may be poor (re)colonisers so that they may be highly dependent on both the scale of sampling and the size of land use units. The life histories and mechanisms of reproduction and dispersal of groups of organisms are obviously of key interest in relation to scale. One might suppose that distances travelled by earthworms may be small relative to termites, whose colonies may cover huge areas and whose reproductives are winged, but earthworms are more aggressive biological invaders of natural and semi-natural habitats (i.e. excluding panglobal termite pests of buildings and other structural timber). In fact, termites have very high endemicity (Eggleton 2000). These relationships and anomalies are of particular interest both in relation to sampling units that may need to differ widely between different groups of organisms, and in relation to the population ecology and conservation of different species groups. Thus the question is raised as to whether having large homogeneous units of particular land use types segregated within the landscape, or having small diverse units integrated within land use units giving rise to meta-populations, is better overall.

To be able to answer that question, a descriptor of land use at the landscape level is needed. The descriptor would need to reflect heterogeneity of the land cover, intensity and fragmentation of the land use, and connectivity and spatial arrangement of the land use and land cover components of the landscape. Application of landscape metrics, which can be related to the dispersal mechanisms and distribution of the soil organisms, to derive an environmental index or indicator of landscape condition (Lausch & Herzog 2002) requires investigation with respect to below-ground biodiversity.

In our attempts to unravel the relationships between above- and below-ground biodiversity and ecosystem functions along gradients of intensification, a full site characterisation will be essential as the gradients may be confounded with other environmental factors. Furthermore, studies to date suggest that simple linear trends between changes in diversity and intensification or land use change may be rare, or else confined to particular faunal indicators, such as termites, in which a high taxonomic resolution is possible and where more than one functional group can be readily recognised. Links between biodiversity and function may differ substantially between different groups of organisms. It is clear that we may have to re-think the basis of ecological theories on relating biodiversity to ecosystem function when it comes to prokaryotes due to the vast opportunities for genetic recombination and horizontal transfer of adaptive and useful functions.

Reflecting on what we know of soil biodiversity in the tropics, we concur with Noti et al. (2003) that our knowledge of the fauna, and indeed of all groups of soil organisms, is 'miserable'! To express this in another way, we cannot even assess the extent of our ignorance of soil biodiversity, which emphasises the need for a major campaign of inventory of soil organisms of all groups. This is a principal goal of a new pantropical project funded by the Global Environment Facility (GEF) entitled 'Conservation and Sustainable Management of Below-ground Biodiversity' through the United Nations Development Programme (UNDP) under the Convention on Biological Diversity (CBD). The project is providing an excellent opportunity for exploring the impacts of land use change on soil biodiversity and soil function across gradients of agricultural intensification at the margins of global biosphere forest reserves in Mexico, Brazil, Côte d'Ivoire, Uganda, Kenya, India and Indonesia.

Science in this area has to date been driven largely by scaling down in terms of pursuing detailed research on individual organisms or processes. We have only just begun to explore the full biodiversity of tropical soils, or the scaling up of such knowledge from scales beyond the soil column or plot, to scales of the broader ecosystem and landscape.

Acknowledgements

We are grateful for the UNEP/GEF grants that have facilitated the close collaboration of scientists from many countries in the development of these ideas. We thank Thom Kuyper, two anonymous referees and the editors for their critical comments which helped us to improve the focus of this chapter.

References

Andrade, D. S., Murphy, P. J. & Giller, K. E. (2002a). The diversity of *Phaseolus*-nodulating rhizobial populations is altered by liming of acid soils planted with *Phaseolus vulgaris* L. in Brazil. *Applied and Environmental Microbiology*, **68**, 4025–4034.

Andrade, D. S., Murphy, P. J. & Giller, K. E. (2002b). Effects of liming and legume/cereal cropping on populations of indigenous rhizobia in an acid Brazilian oxisol. *Soil Biology and Biochemistry*, **34**, 477–485.

Anyango, B., Wilson, K. J., Beynon, J. L. & Giller, K. E. (1995). Diversity of rhizobia nodulating *Phaseolus vulgaris* L. in two Kenyan soils of contrasting pHs. *Applied and Environmental Microbiology*, **61**, 4016–4021.

Anyango, B., Wilson, K. & Giller, K. E. (1998). Competition in Kenyan soils between *Rhizobium leguminosarum* bv. *phaseoli* strain Kim5 and *R. tropici* strain CIAT899 using the gusA marker gene. *Plant and Soil*, **204**, 69–78.

Bala, A. & Giller, K. E. (2001). Symbiotic specificity of tropical tree rhizobia for host legumes. *New Phytologist*, **149**, 495–507.

Bala, A., Murphy, P. & Giller, K. E. (2002). Occurrence and genetic diversity of rhizobia nodulating *Sesbania sesban* in African soils. *Soil Biology and Biochemistry*, **34**, 1759–1768.

Bala, A., Murphy, P. J. & Giller, K. E. (2003a). Distribution and diversity of rhizobia nodulating agroforestry legumes in soils from three continents in the tropics. *Molecular Ecology*, **12**, 917–929.

Bala, A., Murphy, P. J., Osunde, A. O. & Giller, K. E. (2003b). Nodulation of tree legumes and ecology of their native rhizobial populations in tropical soils. *Applied Soil Ecology*, **22**, 211–223.

Barros, E., Curmi, P., Hallaire, V., Chauvel, A. & Lavelle, P. (2001). The role of macrofauna in the transformation and reversibility of soil structure of an oxisol in the process of forest to pasture conversion. *Geoderma*, **100**, 193–213.

Barrios, E., Kwesiga, F., Buresh, R. J., Sprent, J. I. & Coe, R. (1998). Relating preseason soil nitrogen to maize yield in tree legume–maize rotations. *Soil Science Society of America Journal*, **62**, 1604–1609.

Bignell, D. E., Tondoh, J., Dibog, L., *et al.* (2005). Below-ground biodiversity assessment: the ASB rapid, functional group approach. *Alternatives to Slash-and-Burn: A Global Synthesis* (Ed. by P. J. Ericksen, P. A. Sanchez & A. Juo), Special Publication. Madison, WI: American Society for Agronomy.

Cadisch, G. & Giller, K. E. (eds.) (1997). *Driven by Nature: Plant Residue Quality and Decomposition*. Wallingford: CAB International.

Chauvel, A., Grimaldi, M., Barros, E., *et al.* (1999). Pasture degradation by an Amazonian earthworm. *Nature*, **389**, 32–33.

Chen, W.-H., Laevens, S., Lee, T.-M., *et al.* (2001). *Ralstonia taiwanensis* sp. nov., isolated from root nodules of *Mimosa* species and sputum of a cystic fibrosis patient. *International Journal of*

Systematic and Evolutionary Microbiology, **51**, 1729–1735.

Chikowo, R., Mapfumo, P., Nyamugafata, P. & Giller, K. E. (2004). Maize productivity and mineral N dynamics following different soil fertility management practices on a depleted sandy soil in Zimbabwe. *Agriculture, Ecosystems and Environment*, **102**, 109–131.

Coventry, R. J., Holt, J. A. & Sinclair, D. F. (1988). Nutrient cycling by mound-building termites in low-fertility soils of semi-arid tropical Australia. *Australian Journal of Soil Research*, **26**, 375–390.

Daniel, R. M., Limmer, A. W., Steele, K. W. & Smith, I. M. (1982). Anaerobic growth, nitrate reduction and denitrification in 46 rhizobial strains. *Journal of General Microbiology*, **128**, 1811–1815.

Decaëns, T., Galvis, J. H. & Amezquita, E. (2001a). Properties of the structures created by ecological engineers at the soil surface of a Colombian savanna. *Comptes Rendus de L'Academie Des Sciences Serie IIII Sciences De La Vie Life Sciences*, **324**, 465–478.

Decaëns, T., Lavelle, P., Jiménez, J. J., *et al.* (2001b). Impact of land management on soil macrofauna in the eastern plains of Colombia. *Nature's Plow: Soil Macroinvertebrate Communities in the Neotropical Savannas of Colombia* (Ed. by J. J. Jiménez & R. J. Thomas), pp. 19–41. Colombia: CIAT.

Dobson, A. P., Bradshaw, A. D. & Baker, A. J. M. (1997). Hopes for the future: restoration ecology and conservation biology. *Science*, **277**, 515–521.

Eggleton, P. (2000). Global patterns of termite diversity. *Termites: Evolution, Sociality, Symbioses, Ecology* (Ed. by T. Abe, D. E. Bignell & M. Higashi), pp. 25–51. Dordrecht: Kluwer Academic.

Eggleton, P., Bignell, D. E., Hauser, S., *et al.* (2002). Termite diversity across an anthropogenic disturbance gradient in the humid forest zone of West Africa. *Agriculture, Ecosystems and Environment*, **90**, 189–202.

Feijoo, A., Knapp, E. B., Lavelle, P. & Moreno, A. G. (2001). Quantifying soil macrofauna in a Colombian watershed. *Nature's Plow: Soil Macroinvertebrate Communities in the Neotropical Savannas of Colombia* (Ed. by J. J. Jiménez & R. J. Thomas), pp. 42–48. Colombia: CIAT.

Feller, C. & Beare, M. H. (1997). Physical control of soil organic matter dynamics in the tropics. *Geoderma*, **79**, 69–116.

Finlay, J. F. (2002). Global dispersal of free-living microbial eukaryote species. *Science*, **296**, 1061–1063.

Finlay, J. F. & Clarke, K. J. (1999). Ubiquitous dispersal of microbial species. *Nature*, **400**, 828.

Fragoso, C., Lavelle, P., Blanchart, E., *et al.* (1999). Earthworm communities of tropical agroecosystems: origin, structure and influence of management practices. *Earthworm Management in Tropical Agroecosystems* (Ed. by P. Lavelle, L. Brussaard & P. Hendrix), pp. 27–55. Wallingford: CAB International.

Geist, H. J. & Lambin, E. F. (2002). Proximate causes and underlying driving forces of tropical deforestation. *BioScience*, **52**, 143–150.

Giller, K. E. (2000). Translating science into action for agricultural development in the tropics: an example from decomposition studies. *Applied Soil Ecology*, **14**, 1–3.

Giller, K. E. (2001). *Nitrogen Fixation in Tropical Cropping Systems*, 2nd edition. Wallingford: CAB International.

Giller, K. E., Beare, M. H., Lavelle, P., Izac, A.-M. N. & Swift, M. J. (1997). Agricultural intensification, soil biodiversity and ecosystem function. *Applied Soil Ecology*, **6**, 3–16.

Giller, K. E., Witter, E. & McGrath, S. P. (1998). Toxicity of heavy metals to microorganisms and microbial processes in agricultural soils: a review. *Soil Biology and Biochemistry*, **30**, 1389–1414.

Gillison, A. N., Jones, D. T., Susilo, F.-X. & Bignell, D. E. (2003). Vegetation indicates diversity of soil macroinvertebrates: a case study with termites along a land-use intensification gradient in lowland Sumatra. *Organisms, Diversity and Evolution,* **3**, 111–126.

Graham, P. H., Draeger, K. J., Ferrey, M. L., *et al.* (1994). Acid pH tolerance in strains of *Rhizobium* and *Bradyrhizobium,* and initial studies on the basis for acid tolerance of *Rhizobium tropici* UMR1899. *Canadian Journal of Microbiology,* **40**, 198–207.

Heal, O. W., Anderson, J. W. & Swift, M. J. (1997). Plant litter quality and decomposition: an historical overview. *Driven by Nature: Plant Litter Quality and Decomposition* (Ed. by G. Cadisch & K. E. Giller), pp. 3–30. Wallingford: CAB International.

Hernandez-Lucas, I., Segovia, L., Martínez-Romero, E. & Pueppke, S. (1995). Phylogenetic relationships and host range of *Rhizobium* spp. that nodulate *Phaseolus vulgaris* L. *Applied and Environmental Microbiology,* **61**, 2775–2779.

Holt, J. A. & Lepage, M. (2000). Termites and soil properties. *Termites: Evolution, Sociality, Symbioses, Ecology* (Ed. by T. Abe, D. E. Bignell & M. Higashi), pp. 389–407. Dordrecht: Kluwer Academic.

Jones, D. T., Susilo, F.-X., Bignell, D. E., *et al.* (2003). Termite assemblage collapses along a land-use intensification gradient in lowland central Sumatra, Indonesia. *Journal of Applied Ecology,* **40**, 380–391.

Jordan, D. C. (1984). Rhizobiaceae. *Bergey's Manual of Systematic Bacteriology* (Ed. by N. R. Krieg & J. G. Holt), Vol. 1, pp. 235–244. Baltimore, MD: Williams and Wilkins.

Kandji, S. T., Ogol, C. K. P. O. & Albrecht, A. (2001). Diversity of plant parasitic nematodes and their relationships with some soil physico-chemical characteristics in improved fallows in western Kenya. *Applied Soil Ecology,* **18**, 143–157.

Lambin, E. F., Turner, B. L., Geist, H. J., *et al.* (2001). The causes of land-use and land-cover change: moving beyond the myths. *Global Environmental Change, Human and Policy Dimensions,* **11**, 261–269.

Lapied, E. & Lavelle, P. (2003). The peregrine earthworm *Pontoscolex corethrurus* in the East Coast of Costa Rica. *Pedobiologia,* **47**, 471–474.

Lausch, A. & Herzog, F. (2002). Applicability of landscape metrics for the monitoring of landscape change: issues of scale, resolution and interpretability. *Ecological Indicators,* **2**, 3–15.

Lavelle, P. (1997). Faunal strategies and soil processes: adaptive strategies that determine ecosystem function. *Advances in Ecological Research,* **27**, 93–132.

Lavelle, P., Bignell, D., Lepage, M., *et al.* (1997). Soil function in a changing world: the role of invertebrate ecosystem engineers. *European Journal of Soil Biology,* **33**, 159–193.

Lie, T. A. (1981). Gene centres: a source for genetic variants in symbiotic nitrogen fixation – host-induced ineffectivity in *Pisum sativum* ecotype Fulvum. *Plant and Soil,* **61**, 125–134.

Mando, A. (1997). Effect of termites and mulch on the physical rehabilitation of structurally crusted soils in the Sahel. *Land Degradation and Development,* **8**, 269–278.

Martínez-Romero, E. & Caballero-Mellado, J. (1996). *Rhizobium* phylogenies and bacterial genetic diversity. *Critical Reviews in Plant Sciences,* **15**, 113–140.

Martínez-Romero, E., Segovia, L., Mercante, F. M., *et al.* (1991). *Rhizobium tropici*: a novel species nodulating *Phaseolus vulgaris* L. beans and *Leucaena* sp. trees. *International Journal of Systematic Bacteriology,* **41**, 417–426.

Matson, P. A., Parton, W. J., Power, A. G. & Swift, M. J. (1997). Agricultural intensification and ecosystem properties. *Science,* **277**, 504–509.

Moran, E. (1993). Deforestation and land use in the Brazilian Amazon. *Human Ecology,* **21**, 1–21.

Moreira, F. M. S., Gillis, M., Pot, B., Kersters, K. & Franco, A. A. (1993). Characterisation of rhizobia isolated from different divergence groups of tropical Leguminosae by comparative polyacrylamide gel electrophoresis of their total proteins. *Systematic and Applied Microbiology*, **16**, 135–146.

Moreira, F. M. S., Haukka, K. & Young, J. P. W. (1998). Biodiversity of rhizobia isolated from a wide range of forest legumes in Brazil. *Molecular Ecology*, **7**, 889–895.

Moulin, L., Munive, A., Dreyfus, B. & Boivin-Masson, C. (2001). Nodulation of legumes by members of the beta-subclass of Proteobacteria. *Nature*, **411**, 948–950.

Noble, I. R. & Dirzo, R. (1997). Forests as human-dominated ecosystems. *Science*, **277**, 522–525.

Noti, M.-I., Andre, H. M., Ducarme, X. & Lebrun, P. (2003). Diversity of soil oribatid mites (Acari: Oribatida) from High Katanga (Democratic Republic of Congo): a multiscale and multifactor approach. *Biodiversity and Conservation*, **12**, 767–785.

Palm, C. A., Gachengo, C. N., Delve, R. J., Cadisch, G. & Giller, K. E. (2001). Organic inputs for soil fertility management in tropical agroecosystems: application of an organic resource database. *Agriculture, Ecosystems and Environment*, **83**, 27–42.

Romero, D., Singleton, P. W., Segovia, L., *et al.* (1988). Effect of naturally occurring *nif* reiterations on symbiotic effectiveness in *Rhizobium phaseoli*. *Applied and Environmental Microbiology*, **54**, 848–850.

Ruthenberg, H. (1980). *Farming Systems in the Tropics*, 3rd edition. Oxford: Clarendon Press.

Segovia, L., Young, J. P. W. & Martínez-Romero, E. (1994). Reclassification of American *Rhizobium leguminosarum* biovar phaseoli Type I strains as *Rhizobium etli* sp. nov. *International Journal of Systematic Bacteriology*, **43**, 374–377.

Soberon-Chavez, G., Najera, R., Olivera, H. & Segovia, L. (1986). Genetic rearrangements of a *Rhizobium phaseoli* symbiotic plasmid. *Journal of Bacteriology*, **167**, 487–491.

Sprent, J. I. (2001). *Nodulation in Legumes*. Kew: Royal Botanic Gardens.

Susilo, F.-X., Neutel, A. M., van Noordwijk, M., *et al.* (2004). Soil biodiversity and food web synthesis. *Belowground Interactions in Tropical Agroecosystems* (Ed. by M. van Noordwijk, G. Cadisch & C. K. Ong), pp. 285–307. Wallingford: CAB International.

Swift, M. J. (1976). Species diversity and the structure of microbial communities in terrestrial habitats. *The Role of Aquatic and Terrestrial Organisms in Decomposition Processes* (Ed. by J. M. Anderson & A. MacFadyen), pp. 185–221. Oxford: Blackwell Scientific.

Swift, M. J. (1987). Organisation of assemblages of decomposer fungi in space and time. *Organisation of Communities Past and Present* (Ed. by P. Giller & J. Gee), pp. 229–253. Oxford: Blackwell Scientific.

Swift, M. J. (1997). Agricultural intensification, soil biodiversity and ecosystem function. *Applied Soil Ecology*, **6**, 1–2.

Swift, M. J. (1998). Towards the second paradigm: integrated biological management of soil. *Soil Fertility, Soil Biology and Plant Nutrition Interrelationships* (Ed. by J. O. Siqueira, F. M. S. Moreira, A. S. Lopes, *et al.*), pp. 11–24. Lavras: SBCS/UFLA/DCS.

Swift, M. J., Heal, O. W. & Anderson, J. M. (1979). *Decomposition in Terrestrial Ecosystems*. Oxford: Blackwell Scientific.

Sy, A., Giraud, E., Jourand, P., *et al.* (2001a). Methylotrophic *Methylobaterium* bacteria that nodulate and fix nitrogen in symbiosis with legumes. *Journal of Bacteriology*, **183**, 214–220.

Sy, A., Giraud, E., Samba, R., *et al.* (2001b). Certaines légumineuses du genre *Crotalaria* sont spécifiquement nodulées par une nouvelle espèce de *Methylobaterium*. *Canadian Journal of Microbiology*, **47**, 503–508.

Tian, G., Brussaard, L. & Kang, B. T. (1993). Biological effects of plant residues with contrasting chemical compositions under humid tropical conditions: effect on soil fauna. *Soil Biology and Biochemistry*, **25**, 731–737.

Usher, M. B. (1985). Population and community dynamics in the soil ecosystem. *Ecological Interactions in Soil: Plants, Microbes and Animals* (Ed. by A. H. Fitter, D. Atkinson, D. J. Read & M. B. Usher), pp. 243–265. Oxford: Blackwell Scientific.

Wardle, D. A. (2002). *Communities and Ecosystems: Linking the Aboveground and Belowground Components.* Princeton, NJ/Oxford: Princeton University Press.

Wardle, D. A., Bonner, K. I. & Nicholson, K. S. (1997). Biodiversity and plant litter: experimental evidence which does not support the view that enhanced species richness improves ecosystem function. *Oikos,* **79,** 247–258.

Wilson, J. K. (1944). Over five hundred reasons for abandoning the cross-inoculation groups of the legumes. *Soil Science,* **58,** 61–69.

Wood, T. G. (1996). The agricultural importance of termites in the tropics. *Agricultural Zoology Reviews,* **7,** 117–155.

Yeates, G. W. (1981). Soil nematode population depressed in the presence of earthworms. *Pedobiologia,* **22,** 191–195.

Young, J. P. W. (1994). Sex and the single cell: the population ecology and genetics of microbes. *Beyond the Biomass: Compositional and Functional Analysis of Soil Microbial Communities* (Ed. by K. Ritz, J. Dighton & K. E. Giller), pp. 101–107. Chichester: Wiley.

Young, J. P. W. & Haukka, K. E. (1996). Diversity and phylogeny of rhizobia. *New Phytologist,* **133,** 87–94.

CHAPTER SEVENTEEN

Restoration ecology and the role of soil biodiversity

J. A. HARRIS
Cranfield University
P. GROGAN
Queen's University Kingston
R. J. HOBBS
Murdoch University

SUMMARY

1. Restoration ecology is the ultimate test of our understanding of the way ecosystems work. The question is asked: can we assemble the components to achieve an integrated ecosystem that functions sustainably?
2. The terms commonly used in the field of ecological restoration are defined.
3. The relationship of ecological restoration to successional processes is outlined.
4. The types of below-ground measurements commonly made in reclamation and reclamation programmes are illustrated, and the importance of soil biota in different types of land degradation outlined.
5. The potential mechanisms of plant–soil biota–ecosystem interrelationships drawn from the wider literature on facilitation and inhibition are discussed, and their potential implications for restoration explored.
6. Finally, the principal gaps in our knowledge of the relationship between soil biodiversity and restoration ecology are identified.

Introduction

Challenges to the way in which the Earth system deals with anthropogenic pressures are increasing annually. These may be short term and dramatic, such as peat bog fires in Malaysia, or long term and persistent, such as habitat and diversity loss. Much attention and effort has been directed at protecting and enhancing existing 'pristine' systems, but there is a growing realisation that this is not enough (Sutherland 2002). The International Union for the Conservation of Nature (IUCN) has recognised that ecological restoration of systems has a major role to play in re-establishing both structure and function, and there are frequent suggestions that ecological restoration is the critical process (Hobbs &

Biological Diversity and Function in Soils, eds. Richard D. Bardgett, Michael B. Usher and David W. Hopkins. Published by Cambridge University Press. © British Ecological Society 2005.

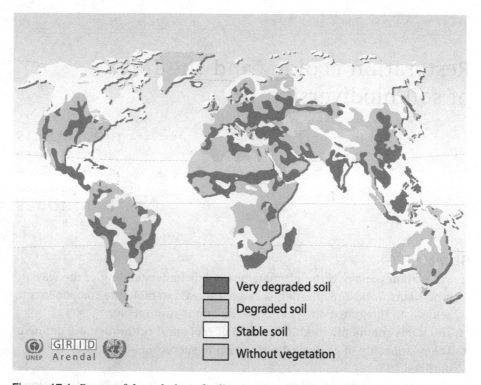

Very degraded soil

Degraded soil

Stable soil

Without vegetation

Figure 17.1. Extent of degradation of soil resources globally (see http://www.grida.no/db/ maps/prod/level3/id_1238.htm). Copyright © UNEP GRID-Arendal.

Harris 2001). One of the challenges in getting effective action is setting the threats and opportunities in a context that decision makers can act upon.

One increasingly common way to set ecosystem function in socio-economic terms is by use of the concept of 'ecosystem services'. These services are regarded as having four components or aspects (de Groot *et al.* 2002). There are regulation functions, providing maintenance of essential ecological processes and life support systems. Second, there are habitat functions, providing suitable living space for 'wild' plant and animal species. Third, there are production functions, providing natural resources from which to make goods (both consumable and structural). Finally, there are information functions, providing opportunities for cognitive development. An idea of how much of each of these functions is being lost, and what might need to be done to re-dress the losses, is illustrated by the extent of soil degradation globally (Fig. 17.1).

This degradation has come about by a variety of pressures (Table 17.1). The uses each come with their own set of challenges to the successful restoration of ecosystems, in terms of both their structure and function. However, in all cases it is important first to identify the opportunities for understanding the importance of soil biodiversity in the degraded land context, which will have wider implications for understanding ecosystem function: second, there may

Table 17.1. *Types of degradative land use. Redrawn from Harris* et al. *(1996).*

Type of use	Examples
Temporary on site	Strip or opencast mining
	Quarrying
	Civil engineering temporary storage and trafficking areas
Permanent on site	Landfill
	Agriculture and forestry
	Built structures
	Structured amenity and recreation
	Uncontrolled recreation
	Hunting
Pressures generated off site	Water and airborne pollution
	Changes in hydrology

also be opportunities for testing established and emerging theory. Bradshaw (1997) clearly identified the role of natural processes in restoring mined lands. However, we need to examine whether manipulation of such land offers us the opportunity to enhance function and structure over shorter time scales than successional ones.

Many of the features of degraded and restored sites offer us the potential for investigating seven ecological concepts: resilience, resistance, patch dynamics, successional theory, life-style strategy theory, symbioses, and the relationship between diversity and function. What are the unique features of degraded sites that allow us to carry out these investigations? Recall, however, one of the difficulties in identifying progress in understanding the role of soil biodiversity in restored systems is the great confusion in the use of terminology. Therefore, before examining the effects of land degradation and restoration, it is necessary to come to some clear definitions of the terminology used.

The Society for Ecological Restoration's (2002) definition of ecological restoration is 'the process of assisting the recovery of an ecosystem that has been degraded, damaged, or destroyed'. This suggests that the goal of ecological restoration is to return ecosystems to a pre-existing condition, although there is much latitude in determining what that state may have been or, more crucially, what the state would be now if no degradation had occurred. This gives us something to work with in terms of auditing what we need to be putting back in place. Ecological restoration is the ultimate test for ecologists: can we repair a system through knowledge of its component parts and how they interact? There are several definitions of management types often associated with the overall goal of fully restoring ecosystems (Aronson *et al.* 1993; adapted by Walker & del Moral 2003). 'Reclamation' relates to actions that stabilise a landscape and increase its utility or economic value. This usually involves amelioration that permits vegetation to establish and become self-sustaining, and rarely uses historical or indigenous ecosystems as a model or target. 'Reallocation' is the management or development actions that deflect succession of a site from one land use to another, with the goal of increased functionality. 'Rehabilitation' consists of actions that seek to repair damaged ecosystem functions rapidly, particularly productivity. 'Restoration *sensu stricto*' aims at carrying out actions that lead to a full recovery of an ecosystem to its pre-disturbance structure and function. Finally, 'restoration *senso lato*' is a series of actions that seek to reverse degradation and to direct the trajectory in the general direction of one aspect of an ecosystem (especially function) that previously existed on the site.

These definitions will have little utility in establishing sustainable and stable end-points if the impact of changing conditions is not taken into account, e.g. extinction of species, making the restoration of a past ecosystem in the strict sense impossible; or climate change, where restoration of a historic ecosystem composition is inherently unstable due to the limits of adaptation of the biotic components.

Restoration ecology is the study and investigation of the ecological restoration process. Linked closely to this definition is the concept that we need targets, goals and conditions to 'aim' at if we are to identify success or failure in restoration schemes. Definition of the restoration 'target' is immensely important. Restoring functions (e.g. major biogeochemical cycles) as well as integrity (e.g. trophic networks that provide resilience capacities) are two possibilities. Ehrenfeld (2000) suggested that the parameters that we might consider are the biotic and abiotic components of a system (their form and arrangement, i.e. the ecosystem architecture) and their function.

It is possible to relate the terms 'reclamation' and 'restoration' to the conceptual arrangement of degraded, recovering and restored ecosystems in a schema (Fig. 17.2) initially defined by Whisenant (1999) and further developed by Hobbs and Harris (2001). Figure 17.2 shows a number of putative stable ecosystem states, from degraded to intact, in relation to ecosystem function. There are two principal barriers between the degraded and restored, or intact, systems. The first are abiotic barriers, which could be lack of an appropriate topology, a contaminated substrate, or little or no organic matter. These barriers all require physical modification to bring the systems to a new level of stability associated with a new 'higher' level of function. The second barrier is biotic; this may be as a result of a lack of appropriate species or a failure of the interaction between them and the abiotic components. Again, active modification allows these barriers to be overcome. The first transition involving the abiotic barriers is the *reclamation* phase, and the second transition involving the biotic barriers is the *restoration* phase of a programme designed to restore both ecosystem function and ecosystem structure.

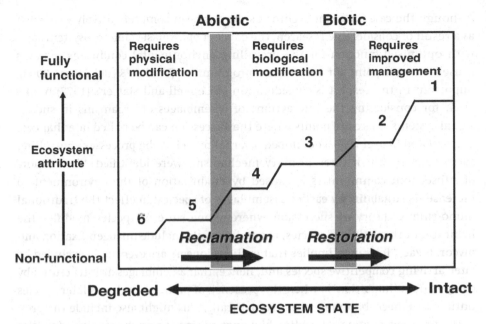

Figure 17.2. Biotic and abiotic barriers to ecosystem restoration. Redrawn from Whisenant (1999).

Relationship to ecological succession

Classical theory has six processes driving succession, neatly summarised by Walker and del Moral (2003), and modified in the following definitions.

- *Nudation*: this creates bare land and a clear analogue for extreme disturbances caused by such activities as opencast coal mining or heavy trafficking during civil engineering procedures.
- *Migration*: the arrival of organisms on that site – something that may have to be accomplished artificially when suitable propagule sources do not occur locally, or site conditions present barriers to their arrival, and in some cases prevented when invasive species are likely to establish and dominate.
- *Ecesis*: the establishment of those newly arrived organisms – offers another potential restoration manipulation, and is one that may be critically depen- dent on the presence of important symbiont propagules, such as mycorrhizae.
- *Competition*: the interactions of the organisms – particularly pertinent when considering soil biota–plant interactions, with pathogens important in mod- ifying direct interactions.
- *Reaction*: modification of the site by the organisms – a process with some potential for management intervention, including the establishment of legumes to increase nitrogen capital in the system, or direct fertilisation.
- *Stabilisation*: the development of a stable end point or 'climax' stage – in effect the restoration 'target' state.

Although the case for a single climax state has been comprehensively weakened as a result of considerable research, the concept of a 'quasi-stable' ecosystem state with optimised function for the prevailing environmental conditions remains a useful target point for restoration programmes (Walker & del Moral 2003). Important in this respect is consideration of Connell and Slatyer's (1977) framework for considering the interactions of assemblages of organisms in successional systems. There are points where the succession can be halted or enhanced. Connell and Slatyer (1977) developed a verbal model of the processes controlling succession, in which three primary mechanisms were identified. 'Facilitation' of subsequent communities is caused by modification of the environment to increase its suitability by earlier assemblages of species; in effect the traditional mono-climax theory of succession whereby one suite of species modifies the habitat for other suites of species, examples would include nitrogen fixation and mycorrhizae. 'Tolerance' implies that earlier pioneers are eventually excluded by later arriving competitive species and, hence, that assemblages are structured by competition. 'Inhibition' is when the presence of pioneers excludes later species until the pioneers become extirpated naturally. This might also include the presence of pathogens and root-feeding biota preventing the establishment of species typical of mature systems. The interrelationships between these three verbal models of succession are represented diagrammatically in Fig. 17.3. Another mechanism that must be considered is termed 'random', whereby the survival of any given species assemblage, especially in conditions of low soil adversity, is determined by chance (Lawton 1987).

Restoration and reclamation projects offer unique opportunities in testing these models of succession. They often provide easy access to nascent substrates on the doorstep of large conurbations in industrialised countries, within easy reach of sophisticated laboratory facilities. If the degraded land is left untreated, it could be expected that one of these models of succession might describe the observed process of succession, but that the time scale might be very much extended. Reclamation and restoration are aiming to work with ecological succession, but to shorten the time scale of recovery very considerably.

What to measure?

Unambiguous measurements are required if effective interventions are to be made so as both to restore ecosystems and to avoid failure (Fig. 17.4). Here the reclamation would allow spontaneous succession to occur by removing abiotic adversity, but restoration facilitates re-instatement of attributes in a shorter time scale. We need to be able to monitor such attributes consistently and effectively if we are to avoid failures, as illustrated. In order to achieve this, we need ecosystem attributes that address five questions (Andreasen *et al.* 2001). (1) Can the attributes relevant to the ecosystem under study and to the objectives of the assessment programmes be defined? (2) Are the attributes sensitive

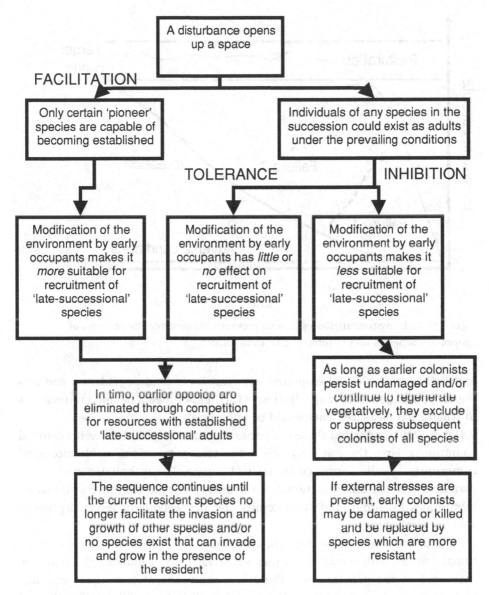

Figure 17.3. Facilitation, tolerance and inhibition in relation to successional events. Redrawn from Connell and Slatyer (1977).

to anthropogenic changes? (3) Can the attributes provide a response that can be differentiated from natural variation? (4) Are the attributes environmentally benign? (5) Are the attributes cost effective to measure?

Currently many metrics are employed, but measurement of the soil biota is emerging as satisfying the requirements of these five questions. Once key attributes are identified and readily measured, they can be used to plan, monitor and adapt management regimes. Harris (2003) provided a review of the

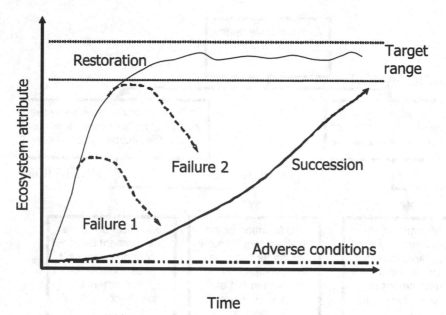

Figure 17.4. Ecosystem attributes may be measured to monitor the progress of restoration schemes and to intervene to avoid failure.

approaches available and employed with respect to land degradation and ecological restoration. Ritz *et al.* (2004) have further outlined a detailed rationale as to when such measurements would be appropriate.

Harris (2003) suggested three categories of measurement of the soil microbial community. First, there is size. This is the total mass of the viable biological community, usually expressed in units of carbon. The techniques used are as varied as fumigation extraction for soil microbial biomass, microscopy-based biovolumes for fungi, and hand counts and dry weights of larger organisms such as earthworms.

Second, there is composition. The abundances of particular species or functional groups, which may be further subdivided into genotypic diversity, are measured at the genetic level. There is a multitude of techniques available including terminal restriction fragment length polymorphisms (T-RFLP) and denaturing gradient gel electrophoresis (DGGE). Phenotypically expressed diversity often uses phospholipid fatty acids (PLFAs).

Third, there is activity, including both *in-situ* transformations, such as respiration, nitrogen fixation and ammonification, and *ex-situ* determination of potential catabolic capabilities. Degens and Harris (1997) provided a protocol for determining the potential of soil communities to respire a wide variety of carbon substrates within two–six hours of being supplemented with them. This is superior to the widely employed BIOLOG technique, which is dependent on the culturable fraction of the community responding to immersion in the substrate solution. It is of limited use due to a difficulty in interpretation and, therefore, in assessing the capabilities of the soil microbial community *in toto* (Preston-Mafham *et al.* 2002).

Measurements of the soil microbial community, particularly in circumstances where the degradation is initially physical, as found in the majority of civil engineering programmes, have proven highly effective in assessing the impact of management regimes for restoring ecosystems. Impacts of such practices as land abandonment are better detected at the community rather than at the species level (Zeller *et al.* 2001).

The use of PLFA profiles is now well established in the study of soil microbial community composition, and has the advantage of showing promise by indicating the sizes of widely differing groups such as fungi, bacteria and nematodes (Zelles 1999). It can also be used to calculate the total size of the biomass. Its ability to discriminate between different land uses, ecosystem types, and management and treatment effects has been amply demonstrated; it should be adopted as a standard approach. Similarly, the functional profile approach, employing Degens and Harris' (1997) technique, has been shown to be robust and interpretable, but suffers from the moderately large effort in obtaining data. This method is, however, amenable to modification for routine analysis. Individual process measurements such as nitrogen transformations will continue to be useful in particular circumstances, and may be employed as indices, but to be of real value they need to be used in conjunction with other techniques. Nucleic-acid-based methods will only be universally applicable as indicators when all sequences are detected with high fidelity, revealing which organisms are viable, and how many copies an individual may contain. This latter difficulty may make the general application of these techniques almost impossible, as many organisms remain to be sequenced. The use of the soil microbial community characteristics in a number of different degradation and restoration situations is exemplified in the following section.

The importance of soil biodiversity in restoration schemes
Opencast (strip) mining

It is commonly observed on opencast sites or strip mines that, when soil is simply re-instated after storage, without further intervention, the changes in the microbial community follow a similar trajectory to those recorded on primary successions, e.g. retreating glacier fronts and volcanic ashes. Early work indicated that increases in soil microbial biomass were correlated with age (Insam & Domsch 1988; Harris & Birch 1989). Studies by Ruzek *et al.* (2001), working on re-instated and abandoned soil-forming materials in the Czech Republic and Germany, found that there were clear, predictable relationships between time since soil re-instatement and soil microbial biomass. These relationships, however, appeared to have a regional dimension, echoing earlier work on retreating glacier fronts

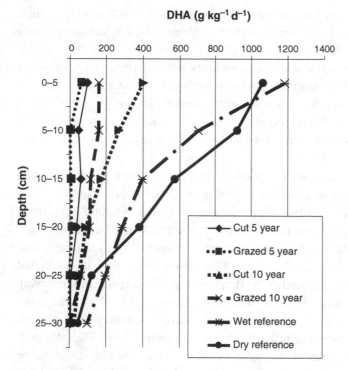

Figure 17.5. Microbial activity, as measured by dehydrogenase activity (DHA), profile development as affected by management regimes. Redrawn from Harris and Birch (1990).

carried out by Insam and Haselwandter (1989). This may be explained by the arrival of primary producers on a site, providing resources for the decomposer community, but it says little as to whether there is a feedback role from the microbial community to the plant community.

The tendency of natural and disturbed systems to converge with time is not universal. Davis *et al.* (2003) investigated dung beetle assemblages in dune forests in Maputuland. A 23-year vegetational chronosequence was available on mined dunes that were initially colonised by grasses, succeeded by open *Acacia* shrub-land thicket, and then developing to *Acacia karroo*-dominated woodland after nine years. In the early stages of succession there was a convergence of dung beetle assemblages with the reference sites in the unmined forest, but the community assemblages diverged again when older *Acacia* woodland was compared with the natural forest. The authors ascribed this to the similarities between microclimates in early mine successions and the reference sites. This suggested that only by interventions, i.e. alteration of the microtopography and woodland density, could a true restoration be achieved. It would not be sufficient to leave the restoration to the natural processes of ecological succession.

The effects of the management regime on the recovery of the function of the microbial community on former opencast sites are illustrated in Fig. 17.5. Two reference sites were available, 'wet' and 'dry', and four adjacent, re-instated

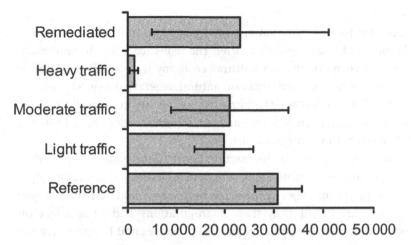

Figure 17.6. The effect of military traffic and remediation on soil microbial biomass as determined by PLFA (pmol g^{-1}). The bars indicate standard deviations. Redrawn from Peacock *et al.* (2001).

fields (two five-year fields and two ten-year fields, after re-instatement). The five-year re-instated areas had little microbial activity, as indicated by dehydrogenase activity (DHA) down the soil profile, but there were higher activities in the ten-year re-instated areas. The results also clearly indicate that cutting and leaving the aftermath was more effective than grazing in encouraging microbial activity. This may be linked to earthworm activity (Scullion & Malik 2000). Apart from earthworms, another group that shows a clear relationship with soil microbial biomass is the size of the ant populations, but this is not true of undisturbed sites where ant population sizes are negatively related to total microbial biomass carbon (Anderson & Sparling 1997).

Military training areas

Many of the pressures on soil systems experienced in opencast mining, and other civil engineering programmes, have analogues in areas used for military training. Figure 17.6 shows the effects of varying intensities of military traffic and training on the soil microbial biomass, as determined by PLFAs (Peacock *et al.* 2001). All of the areas subject to training showed decreases in microbial biomass as compared with the reference site, though not all decreases were statistically significant. Those with the heaviest training loads (those used for main battle tank manoeuvres), showed a decrease of over an order of magnitude with respect to the reference site.

In an area which had been restored by planting tree cover some 10 years previously, the measurement of the microbial biomass had returned to the levels found in the light and moderately trafficked areas. It is important to note the high variability of the restored site, indicating the patchy nature of the recovery, probably related to tree establishment heterogeneity.

Agriculture

There have been numerous programmes taking agricultural fields out of production in Europe and North America. Often the fields have simply been abandoned due to downturns in the agricultural economy (e.g. North America) or as a deliberate policy to increase areas of natural vegetation (e.g. set-aside in Europe) and therefore to further the aims of biodiversity conservation. Some of the fields have simply been left unmanaged, whereas others have been the subject of active restoration interventions.

Maly *et al.* (2000) reported on the interaction between the plant and microbial communities in abandoned arable fields, which included experimental manipulation of vegetation assemblages to produce both early- and late-successional species diversity. They found that these manipulations had little effect on microbial community composition or microbial processes and biomass over the short term (two years). Unfortunately, the only attempt to characterise species diversity was by plate counts, a common practice in some laboratories and one that is severely limited in its interpretation. No effects were determined by this method, probably due to its low sensitivity.

Hemerik and Brussaard (2002) examined the soil invertebrate community in four grasslands undergoing restoration succession (see also Brussaard this volume). The times since the last application of fertilisers were 7, 11, 24 and 29 years. The hypothesis being tested was that decreasing nutrient status would be accompanied by decreasing numbers of macroinvertebrates, simply on a food availability basis. This proved to be the case, particularly for weevil (Coleoptera: Curculionidae) populations. The weevils appeared to be trend followers and not trend leaders. The polyphagous predatory ground beetles (Coleoptera: Carabidae), however, did not respond to impoverishment with changed species composition, but only in numbers due to lower nutrient availability. Again, natural succession will not necessarily result in what might be the desired assemblage of species of plants and animals, and restoration may be required to re-direct the succession.

Interactions between biology and physical structure

Figure 17.7 shows the relationship between water-stable aggregates at five open-cast reclamation sites (Edgerton *et al.* 1995). The relationship appears linear over the first 11 years; longer than this and the relationship becomes a log-linear one. The effects of compaction and waterlogging are also clearly detectable. This indicates the importance of restoring not only the obvious features of a site, but also the below-ground biological component. Without the development of aggregates, reclaimed soils will always remain shedding water, and not infiltrating water, and will tend to be droughty in the summer and waterlogged in winter. The useful spin-off is that measurement of the microbial community can be used as a surrogate for characteristics such as aggregation. This again, however,

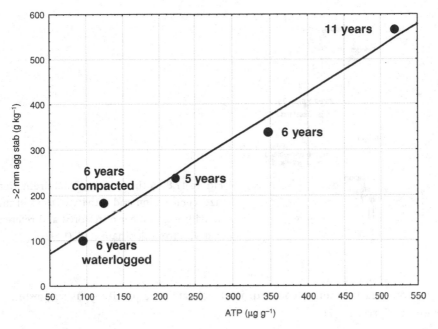

Figure 17.7. Relationship between water-stable aggregates (>2 mm g kg^{-1}) and microbial biomass (ATP µg g^{-1}) at five restored sites. Redrawn from Edgerton *et al.* (1995).

indicates that the total amount of microorganisms in the soil is important, but possibly not their diversity.

Scullion and Malik (2000) clearly demonstrated the effect of earthworms in increasing stable aggregation, soil microbial biomass and re-distribution of organic matter throughout the profile. This observation offers a point of intervention because civil engineering activities have a particularly adverse effect on earthworm populations (Scullion *et al.* 1988).

Semi-natural sites

By using a three-dimensional ordination, Bentham *et al.* (1992) demonstrated the effects of time and management type on three characteristics of the soil microbial community (Fig. 17.8). Here sites reclaimed after opencast coal mining are shown as restored. There are two points to note. First, the ten-year restored grassland is now clustering with a rough (improved) grassland. Second, the restored woodlands are diverging from the five-year grassland restoration in the direction of woodland reference sites.

More comprehensive data would facilitate interpretation of observations of these types, allowing the effects of management interventions to be clearly identified. This is a point re-iterated by Coleman (1994). However, Bentham *et al.* (1992) used a very simple measurement of diversity. The experiment therefore needs to be repeated using both phenotypic diversity measurements (e.g. PLFA profiles)

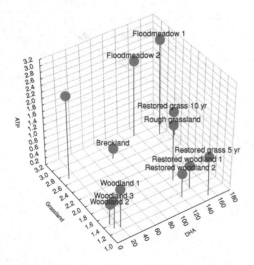

Figure 17.8. A three-dimensional ordination of the size, composition and activity of the soil microbial activity at a number of restored and reference sites. Redrawn from Bentham *et al.* (1992).

and function measurements (community level catabolic profiling, as suggested by Degens & Harris 1997).

Facilitators or followers?

One question central to the role of biodiversity in soils with respect to ecological restoration, and ecosystem succession generally, is whether the soil microbial community is facilitating (in the sense of Connell & Slatyer 1977) or responding to primary producer community establishment? This is a difficult question to answer by simple site observation and the available evidence is conflicting. The studies of fundamental succession and other ecosystem processes carried out on degraded and disturbed sites are few, but we can draw inferences, and a programme of potential future work, from studies of natural systems. There are, however, several potential mechanisms that may be proposed by which the soil biotic community could be critical in the restoration of ecosystems; these are predicated on a desire to establish a 'climax' or stable 'late-successional' ecosystem. First, late-successional plant species may not establish until the soil biota, as a consequence of early successional processes, provide sufficient organic matter reserves and nutrient cycling capacity, nitrogen fixation being clearly important here; this is a *facilitation* by modifying soil conditions. Second, the presence of appropriate symbiont propagules could *facilitate* the establishment of later successional plant species. Third, symbionts of earlier successional species could give them a continuing competitive advantage over incoming plants, therefore *inhibiting* the establishment of late-successional species. Fourth, pathogens and below-ground herbivores, once they have reached a critical population size, could significantly reduce the numbers of earlier plant species, thereby opening gaps for colonisation and therefore *facilitating* the establishment of late-successional species. Finally, pathogens and below-ground herbivores may preferentially infect or feed on late-arriving successional seeds or saplings, thereby *inhibiting* their establishment. In all of these cases, the soil biota contributes to the stability of late-successional community vegetation types.

Facilitation by modifying soil conditions

There is definite evidence for clear switches in community composition during successional processes. It is a well-established observation that heterotrophic organisms are often the primary colonisers of newly exposed substrates, such as glacial moraines, volcanic ejecta and wind-blown sands. Hodkinson *et al.* (2002) have suggested that these communities conserve nutrients and facilitate the establishment of green plants. This mechanism involves the arrival of dead organic matter and living invertebrates from outside of the system, which inter-act to produce a functioning community of detritivores and predators. Coupled with non-symbiotic nitrogen fixation and nitrogen conservation, this allows pioneer plants to become established. Ohtonen *et al.* (1999) showed that there was a change from bacterial to fungal dominance of the microbial biomass on a chronosequence along a retreating glacier front. There was, as might be expected, greater microbial biomass under plant canopies than in the non-vegetated soils. Bardgett and McAllister (1999) suggested that this switch in dom-inance was a good indicator of successful restoration of self-sustaining grassland, previously dependent on fertiliser inputs (see also Yeates *et al.* 1997), in relation to the differences between organic and inorganic management.

Work by Yin *et al.* (2000) showed that the total number of bacterial species on a site recovering from mining activity increased fairly quickly around pioneer vegetation, but the proportion of these species that were active (as determined by a radio-label incorporation method) took much longer to increase, and at no point matched the undisturbed reference forest site (Fig. 17.9). It is important to note that the material used to re-instate the contours was unaltered, uncon-solidated overburden, devoid of organic matter when shaped to final contour. This suggests that the bacterial component of the soil microbial community arrives ahead of the primary producers, but cannot be activated until signifi-cant primary production occurs. Whether the presence of these bacterial species is a prerequisite of successful plant species establishment is unclear; this would have to be tested experimentally. There is also evidence that this may be the case when examining functional profiles. Schipper *et al.* (2001) applied the method of Degens and Harris (1997) to determine the response of soil to a wide variety of carbon substrates from several successional sequences of plant communities. The results suggested that the evenness of response of the heterotrophic commu-nity to this range of carbon substrates re-established quickly after disturbance, once significant organic matter inputs occurred, but subsequently declined. This

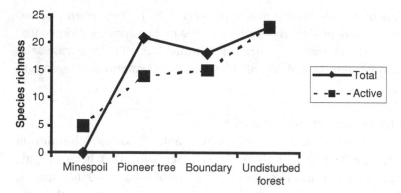

Figure 17.9. Total and active species along an artificial chronosequence on a mined site. At low concentrations, noise signals can be considerable and hence it is possible to find that there are more active species than the total number of species (cf. the minespoil results). Redrawn from Yin *et al.* (2000).

is suggestive of the hump-back model of species diversity, with respect to the availability of resources (Grime 1979). There are also clear relationships between re-establishment of soil microbial communities and soil physical structure. This would be readily amenable to testing further on anthropogenically disturbed sites, particularly where organic input records are available for agroforestry reclamation programmes.

Building sufficient nitrogen capital on degraded and disturbed sites has long been recognised as a prerequisite for establishing stable late-successional communities; figures of between 2000 and 2500 kg N ha^{-1} are often quoted as target amounts for temperate ecosystems (Bradshaw & Chadwick 1980). The approaches usually employed for this purpose are: re-instatement of stored topsoil materials (although this has its own set of challenges; Harris *et al.* 1996); nitrogen addition using inorganic or organic fertilisers; and introduction of legumes and other nitrogen-fixing plants. The last category is dependent on the appropriate symbionts being present in substrate, or being present on seed/planted material. There are few reports of this being a problem. But it is not always clear whether, once established, nitrogen-fixing plants facilitate or inhibit late-successional species.

Mummey *et al.* (2002) working on 20-year-old reclamations after opencast mining and undisturbed reference areas, demonstrated strong spatial correlation of microbial fatty acid methyl ester (FAME) markers around the roots of two pioneer species, *Agropyron smithii* and *Artemesia tridentate*. They concluded that the rate of soil recovery was rather low, and this could be ascribed to poor establishment of shrub species. These results are rather suggestive of the hypothesis that the microbial community is dependent upon the establishment of plant species, but again these observations cannot be used to rule out the potential for microbial species facilitating the establishment of plant species. This could be as simple as facilitating the germination of seeds arriving at the site, but could also be associated with providing co-factors and other precursors for closer symbionts. Of more general interest are the observations that the use of stones to mimic shrubs, by lessening wind flow, could act to cross one of the abiotic barriers to plant establishment. Scullion and Malik (2000) provide clear evidence for the role of earthworm populations in developing stable soil structures, thus facilitating establishment of moisture-sensitive species.

Facilitation by symbionts

One area in which the importance of particular species or groups has been clearly demonstrated is that of symbioses. Mycorrhizal spores are generally dispersed into new substrates readily by wind, water, dusts and animals (Walker & del Moral 2003). Their importance in establishing plants on mined lands has been demonstrated by Allen and Allen (1980), and the artificial re-introduction of mycorrhizae onto reclamation/restoration sites has been advocated, and practised, for some time (Allen & Allen 1988). Johnson (1998) investigated the impact of adding mycorrhizae with sludge to taconite mine tailings and found that the mycorrhizae avoided a large flush of nutrients, promoting late-successional species and suppressed weeds. Requena et al. (2001) experimentally manipulated communities, in degraded semi-arid Mediterranean ecosystems, by inoculating them with combinations of arbuscular mycorrhizal fungi (AMF) and rhizobial nitrogen fixers. They demonstrated that the dual symbioses caused a cascade of biotic and abiotic barriers to be crossed. Increased soil nitrogen and organic matter, production of hydrostable soil aggregates and enhanced transfer of nitrogen from fixing to non-fixing plant species, formed a crucial threshold in further ecosystem development.

Inhibition by symbionts

What effects could an 'inappropriate' pre-existing community have on the establishment of these important symbionts, when they are demonstrably absent in the control plots even though the appropriate plant species are present? There is some tentative evidence for such a mechanism. De Boer et al. (2003) demonstrated that fungistasis in dune soils could be removed by sterilisation; this was attributed to the elimination of pseudomonads. Pseudomonads appear not only to inhibit certain phytopathic species of fungi, but to promote the establishment and function of mycorrhizal fungi (Founone et al. 2002), the so-called 'mycorrhiza helper bacteria'. The pseudomonad community is itself dependent on organic matter lability. This offers the intriguing possibility of manipulating both biotic and abiotic conditions in sites being restored in an effort to overcome restoration barriers, and is certainly worthy of further experimental inquiry.

Facilitation by pathogens and herbivores

Van der Putten *et al.* (1993) investigated the impact of soil-borne diseases on succession in fore dune vegetation. By means of reciprocal transplantation experiments, they demonstrated that later successional species were tolerant of diseases of earlier species, namely *Ammophila arenaria* (marram grass) and *Hippopaë rhamnoides* (sea buckthorn). In effect, once the soil-borne pathogens had reached a critical mass related to the density of the pioneer plant species, they reduced the populations of these plants, opening gaps and thereby facilitating the establishment of later species. Further evidence of this sort, with some indication of a possible mechanism, has been provided by Klironomos (2002) who grew native rare and non-native invasive species in soils that had been used to grow either the same type of plant (home soil) or plants from the other group (foreign soil). He found that the native plants exhibited a relative decrease in growth on home soil as compared with growth of those same species on 'foreign' soil (a similar effect had been reported previously by Bever 1994), and suggested that this was due to the accumulation of pathogens. The invasive species were not only immune to the native species pathogens, but some benefited from interactions with mycorrhizal fungi taken from the native soil. This suggests that the soil biota is critical in facilitating incoming species, particularly invasives; the effect may be less clear when the late-successional species are also natives, but would certainly help to explain why non-native invasives are so successful in establishment, and why some plants are rare. This may also have implications for restoration projects where non-native invasives are a particular problem for site managers – perhaps use of 'home' soil should be discouraged when trying to establish rare native species?

Soil-dwelling herbivores have also been demonstrated as having a key role in promoting late-successional plant species establishment. Brown and Gange (1992) showed that root-feeding insects could accelerate succession by reducing the persistence and colonisation of herbaceous plants. De Deyn *et al.* (2003) demonstrated that the soil invertebrate fauna fed on roots of dominant plant species, opening gaps for subdominants and later species, in grassland ecosystems, promoting the development of 'target' plant species assemblages. This evidence would suggest that simply reducing soil fertility and introducing above-ground grazing may be insufficient to realise the establishment of target ecosystem composition.

Inhibition by pathogens and herbivores

Fagan and Bishop (2000) manipulated the insect populations at various sites on the Mount St Helens eruption zone. They used a topical insecticide, a synthetic pyrethroid called esfenvalerate, on lupin populations. They found that release from herbivory facilitated greater growth and establishment of lupins, allowing them to spread into suitable habitat. Although specifically applied in this instance to reduce above-ground herbivory this pesticide has a half-life in soil of three months, and there may have been impacts on below-ground insect herbivores, which were not controlled for in this experiment. The above-ground mechanism suggested by Fagan and Bishop could equally well apply to below-ground herbivores. Maron (1998) found that there was a positive effect of reducing root-boring ghost moth (*Hepialus californicus*) larvae on the seed productivity of bush lupine (*Lupinus arboreus*) but this took three years to appear as a statistically significant effect. The role played by nitrogen-fixing plants in facilitating the establishment of other species is critical, and it may be that manipulation of below-ground insect populations enhances the rate of restoration. Much more experimental work would need to be done before this could be recommended as a standard treatment, however, in the light of the evidence of the positive effects of herbivores outlined in the previous section.

Maintenance of stability in late-successional assemblages

Several authors have reported on experimental approaches to determine the role of the soil biota in stabilising late-successional plant species assemblages. Mills and Bever (1998) showed that plant pathogenic fungi, specifically from the genus *Pythium*, reduced the biomass and root : shoot ratios of *Danthonia spicata* and *Panicum sphaerocarpon*. The authors suggest that this mechanism limits the growth of individual species in late-successional stages and thereby promotes diversity. This observation offers the opportunity for testing the hypothesis that introduction of late-successional pathogen spores could promote diversity in restoration programmes, whereas simple manipulation of organic matter, fertility and soil physical conditions would not be sufficient. Working in experimental microcosm systems, van der Heijden *et al.* (1998) demonstrated that reducing the diversity of AMF led to greater fluctuations in species composition and community structure, and reduced biomass of plants. These investigations point to the critical role the soil biota has in stabilising late-successional ecosystem composition and function.

Principal gaps and future work

Clearly, the null hypothesis that soil biodiversity does not influence the success of ecological restoration schemes is interesting and needs to be explored. We need to disentangle the effects of biology (i.e. soil organism biomass) from those of biodiversity (i.e. the numbers, relative distributions, 'species' composition, and interactions between soil organisms and the abiotic substrate). We also need to identify the role that the soil biota might have as part of the ecological restoration 'toolkit'. At this stage of research in this applied area, very little has been done to address these issues. Most investigations are, in effect, reports of successional processes on sites variously affected by degradative land use, and in some cases the effects of treatments designed to ameliorate the adverse conditions. These interventions tend to be physical (e.g. to relieve compaction), chemical (to provide inorganic nutrients) or biological (e.g. sowing grass seed or tree planting). The last might be regarded as a means of increasing biological diversity, but this is usually employed on a simplistic level. Sometimes animals are re-introduced, but principally with a focus on restoring the species, rather than as a means of altering overall ecosystem function, or to affect other components. Since there have been few convincing demonstrations of the link between biological diversity and ecosystem function, it is hardly surprising that restoration practitioners have also made little progress in demonstrating such a relationship.

What are the principal gaps in our knowledge of the relationship between soil biodiversity and restoration ecology that should be a priority for future research? First, we must address the question: how much genotypic and functional diversity is required to facilitate plant community function? Second, we need to determine if the shifts from one community profile to another, e.g. from bacterial- to fungal-dominated systems, result in any form of ecosystem stability. If such a shift were proved, it would be a very powerful tool in managing degraded systems during the process of restoration. Third, we must determine whether or not the soil biota, other than the symbiont inoculum, is an essential part of the facilitation mechanism. Is the biota on a particular site actually inhibiting the establishment of desirable species? What is the relative importance of decomposers, mycorrhizae, nitrogen fixers, pathogens and herbivores on sites that we are attempting to reclaim or restore? This may be equally applicable to soils damaged by civil engineering operations, as to those raw substrates used as 'soil-forming' materials on opencast and landfill sites. Fourth, a large-scale survey of reference sites, to provide target end points for restoration programmes, is needed. The microbial community might be uniquely sensitive in this respect, and amenable to large-scale analysis. This would provide comparisons between differing habitats and further our understanding of the interactions between microbial communities, plant communities, the soil fauna and the restoration process.

The soil biological community, other than the microflora, has been the subject of relatively few contemporary studies (other than the research at Sourhope), and rarely on degraded and disturbed sites with a view to elucidating fundamental mechanisms. This work must be broadened and integrated with both plant and microbial community studies. The macroinvertebrates are often the key to step changes in the organisation of soil systems, e.g. earthworms and the development of structural stability. The changes in cycling times brought about by soil herbivores will have significant effects on the soil microbial community composition and function. The role of the soil biota in facilitating, inhibiting or stabilising late-successional 'climax' ecosystems in land reclamation and restoration schemes has been little investigated and has been poorly understood, although work on 'natural' successions suggests that there are strong feedbacks operating. This line of enquiry needs to be vigorously pursued. We need extensive programmes of research to investigate the effect of removal and addition of nitrogen fixers, decomposers and mycorrhizae on degraded and disturbed sites to elucidate clearly their role and application in reclamation and restoration programmes. The opportunities for fundamental advances in our understanding of soil biology and biodiversity are widespread on degraded sites and, as illustrated by Fig. 17.1, are common in many places in the world. The rapid turn-over times of soil microbial communities offer potential for elegant manipulation experiments, with practical outcomes for the restoration of ecosystem services.

References

Allen, E. B. & Allen, M. F. (1980). Natural re-establishment of vesicular–arbuscular mycorrhizae following stripmine reclamation in Wyoming. *Journal of Applied Ecology*, **17**, 139–147.

Allen, E. B. & Allen, M. F. (1988). Facilitation of succession by the monomycotrophic colonizer *Salsola kali* (Chenopodiaceae) on a harsh site: effects of mycorrhizal fungi. *American Journal of Botany*, **75**, 257–266.

Anderson, A. N. & Sparling, G. P. (1997). Ants as indicators of restoration success: relationship with soil microbial biomass in the Australian seasonal tropics. *Restoration Ecology*, **5**, 109–114.

Andreasen, J. K., O'Neill, R. V., Noss, R. & Slosser, N. C. (2001). Considerations for the development of a terrestrial index of ecosystem integrity. *Ecological Indicators*, **1**, 21–35.

Aronson, J., Floret, C., Le Floc'h, E., Ovalle, C. & Pontanier, R. (1993). Restoration and rehabilitation of degraded ecosystems in arid and semi-arid regions: I. A view from the south. *Restoration Ecology*, **1**, 8–17.

Bardgett, R. D. & McAllister, E. (1999). The measurement of soil fungal : bacterial biomass ratios as an indicator of ecosystem self-regulation in temperate meadow grasslands. *Biology and Fertility of Soils*, **29**, 282–290.

Bentham, H., Harris, J. A., Birch, P. & Short, K. C. (1992). Habitat classification and soil restoration assessment using analysis of soil microbiological and physico-chemical characteristics. *Journal of Applied Ecology*, **29**, 711–718.

Bever, J. D. (1994). Feedback between plants and their soil communities in an old field community. *Ecology*, **75**, 1965–1977.

Bradshaw, A. (1997). Restoration of mined lands: using natural processes. *Ecological Engineering*, **8**, 255–269.

Bradshaw, A. & Chadwick, M. J. (1980). *The Restoration of Land: The Ecology and Reclamation of Derelict and Degraded Land*. Oxford: Blackwell.

Brown, V. K. & Gange, A. C. (1992). Secondary plant succession: how is it modified by insect herbivory? *Vegetatio*, **101**, 3–13.

Coleman, D. C. (1994). Compositional analysis of microbial communities: is there room in the middle? *Beyond the Biomass* (Ed. by K. Ritz, J. Dighton & K. E. Giller), pp. 201–220. Chichester: Wiley-Sayce.

Connell, J. H. & Slatyer, R. O. (1977). Mechanisms of succession in natural communities and their role in community stability and organisation. *American Naturalist*, **111**, 1119–1144.

Davis, A. L. V., van Aarde, R. J., Scholz, C. H. & Delport, J. H. (2003). Convergence between dung beetle assemblages of a post-mining vegetational chronosequence and unmined dune forest. *Restoration Ecology*, **11**, 29–42.

De Boer, W., Verheggen, P., Klein Gunnewiek, P. J. A., Kowalchuk, G. A. & van Veen, J. A. (2003). Microbial community composition affects soil fungistasis. *Applied and Environmental Microbiology*, **69**, 835–844.

De Deyn, G. B., Raaijmakers, C. E., Zoomer, H. R., *et al.* (2003). Soil invertebrate fauna enhances grassland succession and diversity. *Nature*, **422**, 711–713.

Degens, B. P. & Harris J. A. (1997). Development of a physiological approach to measuring the catabolic diversity of soil microbial communities. *Soil Biology and Biochemistry*, **29**, 1309–1320.

De Groot, R. S., Wilson, M. & Boumans, R. M. J. (2002). A typology for the classification, description and valuation of ecosystem functions, goods and services. *Ecological Economics*, **41**, 393–408.

Edgerton, D. L., Harris, J. A., Birch, P. & Bullock, P. (1995). Linear relationship between aggregate stability and microbial biomass in three restored soils. *Soil Biology and Biochemistry*, **27**, 1499–1501.

Ehrenfeld, J. G. (2000). Defining the limits of restoration: the need for realistic goals. *Restoration Ecology*, **8**, 2–9.

Fagan, W. F. & Bishop, J. G. (2000). Trophic interactions during primary succession: herbivores slow a plant re-invasion of Mount St. Helens. *The American Naturalist*, **155**, 238–251.

Founone, H., Duponnois, R., Meyer, J. M., *et al.* (2002). Interaction between ectomycorrhizal symbiosis and fluorescent pseudomonads on *Acacia holosericea*: isolation of mycorrhizal helper bacteria (MHB) from a Soudano–Sahelian soil. *FEMS Microbiology Ecology*, **41**, 37–46.

Grime, J. P. (1979). *Plant Strategies and Vegetation Processes*. Chichester: Wiley.

Harris, J. A. (2003). Measurements of the soil microbial community for estimating the success of restoration. *European Journal of Soil Science*, **54**, 801–808.

Harris, J. A. & Birch, P. (1989). Soil microbial activity in opencast coal mine restorations. *Soil Use and Management*, **5**, 155–160.

Harris, J. A. & Birch, P. (1990). Application of the principles of microbial ecology to the assessment of surface mine reclamation. *Proceedings of the 1990 Mining and Reclamation Conference, American Society of Surface Mining and Reclamation*, Charleston, WV, April 1990. (Ed. by J. Skousen, J. Sencindiver & D. Samuel), Part I, pp. 111–120. Morgantown, WV: University of West Virginia.

Harris, J. A., Birch, P. & Palmer, J. (1996). *Land Restoration and Reclamation; Principles and Practice*. Harlow: Longman.

Hemerik, L. & Brussaard, L. (2002). Diversity of soil macro-invertebrates in grasslands under restoration succession. *European Journal of Soil Biology*, **38**, 145–150.

Hobbs, R. J. & Harris, J. A. (2001). Restoration ecology: repairing the Earth's ecosystems in the new millennium. *Restoration Ecology*, **9**, 239–246.

Hodkinson, I. D., Webb, N. R. & Coulson, S. J. (2002). Primary community assembly on land: the missing stages. Why are the heterotrophic organisms always there first? *Journal of Ecology*, **90**, 569–577.

Insam, H. & Domsch, K. H. (1988). Relationship between soil organic carbon and microbial biomass on chronosequences of reclamation sites. *Microbial Ecology*, **15**, 177–188.

Insam, H. & Haselwandter, K. (1989). Metabolic quotient of the soil microflora in relation to plant succession. *Oecologia*, **79**, 174–178.

Johnson, N. C. (1998). Responses of *Salsola kali* and *Panicum virgatum* to mycorrhizal fungi, phosphorus and soil organic matter: implications for reclamation. *Journal of Applied Ecology*, **35**, 86–94.

Klironomos, J. N. (2002). Feedback with soil biota contributes to plant rarity and invasiveness in communities. *Nature*, **417**, 67–70.

Lawton, J. H. (1987). Are there assembly rules for successional communities? *Colonization, Succession and Stability* (Ed. by A. J. Gray, M. J. Crawley & P. J. Edwards), pp. 225–244. Oxford: Blackwell.

Maly, S., Korthals, G. W., van Dijk, C., van der Putten, W. H. & de Boer, W. (2000). Effect of vegetation manipulation of abandoned arable land on soil microbial properties. *Biology and Fertility of Soils*, 31, 121–127.

Maron, J. L. (1998). Insect herbivory above- and belowground: individual and joint effects on plant fitness. *Ecology*, 79, 1281–1293.

Mills, K. E. & Bever, J. D. (1998). Maintenance of diversity within plant communities: soil pathogens as agents of negative feedback. *Ecology*, 75, 1595–1601.

Mummey, D. L., Stahl, P. D. & Buyer, J. S. (2002). Soil microbiological properties 20 years after surface mine reclamation: spatial analysis of reclaimed and undisturbed sites. *Soil Biology and Biochemistry*, 34, 1717–1725.

Ohtonen, R., Fritze, H., Pennanen, T., Jumpponen, A. & Trappe, J. (1999). Ecosystem properties and microbial community changes in primary succession on a glacier forefront. *Oecologia*, 119, 239–246.

Peacock, A. D., McNaughton, S. J., Cantu, J. M., Dale, V. H. & White, D. C. (2001). Soil microbial biomass and community composition along an anthropogenic disturbance gradient within a long-leaf pine habitat. *Ecological Indicators*, 12, 1–9.

Preston-Mafham, J., Boddy, L. & Randerson, P. F. (2002). Analysis of microbial community functional diversity using sole-carbon-source utilisation profiles: a critique. *FEMS Microbiology Ecology*, 42, 1–14.

Requena, N., Perez-Solis, E., Azcon-Aguilar, C., Jeffries, P. & Barea, J.-M. (2001). Management of indigenous plant–microbe symbiosis aids restoration of desertified ecosystems. *Applied and Environmental Microbiology*, 67, 495–498.

Ritz, K., McHugh, M. & Harris, J. (2004). Biological diversity and function in soils: contemporary perspectives and implications in relation to the formulation of effective indicators. *Agricultural Soil Erosion and Soil Biodiversity: Developing Indicators for Policy Analyses* (Ed. by R. Francaviglial), pp. 563–572. OECD: Paris.

Ruzek, L., Vorisek, K. & Sixta, J. (2001). Microbial biomass-C in reclaimed soil of the Rhineland (Germany) and the north Bohemian lignite mining areas (Czech republic): measured and predicted values. *Restoration Ecology*, 9, 370–377.

Schipper, L. A., Degens, B. P., Sparling, G. P. & Duncan, L. C. (2001). Changes in microbial heterotrophic diversity along five plant successional sequences. *Soil Biology and Biochemistry*, 33, 2093–2103.

Scullion, J. & Malik, A. (2000). Earthworm activity affecting organic matter, aggregation and microbial activity in soils restored after opencast mining for coal. *Soil Biology and Biochemistry*, 32, 119–126.

Scullion, J., Mohammed, A. R. A. & Richardson, H. (1988). Changes in earthworm populations following cultivation of undisturbed and former opencast coal mining land. *Agriculture, Ecosystems and Environment*, 20, 289–302.

Society for Ecological Restoration (2002). *The SER Primer on Ecological Restoration.* http://www.ser.org.

Sutherland, W. J. (2002). Restoring a sustainable countryside. *Trends in Ecology and Evolution*, 17, 148–150.

Van der Heijden M. G. A., Klironomos, J. N., Ursic, M., *et al.* (1998). Mycorrhizal fungal diversity determines plant biodiversity, ecosystem variability and productivity. *Nature*, 396, 69–72.

Van der Putten, W. H., van Dijk, C. & Peters, B. A. M. (1993). Plant-specific soil-borne diseases contribute to succession in foredune vegetation. *Nature*, 362, 53–55.

Walker, L. R. & del Moral, R. (2003). *Primary Succession and Ecosystem Rehabilitation.* Cambridge: Cambridge University Press.

Whisenant, S. G. (1999). *Repairing Damaged Wildlands: A Process-Oriented, Landscape-Scale Approach.* Cambridge: Cambridge University Press.

Yeates, G. W., Bardgett, R. D., Cook, R., *et al.* (1997). Faunal and microbial diversity in three Welsh grassland soils under conventional and organic management regimes. *Journal of Applied Ecology*, 34, 453–471.

Yin, B., Crowley, D., Sparovek, G., de Melo, W. J. & Borneman, J. (2000). Bacterial functional redundancy along a soil reclamation gradient. *Applied and Environmental Microbiology*, 66, 4361–4365.

Zeller, V., Bardgett, R. D. & Tappeiner, U. (2001). Site and management effects on soil microbial properties of sub-alpine meadows: a study of land abandonment along a north–south gradient in the European Alps. *Soil Biology and Biochemistry*, 33, 639–649.

Zelles, L. (1999). Fatty acid patterns of phospholipids and lipo-polysaccharides in characterization of microbial communities in soil: a review. *Biology and Fertility of Soil*, 29, 111–129.

CHAPTER EIGHTEEN

Soil biodiversity: stress and change in grasslands under restoration succession

LIJBERT BRUSSAARD
Wageningen University
RON G. M. DE GOEDE
Wageningen University
LIA HEMERIK
Wageningen University
BART C. VERSCHOOR
Wageningen University and De Groene Vlieg BV

SUMMARY

1. Comparative field research, backed up by field and laboratory experimentation, on the effects of stress on communities is necessary to increase insight into the relationships, if any, between stress, (soil) biodiversity and ecosystem functioning.
2. Such insight is needed as a complement to ecotoxicological research on the effects of contaminants on species populations so as to broaden the base for environmental policies and legislation.
3. From a synthesis of research on reversed succession of plants, nematodes and insects in the Drentse A grasslands in the northern Netherlands, we infer that knowledge of life-history strategies at various levels of taxonomic detail is an important key to understanding stress effects on natural communities.
4. We observe that different approaches in research of stress on communities and ecosystems are beginning to converge in taking life-history strategies into account.

Introduction

Ecosystem structure and functioning are governed by three classes of driving variables: the physical environment, resource quality and organisms (Swift *et al.* 1979). Stress effects on organisms may result from biotic interactions or from changes of the physical environment and resource quality. Stress can be defined to occur when the organisms within an ecosystem are chronically confronted with abiotic conditions, resource quality conditions, new species or abundance of existing species (especially herbivores, predators or parasites) near or beyond

Biological Diversity and Function in Soils, eds. Richard D. Bardgett, Michael B. Usher and David W. Hopkins.
Published by Cambridge University Press. © British Ecological Society 2005.

the range of their ecological amplitude (Grime *et al.* 1988), and when physiological adaptation to such changes is absent (Calow & Forbes 1998). Many such stresses are human induced and may result from contaminants, global warming, acidifying atmospheric deposition, land use change and agricultural measures such as fertilisation, pesticide application and tillage.

Because ecosystem processes in soil, such as decomposition, nutrient cycling, soil (dis)aggregation and water transport, are to a large extent carried out or influenced by organisms, the effects of stress on the soil biota and ecosystem functioning are closely linked. Therefore, it is important to understand how the soil biota is affected by stress in order to understand how ecosystem functioning is influenced. Such understanding is necessary for developing criteria for setting standards for the acceptability of human-induced stress, for developing measures to counteract undesired effects of such stress and especially to manage ecosystems. It is not self-evident, however, at which level of aggregation the soil biota is most meaningfully investigated when considering stress effects: populations of individual species, the number of species, taxonomic or functional groups, or some other measure of diversity?

In the absence of sound taxonomic knowledge of much of the soil biota, it is difficult to argue how much diversity is needed to sustain the ecosystem processes mentioned. In an effort to make this issue more tractable, the soil fauna have, to the best of current knowledge, been assembled into so-called functional groups, mainly based on food preference (Moore *et al.* 1988; Brussaard 1998). Hence, the question 'how much diversity is needed to sustain ecosystem processes' may be narrowed down to *functional diversity*: how many functional groups are needed? Next to the possible relationship between functional group diversity and ecosystem functioning, the diversity of species *within* functional groups may be important for ecosystem functioning. The unique nature of species has entered the discussion in the use of the term *functional composition*, which suggests that the presence of one or more particular species, rather than other species, of a functional group may be decisive for (the intensity of) an ecosystem process (Brussaard *et al.* 1997; Chapin *et al.* 2000).

To understand the possible relationship between biodiversity and ecosystem functioning, it is furthermore important to consider the various temporal and spatial scales at which species or functional groups interact to isolate the 'signal' of individual species populations or functional groups and biological interactions from the 'noise' of all the other factors affecting ecosystem functioning (Beare *et al.* 1995; Ettema & Wardle 2002). Yet, the spatial and temporal patterning of the distribution of populations and species groups, the non-linear nature and the multitude of trophic, competitive and mutualistic interactions among soil organisms and the possible averaging out of their contributions to ecosystem processes across spatial and temporal scales, make it unlikely that a straightforward relationship between biodiversity per se and ecosystem functioning will be found. If some kind of relationship between biodiversity and ecosystem functioning will be found anywhere, it is probably in 'narrow physiology' groups, such as bacteria (Giller *et al.* 1998; Schimel & Gulledge 1998) or in species-poor taxonomic or functional groups, such as earthworms (Hoogerkamp *et al.* 1983).

Evolutionary biology carries an important message when considering effects of stress on communities and ecosystems. Each species has passively, i.e. by differential survival of organisms with favourable genetically encoded properties, evolved to meet the characteristics of a particular environment, consisting of both abiotic factors and other organisms. The degree then to which species can withstand or cope with *human-induced* stress is closely related to the genotypic variation and phenotypic plasticity resulting from the natural factors they adapted to during evolution. This is the evolutionary playground. Admittedly, many nuances can be added; the potential amplitude of a species for a certain environmental factor may be broader than the realised amplitude (Hutchinson 1957; Rorison 1969): 'favourable' behaviour may be acquired by learning (Vet *et al.* 1995); genetic variability, 'unmined' during evolution, that 'helps' them 'handle' a new type of stress, may exist. But this does not alter the basic rule.

So, it should be noted that even within the narrow-physiology or low-diversity groups mentioned above, it will be the ability or inability of each of the constituent species to fill niche space, i.e. the functional composition of the assemblage, that determines the outcome at the process level, rather than diversity per se.

Perhaps due to the specialisation of both taxonomists and ecologists in certain groups of soil biota, research on the effects of stress is in many cases restricted to sets or subsets of taxonomic groups. This is less unfortunate than it may seem, because sets or subsets of these groups constitute meaningful aggregations of the soil biota in view of their functions in nutrient cycling, dynamics of soil structure and rhizosphere processes. Nutrient cycling is largely due to bacteria, fungi, protozoa and nematodes (Hendrix *et al.* 1990). Depending on soil type, either earthworms or enchytraeids, mites and springtails govern the dynamics of soil structure (Hendrix *et al.* 1990), while rhizosphere processes are largely influenced by rhizobia, mycorrhizal fungi, plant-feeding nematodes and root-herbivorous insects. Hence, it would seem that analysing how stress impacts on taxonomic groups might improve our understanding of these effects on ecosystem functioning.

There are different approaches in the research on stress effects on communities and ecosystems. One approach has become prevalent in ecotoxicology, where effects of contaminants are studied at the level of the individual organism, followed by scaling up the effects on growth, reproduction, fecundity and longevity to the population level. Species-sensitivity distributions (SSDs), representing the variation in sensitivity of species to a contaminant by a statistical or empirical distribution function of responses for a sample of species, are then used to derive environmental quality criteria and ecological risk assessments

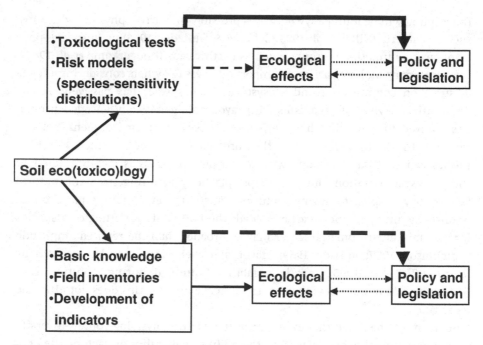

Figure 18.1. Modelled stress effects on ecosystems have had a stronger impact on policy and legislation than observed effects of stress in the field. Based on personal observation by A. J. Schouten in 2002.

(Posthuma *et al.* 2002). Currently, efforts are made to merge the SSD approach with food web modelling so as to incorporate biological interactions.

The SSD approach has been endorsed by the Organisation for Economic Co-operation and Development and adopted in the legislation of several countries in the European Union and the USA, despite the fact that the relationship between contaminant stress and observed effects on communities and ecosystems is weak. In contrast, where such relationships have been observed using different approaches, their generality beyond the site of observation is always an issue. Hence, it has been difficult to bridge the gap between scientists and policy makers and legislators (Fig. 18.1). From both observations it follows that there is still an urgent need for careful analysis of stress effects on real communities and ecosystem functioning in the field.

Because it is necessary to find the appropriate level of community organisation at which the distinction between stress effects and the background noise of natural variation can easily be made, studies about stress factors in natural environments are indispensable. During the last decade, we studied the effect of diminishing the nutrient supply on the soil biota in grasslands under restoration succession. We first summarise how the decrease in nutrient supply caused by annual cutting and hay-making affected the plant species diversity and the availability of certain nutrients. Thereafter we concentrate on plant-feeding nematodes and beetles to illustrate the quest for determining stress effects on these groups of organisms.

The Drentse A area

The nutrient supply to plants and the organisms depending on them is related to soil type and climate. (Over)fertilisation and atmospheric deposition of nutrients, in general, cause a change from a species-rich, nutrient-limited vegetation, to a species-poor, light-limited vegetation (e.g. Marrs 1993). Nutrient supply influences the chemical composition of plant production and, ultimately, plant residue quality, which in its turn affects the nutrient supply from the soil. Nutrient supply and plant residue quality are important aspects of resource quality. In addition, biotic stress may influence vegetation succession (van der Putten 2001).

In an earlier paper (Brussaard et al. 1996), we summarised thesis work on the vegetation succession in grasslands alongside a small stream in the Drentse A area, northern Netherlands (Olff 1992) and, in particular, the parts published by Olff and Bakker (1991), Olff and Pegtel (1994) and Olff et al. (1994). They showed that, when the above-ground production drops to below approximately 6 Mg ha^{-1} a^{-1}, plant-species diversity rises with a characteristic shift from a few, light-limited species (e.g. Lolium perenne L. and Holcus lanatus L.), to many, nutrient-limited species (e.g. Festuca rubra L. and Anthoxanthum odoratum L.). As the reversed succession proceeded, first nitrogen and potassium (after two years without fertilisation), later also phosphorus (after between 6 and 19 years) became limiting. Nutrient limitation was concluded to be an explanation for the increasing species richness in these grasslands.

Plant-feeding nematodes and insects were not considered by Olff (1992). We continued the work in some of the same and some nearby grasslands in the Drentse A area, focusing on nematodes and beetles. Basic soil characteristics are given in Table 18.1. In the past two decades, there has been an increasing interest in the role of below-ground grazers and pathogens in spatio-temporal vegetation processes. Studies on this subject have recently been comprehensively reviewed (Brown & Gange 1990; Mortimer et al. 1999; van der Putten 2001; Kuyper & de Goede 2005). In particular, plant-feeding nematodes are considered major below-ground herbivores affected by, and affecting, vegetation composition (Stanton 1988; Mortimer et al. 1999). Among the beetles, weevils (Curculionidae) and click beetles (Elateridae) would be expected to show close relationships with the plants on which they feed.

Objectives

The objective of this chapter is to analyse how nutrient stress and biotic stress affect plants, nematodes and beetles in a reversed succession of grasslands,

Table 18.1. *Soil characteristics (mean values of three replicates) of the four Drentse A sites studied, at 0–10 cm depth.*

	Field O	Field B	Field C	Field K
Years unfertilised	6	10	23	28
Soil bulk density ($g\,cm^{-3}$)	1.21a	1.12ab	1.03b	1.17a
pH (H_2O)	4.87a	3.89b	4.00b	3.96b
Clay + silt (%)	23	21.8	19.9	21.9
Total C ($g\,cm^{-3}$)	0.53b	0.41c	0.77a	0.79a
Total N ($g\,cm^{-3}$)	0.028ab	0.021c	0.034a	0.026bc
C : N ratio	19.3b	24.4b	23.5b	35.0a
Organic C (%)	3.89bc	3.60c	5.63a	5.31ab

Values with the same letter were not significantly different among the four Drentse A sites ($P < 0.05$). From Verschoor (2001).

where fertilisation was stopped 6–7, 10–11, 23–24 or 28–29 years ago, while nutrient impoverishment by hay-making continued. On the assumption that plants are more important for plant feeders than for organisms that do not have direct trophic interactions with plants, we discriminate between plant-feeding and free-living nematodes, and between herbivorous and polyphagous insects. Details on sampling nematodes are given by Verschoor *et al.* (2001) and on insects by Hemerik and Brussaard (2002). Initially, we also discriminated between species richness of monocotyledonous and dicotyledonous plants, because of known preferences of faunal groups for either one or other of these plant groups. Because the share of monocotyledonous and dicotyledonous species was approximately 50% in all four grasslands, and because plant cover was almost entirely determined by monocotyledons, discriminating between the two groups yielded no extra information in the diversity analysis and hence will not be reported.

We hypothesised that changes in plant species richness or the Shannon index of plant diversity (determined using plant cover data) would be correlated with

- changes in the taxonomic richness or the Shannon index of faunal diversity (determined using proportional numbers) of plant-feeding groups (plant-feeding nematodes, weevils (Curculionidae) and click beetles (Elateridae)), but
- not with changes in taxonomic richness or the Shannon index of faunal diversity of non-plant-feeding groups (free-living nematodes and ground beetles (Carabidae)).

We further analysed the data, using various methods, which will be described later. Because some abiotic soil factors showed a decreasing (pH) or an increasing

Table 18.2. *Taxonomic richness of faunal groups correlated with the taxonomic richness of Angiospermae; and taxonomic richness of Angiospermae and faunal groups correlated with time since fertilisation stopped.*

Taxonomic richness	Field O	Field B	Field C	Field K	Angiospermae vs. faunal groups Probability	Years without fertilisation vs. taxonomic groups Probability
Years without fertilisation	7	11	24	29		
Angiospermae	17	16	13	12		<0.001
Carabidae	43	49	42	27	ns	ns
Elateridae	2	5	5	3	ns	ns
Curculionidae	21	20	28	13	ns	ns
Free-living Nematoda	24	20	19	16	0.089	0.089
Plant-feeding Nematoda	10	6	9	8	ns	ns
Total Nematoda	34	26	28	24	ns	ns

(carbon : nitrogen ratio of soil organic matter) trend with increasing number of years without fertilisation (Table 18.1), we hypothesised that taxonomic richness and Shannon diversity of non-plant-feeding groups would also change, without being able to predict an increase or decrease.

Diversity of plants, nematodes and beetles

The loss of species of nutrient-rich conditions was larger than the gain of species of nutrient-poor conditions. Therefore, contrary to our expectation, the species richness of higher plants declined with increasing number of years without fertilisation. We know from observations in adjacent grasslands that eventually the species richness increases. One reason why we did not observe this in our grasslands may be the limited dispersal ability of many of the rarer plant species, for which we gave evidence in an earlier paper (Brussaard *et al.* 1996).

We used taxonomic richness and Shannon diversity to correlate the different faunal groups with plants and all groups with number of years without fertilisation. Table 18.2 shows taxonomic richness values for the various groups investigated, and correlations of all groups with years without fertilisation and of faunal groups with plants. In view of the low number of observations on fields under restoration succession, we report P values of <0.10. Taxonomic richness of plants and free-living nematodes, but not ground beetles, was significantly negatively correlated with number of years without fertilisation and, hence, taxonomic richness of plants and free-living nematodes was mutually correlated as well. Contrary to our expectation, taxonomic richness of plant-feeding nematodes, weevils or click beetles did not correlate with that of plants.

Table 18.3. *Shannon's index of diversity* H *of faunal groups correlated with* H *(Angiospermae) and* H *of Angiospermae and faunal groups with time since fertilisation stopped.*

Shannon diversity, H	Field O	Field B	Field C	Field K	Angiospermae vs. faunal groups Probability	Years without fertilisation vs. taxonomic groups Probability
Years without fertilisation	7	11	24	29		
Angiospermae	1.38	0.96	0.89	1.08		ns
Carabidae	2.19	2.66	2.64	2.02	ns	ns
Elateridae	0.16	0.78	0.53	0.58	ns	ns
Curculionidae	1.95	2.22	2.96	2.12	ns	ns
Free-living Nematoda	2.88	2.61	2.44	1.98	ns	0.066
Plant-feeding Nematoda	1.9	1.32	1.83	1.48	ns	ns
Total Nematoda	3.14	2.85	2.83	2.45	ns	ns

Table 18.3 shows values of Shannon's diversity index H for the various groups investigated and correlations of all groups with years without fertilisation and of faunal groups with plants. We will further use H(group) for indicating the Shannon diversity index of that group. In accordance with results on taxonomic richness, H(plant-feeding nematodes), H(click beetles) and H(weevils) did not correlate with H(plants). When restoration succession lasted longer, H(free-living nematodes) decreased but not H(ground beetles). Because no correlation was shown between H(free-living nematodes) and H(plants), it is likely that the positive correlation between the taxonomic diversities of both groups (Table 18.2) was a side effect of the positive correlations of taxonomic diversity of each of the two groups with number of years without fertilisation.

Further analysis of stress effects on nematodes

The nematodes most likely to affect, and to be affected by, the vegetation are plant-feeding nematodes. We performed correspondence analysis to describe changes in the plant-feeding nematode community structure with explanatory variables grouped into five classes: field (fields labelled O, B, C and K representing the treatment of non-fertilisation for 6, 10, 23 or 28 years, respectively), plant species (*L. perenne, H. lanatus, F. rubra* and *A. odoratum* characteristic for the various stages of reversed succession), root biomass, sampling date and abiotic factors. Figure 18.2 clearly shows the distinct nature of the plant-feeding nematode communities of fields O, B and C/K.

Statistical analyses revealed significant but small differences in plant-feeding nematode abundance between the rhizospheres of co-existing plant species

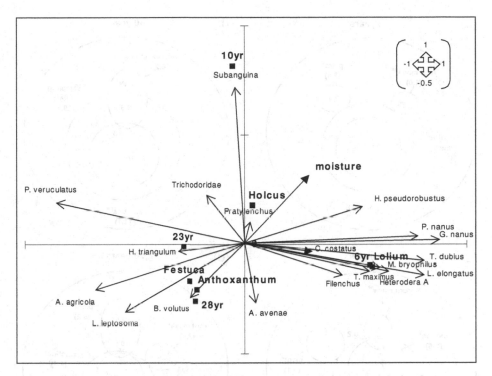

Figure 18.2. Correlation triplot (axes 1 and 2) based on redundancy analysis of nematode species and environmental variables (axes 1 and 2) in fields O (6 years without fertilisation), B (10 years), C (23 years) and K (28 years) of the Drentse A grasslands. Only significant explanatory variables (Monte Carlo permutation test, P < 0.005) are presented. From Verschoor *et al.* (2001).

('plant' in Fig. 18.3). In contrast to these small differences, the plant-feeding nematode community in the rhizosphere of one species, which occurs in all but the poorest stages of succession, *H. lanatus*, markedly changed as the period of non-fertilisation increased. Moreover, the results of redundancy analysis performed on the plant-feeding nematode data collected under *H. lanatus* closely resembled those of the redundancy analysis performed on all plant-feeding nematode data. Almost all explained variance (49 out of 55%) could be ascribed to time of non-fertilisation (Fig. 18.2).

This indicates that *intra*specific rather than *inter*specific differences in food quality affected the plant-feeding nematode succession in the grasslands. This is further exemplified by clear shifts in nematode feeding types and biomass from large to small plant-feeding nematodes, with mean individual biomasses of 0.051, 0.028, 0.020 and 0.024 µg in fields O, B, C and K, respectively (Table 18.4). Abiotic factors and sampling date also significantly contributed to the explained variance in plant-feeding nematode abundance. Differences in abiotic factors, such as pH, however, will also bring about differences in amount and quality of

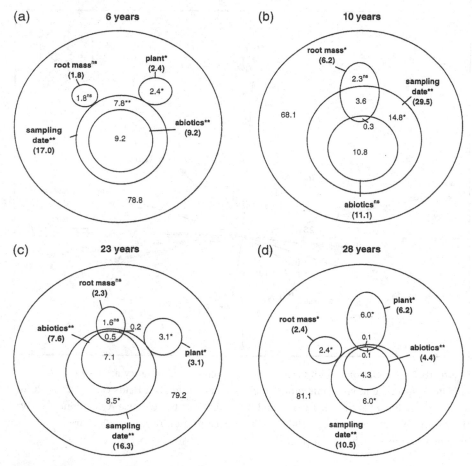

Figure 18.3. Variance diagrams based on the nematode data in the redundancy analysis of Fig. 18.2. For each of the explanatory variables, the percentage of the total variance of the nematode data explained is indicated. The outer circles represent the total variance (100%) in the nematode data. The inner circles represent the variance explained by the different combinations of variables (bold letters): ns, not significant; *, $P \leq 0.05$; **, $P \leq 0.01$. From Verschoor *et al.* (2001).

food plants and, hence, are difficult to interpret separately. Root biomass did not significantly explain changes in plant-feeding nematode abundance (Fig. 18.3).

Greenhouse and field experiments (Verschoor *et al.* 2002a, b) confirmed that root biomass was not correlated with plant-feeding nematode numbers. Root nutrient concentration, however, strongly affected plant-feeding nematode numbers and community composition. In a full factorial field manipulation experiment, with three replicate plots consisting of eight quadrats of 1 m^2 in each field, in which eight different treatments were established, one for each quadrat (control (C), nematicide (N), fertilisation (F), liming (L), nematicide and fertilisation (NF), nematicide and liming (NL), fertilisation and liming (FL), and

Table 18.4. *Mean annual biomass (μg) 100 cm^{-3} of soil of plant-feeding nematodes extracted from the rhizosphere soil or the roots of dominant plant species in each field studied.*

Field	O	O	B	C	C	K	K	K
Years unfertilised	6	6	10	23	23	28	28	28
Plant species	Lp	Hl	Hl	Hl	Fr	Hl	Fr	Ao

Mean biomass (μg) of nematodes 100 cm^{-3} extracted from soil fraction

Plant feeders	244 a	284 a	96 bc	96 b	81 bc	83 bc	62 c	79 bc
Sedentary endoparasites	3 b	2 b	15 a	+ c	+ c	0 c	0 c	+ c
Migratory endoparasites	4	6	2	3	2	4	2	2
Semi-endoparasites	27 a	34 a	36 a	2 b	1 b	7 b	3 b	2 b
Ectoparasites	173 a	195 a	29 c	54 b	57 b	20 cd	8 d	11 cd
Epidermis/root hair feeders	37 bd	47 bcd	14 f	37 de	21 e	52 ab	49 ac	64 a
Fungal feeders	10 bc	10 bc	15 b	39 a	27 a	7 bc	6 c	9 bc

Mean biomass (μg) of nematodes 100 cm^{-3} extracted from root fraction

Plant feeders	0.25 b	23 b	123 a	20 b	15 b	21 b	14 b	20 b
Sedentary endoparasites	0.01 b	+ c	112 a	+ c	+ c	+ c	+ c	+ c
Migratory endoparasites	0.07	9	8	13	10	13	9	11
Semi-endoparasites	0.01 bc	2 ab	+ c	+ c	+ c	2 a	2 ab	1 ab
Ectoparasites	0.05 a	3 a	+ b	1 b	1 b	1 b	+ b	+ b
Epidermis/root hair feeders	0.12 a	9 b	2 e	5 cd	3 de	4 cde	4 de	7 bc
Fungal feeders	0.02 a	1 a	1 bc	1 b	+ c	1 bc	1 b	1 b

Abbreviations: Lp, *Lolium perenne*; Hl, *Holcus lanatus*; Fr, *Festuca rubra*; Ao, *Anthoxanthum odoratum*; +, present, but with abundance of <0.5 100 cm^{-3}. Mean values with the same letter were not significantly different ($P < 0.05$). From Verschoor *et al.* (2001).

nematicide, fertilisation and liming (NFL)), fertilisation did not increase the plant-feeding nematode numbers in the high-production fields O and B, but it significantly did in the low-production fields C and K (Fig. 18.4).

These results further confirm that qualitative differences within, rather than between, plant species affected the succession of plant-feeding nematodes. This would explain the lack of correlation between taxonomic and Shannon diversities of plants and plant-feeding nematodes reported above. It is beyond the scope of this chapter to discuss the effects of stress by plant feeders on plant performance and diversity. We did, however, observe such effects of nematodes; they were statistically significant, but small in comparison with effects of nutrient limitation, and they were additive, not synergistic (Verschoor *et al.* 2002b).

The number of taxa of free-living nematodes from samples taken in the four grasslands is given in Table 18.5, subdivided into life-history group (according to Bongers 1990) or trophic group (according to Yeates *et al.* 1993). The nematode Maturity Index (MI; Bongers 1990) for free-living nematodes is known to

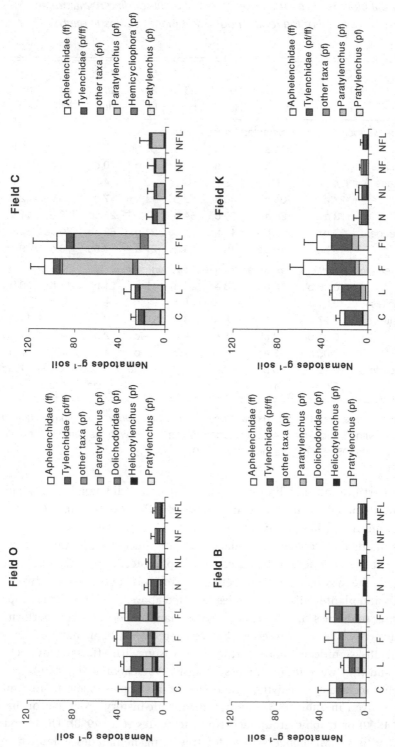

Figure 18.4. Mean numbers (±s.e.) of plant-feeding (pf) and fungal-feeding (ff) nematodes in experimental quadrats of fields O, B, C and K at different treatments. The quadrats were treated by different combinations of lime (L), fertilisers (F) and nematicides (N), or remained untreated (C). From Verschoor (2001).

Table 18.5. *Number of nematode taxa (trophic and life-history groups) and nematode indices for the four Drentse A grassland fields O, B, C and K and means and ranges for 15 nutrient-rich reference fields (Friesland) and 10 nutrient-poor reference fields (Junner Koeland).*

Field	Friesland	O	B	C	K	Junner Koeland
Trophic groups						
Endoparasites		2	2	2	2	
Ectoparasites		4	2	5	3	
Root hair/ epidermal feeding		4	2	2	3	
Bacterial feeding		18	13	11	10	
Fungal feeding		2	2	3	2	
Predatory		1	1	2	0	
Omnivorous		4	3	2	4	
Total		33	25	27	24	
Life-history groups						
c-p 1		3	3	2	2	
c-p 2		7	7	5	4	
c-p 3		6	4	6	4	
c-p 4		5	4	4	4	
c-p 5		2	1	1	2	
Total		23	19	18	16	
Indices						
Maturity Index (MI)	1.85 (1.44 2.26)	2.78	2.63	2.82	2.33	2.49 (2.24–2.65)
MI 1–2	1.59 (1.28–1.74)	1.81	1.90	1.88	1.95	1.96 (1.84–2.00)
MI 3–5	4.30 (3.33–5.00)	3.62	3.29	3.25	3.15	3.34 (3.03–3.65)
Enrichment Index (EI)	71.51 (55.44–91.43)	49.13	32.40	35.21	22.31	16.59 (2.20–45.5)

decrease with increasing fertilisation and increasing stress. Because fertilisation decreased and all other abiotically caused stresses increased with succession, the change in the MI is not straightforward (fluctuating during succession: Table 18.5). We therefore split the MI into MI 1–2, comprising the c-p groups 1 and 2, mainly consisting of bacterivores and fungivores, which are the most sensitive to fertilisation, and MI 3–5, comprising the c-p groups 3–5, mainly consisting of omnivores and predators, which are the most sensitive to all other stress factors except nutrient availability. Following Ferris *et al.* (2001), we also calculated the Enrichment Index (EI), for which certain taxa from c-p 1 and 2 groups are selected and weighed according to their life history. As expected, the MI 1–2 increased and the EI decreased during succession, meaning that the share of 'enrichment opportunists' decreased and indicating a decrease in easily available nutrients. Also as expected, the MI 3–5 decreased, meaning a relative decrease of the higher c-p groups and indicating increasing environmental stress. In fields O + B, 19 c-p 3–5 taxa occurred, of which 6 did not occur in fields C + K; in fields C + K, 18 c-p 3–5 taxa occurred, of which 5 did not occur in fields O + B. Given the clear change in environmental conditions during succession (Table 18.1), we consider the change in the taxonomic composition of c-p 3–5 nematodes and the change in MI 3–5 to be small.

As a reference for the MI and EI values we obtained, we used values from 15 grasslands in Friesland in the study of de Goede *et al.* (2003), which have similar soil characteristics and are still being fertilised (nutrient-rich reference to field O). We also used unpublished values obtained from 10 sampling sites, in the study of Blomqvist *et al.* (2000) in grasslands on loamy sand in the Junner Koeland, which have never been fertilised (nutrient-poor reference to field K). As expected, the MI 1–2 values in Friesland were (much) lower and the EI values (much) higher than those from the Drentse A grasslands and the Junner Koeland (Table 18.5). The MI 1–2 value of 1.95 in field K, although close to the maximum value of 2.00, was still at the lower end of the range of the 10 reference values in Junner Koeland (between rank numbers 2 and 3), while the EI value of 22.31 was at the higher end of the range of the 10 reference values (between rank numbers 8 and 9). As also expected, most MI 3–5 values in Friesland were higher than those from the Drentse A grasslands and Junner Koeland (Table 18.5). The MI 3–5 value of field K was at the lower end of the 10 reference values in Junner Koeland (between rank numbers 1 and 2). This suggests that stress in this field is high. This may either mean that the Drentse A grasslands are close to the final stage of succession as regards nematodes or that nematodes adapted to such stress have been unable to colonise the nutrient-poor sites. The latter explanation fits the finding of rather little change in the nematode composition during succession. In an earlier paper we gave evidence that dispersal ability is hampering the colonisation of the nutrient-poor sites in some plant species (Brussaard *et al.* 1996) and we suggest that this may also apply to nematode species.

Further analysis of stress effects on beetles

We performed Canonical Correspondence Analysis (CANOCO) of catches of Carabidae and Curculionidae with day number, mean air temperature and total precipitation in the previous week, soil moisture content and 7, 11, 24 and 29 years without fertilisation as environmental variables (Fig. 18.5). Due to the low number of species (five) and dates at which adults were caught, such an analysis was not possible for Elateridae. In both cases the explained variance was rather low (less than 25%), but significant for the first two axes. Years without fertilisation explained all the variance of numbers of individuals of weevil species (first axis: 7 and 11 years; second axis: 29 and 24 years), whereas abiotic variables did not contribute to the explained variance at all. In the ground beetle

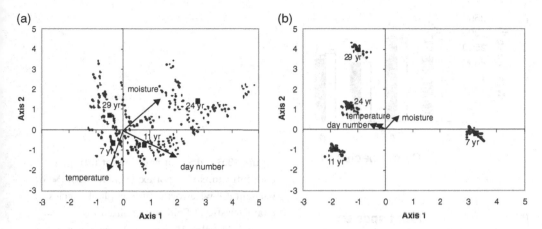

Figure 18.5. Canonical correspondence analysis of beetle species and environmental variables in fields unfertilised for 7, 11, 24 or 29 years of the Drentse A grasslands: (a) Carabidae and (b) Curculionidae. From Hemerik and Brussaard (2002).

species, the variance in numbers of individuals was explained by day number and 29 years (first axis) and temperature and precipitation (second axis). These results are consistent with our hypothesis that the distribution of numbers in weevil species is associated with plant successional stage, whereas the distribution of numbers in ground beetle species is largely associated with abiotic variables.

Because it is not the taxonomic or Shannon diversities of plants and weevils that are associated, we tested in successive experiments whether intraspecific rather than interspecific differences between food plants would affect weevil numbers. These follow-up experiments were performed with wireworms (click-beetle larvae), because they were by far the most numerous group in our soil samples. A food choice experiment was done with larvae of the click beetle *Agriotes obscurus* (L.), which comprised 83% of the Elateridae we caught in the four grasslands during nine months of combined soil, pitfall and light fall sampling. *Agriotes obscurus* comprised 70% of the wireworms we collected for the food choice experiment; the other 30% belonged to *Athous haemorrhoidalis* (F.), which we also tested in the experiment following storage at 5 °C. Of the four grass species characteristic for each of the stages of reversed succession, *L. perenne*, *H. lanatus*, *F. rubra* and *A. odoratum*, four individuals of each were planted in pots with potting soil after six or nine weeks of culturing from seed. Three days later, 40 or 60 wireworms of one species were added. All six possible spatial arrangements of the plants around the central release point were duly replicated. The spatial arrangement of the pots was randomised and the pots were kept at constant weight. Wireworms were recaptured after one week from the vicinity of each of the plants (94% recovery). The data were analysed with multi-nomial logit models, revealing a statistically significant preference for the plant species from the relatively nutrient-rich sites, *L. perenne* and *H. lanatus*, in *A. obscurus*,

(a)

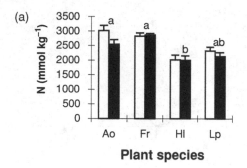

(b)   **Plant species**

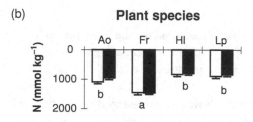

Figure 18.6. (a) Shoot and (b) root nitrogen concentrations in a pot experiment with (■) or without (□) wireworms in four plant species characteristic of fields unfertilised for 7 (Lp = *L. perenne*), 11 (Hl = *H. lanatus*), 24 (Fr = *F. rubra*) or 29 (Ao = *A. odoratum*) years. Bar pairs with the same letters are not significantly different at $P < 0.05$. Only in Ao shoot was the difference between + or − wireworms marginally significant ($0.10 > P > 0.05$).

with no preference in *Athous haemorrhoidalis*, but avoidance of *F. rubra* in both species (Hemerik *et al.* 2003). We then proceeded by assessing whether *A. obscurus* could grow in pots containing one of each of the plant species. Starting from four batches of 45 pots with one individual plant and one wireworm and a batch with only one wireworm, all wireworms significantly gained in weight on all four plant species (recovery at least 82% or more) during five weeks; whereas in the control pots, weights diminished slightly (not significantly) and recovery was low (38%). The shoot-to-root biomass ratios were significantly lowered in *L. perenne*, *H. lanatus* and *A. odoratum* (Hemerik *et al.* 2003).

The preference for *L. perenne* and *H. lanatus* in *A. obscurus*, although it could grow on all four grass species, may not be related to the plant species as such, but may be caused by the food quality irrespective of plant species. We thus proceeded by determining the nitrogen concentrations in shoots and roots of each of the grass species after growth of a plant of one species in pots with or without wireworms. The results are given in Fig. 18.6. Wherever the treatments *with* wireworms showed significant differences in nitrogen concentrations between plant species, such significant differences between plant species were also found in the treatments *without* wireworms. This shows that such differences had not been caused by the wireworms, but were plant species-specific under the prevalent growth conditions (plants in potting soil). In the control treatments, *L. perenne* and *H. lanatus* showed significantly lower nitrogen shoot concentrations than *F. rubra* and *A. odoratum*, and lower nitrogen root concentrations than *F. rubra*. Although the plants in the latter experiment were three weeks old at the start of the experiment, whereas those in the food choice experiment mentioned above were six or nine weeks old at the start, we tentatively conclude that the preference for L. perenne and H. lanatus in the food choice experiment was not determined by the nitrogen concentrations in shoot and root, because these were lower than in non-preferred plant species. Neither was avoidance of F. rubra caused by nitrogen concentrations, because they were higher than in preferred plant species. Hence, our hypothesis that, as in plant-feeding nematodes, intraspecific rather than interspecific differences between plant species are more important determinants of wireworm abundance was not confirmed. Insect-deterrent compounds in the rhizosphere may be a more likely cause in this species, and this warrants further research.

Conclusions

Our studies fully reflect the complexity of the biotic and abiotic environment in which plant and faunal communities occur and develop. By comparing faunal groups that show direct trophic links with plants, i.e. plant-feeding nematodes and insects, with those that do not show such direct links, i.e. free-living nematodes and insects, we attempted to shed light on supposedly meaningful differential biotic and abiotic stress effects of nutrient impoverishment of the soil on plant and faunal communities. We found evidence that, in the case of plant-feeding nematodes, the quality of their food rather than the nature of the plant species is decisive in shaping the plant-feeding nematode community towards prevalence of groups with distinctive life-history characteristics (functional groups). The only group of plant-feeding insects that was speciose and numerous enough to perform analysis of distribution and abundance, i.e. the weevils, also showed a clear association with the stage of successional development.

However, in the one wireworm species that we experimentally tested, it was not the quality of the food that determined its preference for certain plant species. These results are consistent with the hypothesis that under the conditions studied plants tolerate nematodes rather than deter them. In the case of root-herbivorous insects, tolerance may prevail in plant species that occur under nutrient-rich conditions, whereas deterrence may prevail in species of nutrient-poor conditions. Although the available evidence should serve more as an encouragement for further research and not as a basis for general conclusions, such a difference may be considered functional, because flying insects can actively select plant species. In contrast, dispersal ability will be much more limited in nematodes and, hence, the community of nematodes in the rhizosphere of plant species will be much less specific than in the case of insects.

As expected, free-living nematodes and ground beetles were more associated with abiotic than plant (restoration succession) variables. In the case of the plant-feeding nematodes, the community structure could be characterised, based on feeding strategies. In the case of the free-living nematodes various other life-history traits enabled characterisation of the community.

What did we learn? Twenty years ago, Harper (1982) critically reflected on issues such as adaptation, strategy and stress as investigated by population, community and ecosystem ecologists. He concluded that from studies like ours it is not obvious where increased understanding will take us: 'If it is true that natural vegetation is what it is because of the evolution of individualistic selfish traits, it is unlikely that we will find ultimate explanations of ecological phenomena in nature at the area or population level. Both area and ecosystem (. . .) may be inappropriate levels of study at which to seek "ultimate" explanations of ecological behaviour. If this is the case, only superficial generalities about ecosystems may be expected to emerge from systems studies of natural communities' (Harper 1982). In a review of the state-of-the-art on interactions of plants, soil pathogens and their antagonists in natural ecosystems, van der Putten (2001), however, concluded that 'there is a strong need for a more comparative approach when studying soil pathogens in natural ecosystems. Ecosystem development (the stage of successional development) may be one form of stratification that can be used'.

Our studies show that this approach, backed up by experimental work to test hypotheses gained at the community and ecosystem level, gives a better insight into community and ecosystem processes. We believe that this approach will have to meet the SSD approach (Posthuma *et al.* 2002), which has originally started from clear-cut sources of stress at the population level of certain species, but is now in the process of incorporating different reactions of broad taxonomic groups to different sources of stress. Our studies indicate that within such broad taxonomic groups differences in feeding and/or life-history strategies, which lead to the definition of functional groups, can be related to stress effects on communities in a meaningful way. So taking the limitations and potentials of the evolutionary playground into account will be a common feature of both approaches. Only in this way will it be possible to develop a realistic base for policy and legislation for stress effects on communities and ecosystems.

Acknowledgements
We thank Gerda Bijl for help with preparing the illustrations and editing the text and Thom Kuyper for critical comments on the manuscript.

References

Beare, M. H., Coleman, D. C., Crossley, D. A., Jr., Hendrix, P. F. & Odum, E. P. (1995). A hierarchical approach to evaluating the significance of soil biodiversity to biogeochemical cycling. *Plant and Soil*, 170, 5–22.

Blomqvist, M. M., Olff, H., Blaauw, M. B., Bongers, T. & van der Putten, W. H. (2000).

Interactions between above- and belowground biota: importance for small-scale vegetation mosaics in a grassland ecosystem. *Oikos*, **90**, 582–598.

Bongers, T. (1990). The maturity index: an ecological measure of environmental disturbance based on nematode species composition. *Oecologia*, **83**, 14–19.

Brown, V. K. & Gange, A. C. (1990). Insect herbivory below ground. *Advances in Ecological Research*, **20**, 1–58.

Brussaard, L. (1998). Soil: fauna, guilds, functional groups and ecosystem processes. *Applied Soil Ecology*, **9**, 123–135.

Brussaard, L., Bakker, J. P. & Ollf, H. (1996). Biodiversity of soil biota and plants in abandoned arable fields and grasslands under restoration management. *Biodiversity and Conservation*, **5**, 211–221.

Brussaard, L., Behan-Pelletier, V. M., Bignell, D. E., *et al.* (1997). Biodiversity and ecosystem functioning in soil. *Ambio*, **26**, 563–570.

Calow, P. & Forbes, V. E. (1998). How do physiological responses to stress translate into ecological and evolutionary processes? *Comparative Biochemistry and Physiology A*, **120**, 11–16.

Chapin III, F. S., Zavaleta, E. S., Eviner, V. T., *et al.* (2000). Consequences of changing biodiversity. *Nature*, **405**, 234–242.

De Goede, R. G. M., Brussaard, L. & Akkermans, A. D. L. (2003). On-farm impact of cattle slurry management on biological soil quality. *NJAS, Wageningen Journal of Life Sciences*, **51**, 103–133.

Ettema, C. H. & Wardle, D. A. (2002). Spatial soil ecology. *Trends in Ecology and Evolution*, **17**, 177–183.

Ferris, H., Bongers, T. & de Goede, R. G. M. (2001). A framework for soil food web diagnostics: extension of the nematode faunal analysis concept. *Applied Soil Ecology*, **18**, 13–29.

Giller, K. E., Witter, E. & McGrath, S. P. (1998). Toxicity of heavy metals to microorganisms and microbial processes in agricultural soils:

a review. *Soil Biology and Biochemistry*, **30**, 1389–1414.

Grime, J. P., Hodgson, J. G. & Hunt, R. (1988). *Comparative Plant Ecology*. London: Unwin Hyman.

Harper, J. L. (1982). After description. *The Plant Community as Working Mechanism* (Ed. by E. I. Newman), pp. 11–25. Oxford: Blackwell Science.

Hemerik, L. & Brussaard, L. (2003). Diversity of soil macro-invertebrates in grasslands under restoration. *European Journal of Soil Biology*, **38**, 145–150.

Hemerik, L., Gort, G. & Brussaard, L. (2003). Food preference of wireworms analyzed with multinomial Logit models. *Journal of Insect Behaviour*, **16**, 647–665.

Hendrix, P. F., Crossley Jr., D. A., Blair, J. M. & Coleman, D. C. (1990). Soil biota as components of sustainable agroecosystems. *Sustainable Agricultural Systems* (Ed. by C. A. Edwards, R. Lal, P. Madden, R. H. Miller & G. House), pp. 637–654. Ankeny: Soil and Water Conservation Society.

Hoogerkamp, M., Rogaar, H. & Eijsackers, H. J. P. (1983). Effect of earthworms on grassland on recently reclaimed polder soils in the Netherlands. *Earthworm Ecology, from Darwin to Vermiculture* (Ed. by J. E. Satchell), pp. 85–105. London: Chapman and Hall.

Hutchinson, G. E. (1957) Concluding remarks. *Cold Spring Harbour Symposium on Quantitative Biology*, **22**, 415–427.

Kuyper, Th. W. & de Goede, R. G. M. (2005). Interactions between higher plants and soil-dwelling organisms. *Vegetation Ecology* (Ed. by E. van der Maarel), pp. 286–308. Oxford: Blackwell Science.

Marrs, R. H. (1993). Soil fertility and nature conservation in Europe: theoretical considerations and practical management solutions. *Advances in Ecological Research*, **24**, 241–300.

Moore, J. C., Walter, D. E. & Hunt, H. W. (1988). Arthropod regulation of micro- and mesobiota in below-ground detrital food webs. *Annual Review of Entomology*, **33**, 419–439.

Mortimer, S. R., van der Putten, W. H. & Brown, V. K. (1999). Insect and nematode herbivory below ground. *Herbivores: Between Plants and Predators* (Ed. by H. Olff, V. K. Brown & R. H. Drent), pp. 205–238. Oxford: Blackwell Science.

Olff, H. (1992). On the mechanisms of vegetation succession. Unpublished Ph.D. thesis, University of Groningen.

Olff, H. & Bakker, J. P. (1991). Long-term dynamics of standing crop and species composition after the cessation of fertilizer application to mown grassland. *Journal of Applied Ecology*, **28**, 1040–1052.

Olff, H., Berendse, F. & de Visser, W. (1994). Changes in nitrogen mineralization, tissue nutrient concentrations and biomass compartmentation after cessation of fertilizer application to mown grassland. *Journal of Ecology*, **82**, 611–620.

Olff, H. & Pegtel, D. M. (1994). Characterisation of the type and extent of nutrient limitation in grassland vegetation using a bioassay with intact sods. *Plant and Soil*, **163**, 217–224.

Posthuma, L., Suter II, G. W. & Traas, T. P. (Eds.) (2002). *Species Sensitivity Distributions in Ecotoxicology*. Boca Raton, FL: Lewis.

Rorison, I. H. (1969). Ecological inferences from laboratory experiments on mineral nutrition. *Ecological Aspects of the Mineral Nutrition of Plants* (Ed. by I. H. Rorison), pp. 155–175. Oxford: Blackwell Science.

Schimel, J. P. & Gulledge, J. (1998). Microbial community structure and global trace gases. *Global Change Biology*, **4**, 745–758.

Stanton, N. L. (1988). The underground in grasslands. *Annual Review of Ecology and Systematics*, **19**, 573–589.

Swift, M. J., Heal, O. W. & Anderson, J. M. (1979). *Decomposition in Terrestrial Ecosystems*. Oxford: Blackwell Science.

van der Putten, W. H. (2001). Interactions of plants, soil pathogens, and their antagonists in natural ecosystems. *Biotic Interactions in Plant–Pathogen Associations* (Ed. by M. J. Jeger & N. J. Spence), pp. 285–305. Wallingford: CAB International.

Verschoor, B. C. (2001). Nematode–plant interactions in grasslands under restoration management. Unpublished Ph.D. thesis, Wageningen University.

Verschoor, B. C., de Goede, R. G. M. & Brussaard, L. (2002a). Do plant parasitic nematodes have differential effects on the productivity of a fast- and a slow-growing grass species? *Plant and Soil*, **243**, 81–90.

Verschoor, B. C., de Goede, R. G. M., de Vries, F. W. & Brussaard, L. (2001). Changes in the composition of the plant-feeding nematode community in grasslands after cessation of fertiliser application. *Applied Soil Ecology*, **17**, 1–17.

Verschoor, B. C., Pronk, T. E., de Goede, R. G. M. & Brussaard, L. (2002b). Could plant-feeding nematodes affect the competition between grass species during succession in grasslands under restoration management? *Journal of Ecology*, **90**, 753–761.

Vet, L. E. M., Lewis, W. J. & Cardé, R. T. (1995). Parasitoid foraging and learning. *Chemical Ecology of Insects* (Ed. by W. Bell & R. T. Cardé), pp. 65–101. London: Chapman and Hall.

Yeates, G. W., Bongers, T., de Goede, R. G. M., Freckman, D. W. & Georgieva, S. S. (1993). Feeding habits in soil nematode families and genera: an outline for soil ecologists. *Journal of Nematology*, **25**, 315–331.

CHAPTER NINETEEN

Soil biodiversity, nature conservation and sustainability

MICHAEL B. USHER

University of Stirling

SUMMARY

1. Soils have hardly featured in nature conservation thinking. Criteria have been developed for selecting networks of important Earth science sites, but these have not included criteria for soils.
2. Above ground, nature conservation has focused on communities of plants and the animals that they support, and criteria have been developed for selecting the 'best' sites. There has been scant attention to the soils on which those plant communities depend.
3. Although species rich, soils do not contain the charismatic species that have been favoured by conservationists. There is no giant panda, corncrake or lady's slipper orchid.
4. Both the increasing concentration on biodiversity and 'the ecosystem approach' are shifting thinking in relation to soils. Despite limited taxonomic knowledge, some attention is being paid to fungi (e.g. the waxcaps, *Hygrocybe* spp.) and the larger soil-inhabiting invertebrates (e.g. the mole cricket, *Gryllotalpa gryllotalpa*, and earthworms). The 'ecosystem approach' is forcing a more holistic view, focusing on the function of terrestrial ecosystems.
5. Soils are intimately involved in many ecosystem processes that contribute to the sustainable use of the planet's land resources. The contribution of soils and their biota to sustainable development will ultimately be far more important than the protection of either individual soil types or individual species.

Introduction

Although it is true that nature conservationists have generally neglected soils, there is one notable exception, that relating to peat soils (e.g. Heathwaite & Göttlich 1990). The literature explicitly relating soil science to nature

Biological Diversity and Function in Soils, eds. Richard D. Bardgett, Michael B. Usher and David W. Hopkins.
Published by Cambridge University Press. © British Ecological Society 2005.

conservation is small, though there is a more extensive literature about nature conservation that implicitly incorporates ideas about soils. A pioneering publication in the more modern literature that explicitly addressed soils and nature conservation was Ball and Stevens' (1980–81) discussion of ancient woodlands and the conservation of 'undisturbed soils'. They considered that the recognition of undisturbed soil profiles would be a valuable criterion for assessing whether a site was an ancient woodland or whether there were more recent modifications to the woodland's natural character. In many ways this relates to Tansley's (1949) discussion of the complex interactions between climate, vegetation cover and soil profile development, or to Tamm's (1950) descriptions of the parallel developments of soil and coniferous forest types in the Boreal region of Scandinavia.

The wealth of literature on both soil science and nature conservation needs to be brought together so as to develop the relationships between these two fields of knowledge. The advent of an increasing interest in the topic of 'biodiversity' provides an impetus for exploring the value of the diversity of life in the soil, its function and its conservation. This chapter reviews criteria that have been proposed for identifying important aspects of either the Earth sciences or the biological sciences for conservation, and then explores various aspects of soils, such as the conservation of soil organisms and the designation of protected areas. It will also explore two of the three levels of biodiversity mentioned in the Convention on Biological Diversity, namely species and ecosystems (the genetic level is not discussed). Finally, it will focus on the difficult topic of soils in relation to the sustainability of land management, investigating a definition of 'sustainability' and asking about the links between soils and sustainable development.

What should be conserved? The development of criteria

The development of criteria for achieving the aims of biological conservation has had a long history. Ten criteria were proposed by Ratcliffe (1977, pp. 7–10) in his review of the most important nature conservation sites in Great Britain. An international review (Usher 1986a, p. 13) found a series of 26 criteria being used in 17 publications appearing in the decade from 1971 to 1981. In general, if an area of land or water was being evaluated for possible designation for biological conservation purposes, diversity of both species and habitats was considered desirable, as were naturalness, large extent and rarity, again of both species and habitats. Soils were not explicitly included in any of the criteria, but implicitly they might have been considered in some of the lesser-used criteria such as 'typicalness', 'successional stage' or 'archaeological interest'. In terms of the legal requirements for site designation in Great Britain, much of this work on criteria became codified into a series of guidelines for different habitats and different taxonomic groups of species (Anonymous 1989). Apart from peatlands (Anonymous 1994a), soils do not feature in these guidelines.

Rather later, thinking on the conservation of the Earth's heritage developed. Ellis *et al.* (1996, p. 105) said 'The importance of soils, somewhat neglected hitherto in nature conservation, is now gaining increased prominence and their study within the concept of sustainability is likely to become an important task for the future'. Although this statement is encouraging, the analysis of important Earth heritage sites in Great Britain considered stratigraphy, palaeontology, quaternary geology, geomorphology, igneous petrology, structural and metamorphic geology, and mineralogy, but not soils. Ellis *et al.* (1996, p. 45) used three criteria for the recognition of important Earth heritage sites, namely (1) the importance of the site to the international scientific community; (2) the presence of exceptional features that make the site scientifically important; and (3) national importance because the site is representative of a feature, event or process that was fundamental to the shaping of the country as it is known today.

The sets of criteria used for recognising sites of importance for either Earth science or biological conservation both omit explicit mention of soils. There are, however, very close relationships between rocks or glacial debris and the biological communities that they support, mediated through the soil (Usher 2001; Hopkins 2003). There is therefore a question as to why soils have not featured more in nature conservation thinking, at least not until the mid 1990s. In an interesting short essay on soils, Yaalon (2000) asked the question 'Is soil just dirt, too commonplace to mention or study?' Even although attitudes to soil are changing, there remain few links between nature conservation and soils. For example, Mermut and Eswaran (2001) reviewed major developments in soil science during the final third of the twentieth century, and in their list of 15 topics the word 'conservation' only appears as 'conservation tillage'. Heal's (1999) account of current issues in Arctic soil ecology similarly did not consider conservation. Why have soils been forgotten by nature conservationists, or has nature conservation been forgotten by soil scientists?

There are four possible reasons. First, the taxonomy of soil organisms is relatively poorly understood (André *et al.* 2001, 2002), and where it is better known species assemblages appear to be both widely distributed (Finlay & Clarke 1999; Finlay 2002) and without obvious rarities. Second, there are no 'charismatic' species, such as the giant panda, birds of prey or orchids, which capture the public's attention and imagination (and this highlights the sociological driver for much nature conservation work). Third, and related to the lack of taxonomic knowledge, there are no known 'hot spots' of species richness in soils (Wall *et al.* this volume), although there might be 'hot spots' of soil biological activity (Bundt *et al.* 2001). Finally, soils tend to be 'out of sight and out of mind' (Usher 1996), although the importance of below-ground processes within the whole ecosystem is increasingly being stressed (Copley 2000).

Should some criteria for soils be developed and added to those for the biological and Earth sciences? Soils are dynamic and their profiles can change substantially as the plant cover above them changes. Soils can also change as a result of geomorphological processes, such as erosion of existing material or the deposition of new material. The taxonomic uncertainties, concerning many of the groups of soil biota, makes species diversity a poor criterion, and the criterion of species' rarity impossible to use. It therefore seems that only a few of the criteria that have been developed for biological sites can be used for assessing the importance of soils, whereas the three that have been developed in the Earth sciences all have a role. However, analogous to the decline in a species' abundance and/or range, which is used for the IUCN's Red Data Lists (Anonymous 1994b), the rate of decline in extent of a soil type could be used as a criterion. A possible set of ten criteria is given in Table 19.1, but it is recognised that, as with most sets of criteria, these ten criteria are not all independent of each other.

The soils of designated areas

Because little attention has been given to designating sites for their soil interests, very limited data are available on the soils of designated areas. For example, the creation of a series of designated areas throughout the European Union, based on the 1979 Birds Directive and the 1992 Habitats Directive, does not require soil information to be notified. This 'Natura 2000' series of Specially Protected Areas (for birds) and Special Areas of Conservation (for habitats and an eclectic range of species other than birds) will thus not be capable of explicitly addressing soil conservation either generally or for the rarer types of soil. Similarly, the Emerald Network of Areas of Special Conservation Interest, championed by the Council of Europe and extending the Natura 2000 process from the 15 EU member states to a further 23 non-EU nations at the end of 2002, had not provided any focus on soils. Given the diversity of life in soil, it is astonishing that soil has been so neglected by nature conservationists.

An exception to this generalisation is in Scotland. Gauld and Bell (1997) used the 49 classes of soils, recognised in the National Soil Classification at the soil group (12) and subgroup (37) levels. Terminology in the various soil classifications can be difficult; that used in the UK was described and illustrated by Trudgill (1989), but this is comparable with that used in North America (Birkeland 1984) and elsewhere in the world (White 1987). Gauld and Bell (1997) compared their mapped distribution to a stratified random sample of 152 Sites of Special Scientific Interest (SSSIs), which form the statutory basis for designated sites in the UK. All of the major soil groups were represented in this series of SSSIs (Table 19.2), as were 36 of the 37 subgroups. The one soil subgroup not

Table 19.1. *Ten criteria that could be used to recognise soils of nature conservation importance. These criteria are derived from those developed in the biological and Earth sciences, with three additional ones specifically relating to soils.*

Origin	Criterion	Explanation
Biological	Naturalness	Lack of anthropogenic disturbance
	Typicalness	A soil type that is common and is supporting a characteristic assemblage of plants and animals
	Recorded history	Soils, which have been studied for a long period of time, can provide time-series data indicating change, or base-line data from which change can be measured
	Uniqueness	Analogous to 'rarity', indicating that the soil type occurs in only one place or in a very few places, or to 'distinctiveness', because evolutionarily it is very different from other soils
Earth science	International importance	Contributing to soil processes, geological processes that form soils, and biological communities being supported
	Exceptional features	Also analogous to 'rarity', indicating soils that are of particular scientific importance
	Representativeness	Similar to 'typicalness', but also associated with historical events and soil processes
Soils	National importance	Typical of a soil type that is nationally distributed, but which could be monitored to detect future change (cf. Black *et al.* 2003)
	Historical importance	Soils that have been created by a distinctive, historical event or series of events, either natural or anthropogenic (see Foster & Smout 1994)
	Loss of extent	The *IUCN Red List Categories* (Anon. 1994b) are predicated upon the rate of loss of either abundance or area of occupancy of a species; a similar criterion for the rate of loss of extent occupied by a soil type could be constructed

represented was humus podzol (a form of spodosol), which has a very restricted occurrence in Scotland. This poses an interesting question – should at least one site with this soil type be sought and designated as an SSSI? The pattern of occurrence of soil types in this sample of SSSIs has a strong degree of correspondence with the occurrence of the soil types in Scotland as a whole (Spearman's $\rho = 0.881$, $P < 0.001$). Thus about 23% of Scotland is considered to support podzols, 23% peat, 19% surface-water gleys and 15% brown earths, and these account for 20%, 14%, 24% and 13%, respectively, of the combined major and minor occurrences in the SSSI sample. Peat soils are clearly under represented in the SSSI series, whereas the main over representations are the ground-water

Table 19.2. *The occurrence of the major soil groups and subgroups in a sample of 152 designated areas in Scotland (from Gauld and Bell 1997). The table indicates whether the occurrence of a soil group was likely to be a major or a minor feature of the soil of the designated areas, and is discussed further in the text.*

Major soil group	No. of subgroups	No. of major occurrences	No. of minor occurrences
Lithosols	1	3	0
Regosols	2	11	1
Alluvial soils	3	24	9
Rankers	4	32	45
Rendzinas	1	1	0
Calcareous soils	1	6	0
Magnesian soils	1	1	0
Brown earths	2	70	22
Podzols	6	118	27
Surface-water gleys	6	90	81
Ground-water gleys	6	37	42
Peat	4	89	11

gleys (11% in the series compared to 2% nationally), rankers (11% compared to 6% nationally) and calcareous soils (0.8% compared to 0.1% nationally). The reasons for these discrepancies are unknown, especially as ground-water gley soils are not obviously associated with any habitats or species of particular conservation concern, although calcareous soils often support a greater diversity of vascular plants species.

While more use could be made of soil data in conserving nature (Towers *et al.* 2002), there are three particularly important facets of the nature conservation value of soils. One is the value of undisturbed soils whereby sufficient time has elapsed for them to develop both their characteristic profile and an appropriate (and possibly stable) range of biota. In Scotland, a series of ancient woodland sites with undisturbed soils has been identified (Gauld & Bell 1997), with soil subgroups ranging from brown forest soils (supporting deciduous woodland) to humus iron podzols (supporting native pinewoods) and eutrophic peats (supporting alderwoods). Most, if not all, of these have been designated as SSSIs for reasons other than their soils. The second is the importance of the biota, especially the microbial communities, in carrying out the various ecosystem functions such as the decomposition of dead plant materials, animal wastes and pollutants. Whereas these organisms remain relatively unknown, they are never-the-less a facet of the Convention on Biological Diversity and need to be considered in national plans and strategies (Davison *et al.* 1999). At least some of the microbial species are likely to be of importance in the sustainable use of soils. Third, increasing efforts are being made world wide to restore damaged or degraded ecosystems (Harris *et al.* this volume). As soil is a vital component of virtually all terrestrial ecosystems, there is again a nature conservation interest in the biota that live within soils (Puri 2002).

Interestingly, the series of designated areas in Scotland, devised on the basis of criteria for habitats and species, provides a reasonable representation for soil types. This is, however, accidental rather than by design, and there are no data to demonstrate whether or not the series of designated areas really encompasses the range of biodiversity in the soil. As Adderley *et al.* (2001, p. 57) have concluded, 'although soils do not feature in the criteria for identifying SSSIs, never-the-less, where relevant, the notification of an SSSI should include soil attributes as part of the habitat description'. Nature conservationists have been guilty of neglecting soils, the very basis of the terrestrial ecosystems that they wish to protect.

Flagships, keystones and biodiversity action plans

Much conservation activity is driven by concern for species that have an emotional appeal to people. To a considerable extent this is because conservation management requires money, and money is more easily attracted to 'save' the 'flagship' species. This is demonstrated by the fact that these 'flagships' have often become the logos of non-governmental conservation organisations, such as the panda for the Worldwide Fund for Nature (WWF), the avocet for the Royal Society for the Protection of Birds (RSPB), sharks, tigers and a range of other vertebrate animals. What 'flagship' species might there be in the soil? Certainly there are no birds or fish, there are very few soil-inhabiting mammals, and the list of soil-inhabiting herpetofauna would be small. Although vertebrate animals are noticeably absent from soil, some potential 'flagship' plants have their roots in soil. However, 'flagship' species are a social construct rather than a scientific reality.

Similarly the soil is apparently lacking in 'keystone' species, defined as those that exert a controlling influence, out of proportion to their density or biomass, on the overall diversity and/or long-term stability of the ecological community in which they occur (Kapoor-Vijay & Usher 1993). In a series of appendices to this publication, the 'keystone' species for Grenada, Trinidad and Tobago, Nigeria, Zambia and New Zealand are listed. It is instructive to see how few of the hundreds of listed species could be considered as part of the soil's biodiversity, excluding the roots of vascular plants. The only entries that relate to soil-inhabiting species are for 'fungi – important decomposers' (p. 69) on Grenada, and a list of 16 named species of fungi, for Zambia, all of which are edible (pp. 175–176), including four species of *Termitomyces* which are associated with subterranean nests of termites in family Macrotermitidae.

Despite such few listed species, the building of large mounds by termites can influence both the hydrology of ecosystems and the structure of plant communities of tropical savannas by the eventual growth of small thickets on abandoned mounds. These 'termite thickets' provide a refuge for many animal species, especially at times when the savanna grasslands are burnt (Usher 1986b). The role of moles, *Talpa europaea*, in creating mounds of soil which then act as foci for the germination of many plant seeds, creates a succession, and hence the mole could also be considered to have 'keystone' status. Clearly the concept of 'keystone' species has not yet progressed far for the soil biota, though it is becoming increasingly recognised that some of the larger species, either in size, such as the mole, or in biomass, such as a termite colony or a particular strain of fungal mycelium, could have very considerable impacts on the above-ground part of the ecosystem as well as the transport of water and associated materials within ecosystems.

Processes of identifying, listing and using the 'flagship' and 'keystone' species concepts have therefore by-passed the soil's biota. The problem with 'flagship' species is that they are defined on the basis of a social, rather than a scientific, construct. Although they can be used to fund conservation campaigns by rousing public interest and sympathy, they may be of very limited value in wider conservation efforts, in indicating overall biodiversity loss, or in focusing on real conservation priorities. 'Keystone' species are defined on the basis of an ecological construct, but the definition is imprecise and there remains a subjective element in their identification. But is this reliance upon a few 'special' species appropriate? Simberloff (1998) argued that single-species management should be replaced by an ecosystem approach. In an analysis of the protected species of European land and freshwater molluscs, Bouchet *et al.* (1999) demonstrated that the listed species might not necessarily be those of greatest conservation need. It seems that all forms of 'red lists' are likely to suffer from taxonomic and geographic imbalance, reflecting the expertise available, which is a problem facing all biodiversity inventory work.

Despite the growing recognition that single-species management is unlikely to be effective for protecting biodiversity in the long term, action plans for species continue to be written, published and (to a greater or lesser extent) implemented. In the UK, the first collection of Species Action Plans (SAPs) enumerated a series of criteria for the inclusion of species in the lists (Anonymous 1995). However, application of these criteria assumes that enough is known about the size of populations in order to estimate the percentage decline in either range or abundance over the last 25 years, or to know what proportion of the world's population is present in the UK. Not surprisingly, few soil organisms were included in the first and subsequent collections of plans. Exceptions, however, include the mole cricket (*Gryllotalpa gryllotalpa*), a few species of ants (if they can be considered to be soil insects) and a small number of fungi, among which are the waxcaps, *Hygrocybe* spp. (Griffith *et al.* 2002). Internationally, a few species of earthworms have been recognised as being vulnerable to extinction (Wells *et al.* 1983).

It is unlikely that any 'flagship' species will be identified among the soil biota. It is possible that, with further knowledge about soil biodiversity, more 'keystone' species will be identified, as well as some 'umbrella' species (Simberloff 1998). However, there are two major reasons why identification of such species might not be a useful avenue of research to follow. First, it is possible that taxonomic advances will demonstrate that many soil organisms, especially the microbes and protozoa, are widely distributed and abundant (Finlay & Clarke 1999; Finlay 2002) and therefore not easily identifiable as in need of conservation management per se, although this conjecture is currently disputed (T. M. Embley personal communication). Second, thought is moving away from single-species management and towards the management of communities, habitats and ecosystems. For soils, it is likely to be the functions of their biota that are most in need of conservation (Wolters 2001); for above-ground conservation, the soil biota and the functions that they perform should not be forgotten. As Morris (2000) said, 'failure to consider all the components of ecosystems can lead to incomplete, or even erroneous, understanding of their dynamics'. Even with SAPs for single species, or the designation of sites on the basis of one (or a few) species, the nature and function of soil biodiversity should be considered and included within the management prescriptions. This is potentially extremely difficult because we know so little about the biodiversity of virtually all soils, but ignorance is no reason to forget that the ecology of soil is an integral part of the ecology of all terrestrial ecosystems.

Considering the ecosystem

The Convention on Biological Diversity recognised biodiversity at three levels: genetic, species and ecosystem. It is, however, often difficult to move from one level to another. This review has omitted the genetic level, focusing rather on the species level and found it to be lacking in concepts that are useful for conservation of soil biodiversity, and has pointed to the potential importance of the ecosystem level. Eijsackers (2001) also considered these three levels in relation to the science of soil ecology, but none of his five research themes addressed, even implicitly, the conservation of soil biodiversity, nor did they address the use of soil and soil organisms.

These more applied aspects of soil ecology relate closely to what has been termed 'The Ecosystem Approach' (Hadley 2000). This approach sets out a series of 12 principles, some of which are science orientated but all of which form an essentially socio-economic context for conservation. Two of the principles particularly relevant to soil biodiversity are Principle 5 'conservation of ecosystem structure and functioning, in order to maintain ecosystem services, should be a priority target for the ecosystem approach' and Principle 10 'the ecosystem approach should seek the appropriate balance between, and integration of, conservation and use of biological diversity'. How do these apply to the soil?

In relation to Principle 5, much has been written about ecosystem functioning, and a number of chapters in this book explore the subject in detail. Bradford and Newington (2002) enumerated some functions of soil, such as it being a carbon sink, a nutrient recycler and a pollutant remover, and they linked these functions to the activities of soil animals and microbes. Their key question was whether we have enough knowledge of the soil organisms to elucidate their functions and thus prevent permanent damage to the ability of soils to perform their beneficial actions. Similarly, Loreau *et al.* (2001) queried whether we have sufficient knowledge about the function of organisms in the soil, and they postulated that large numbers of species are required to reduce temporal variability in ecosystem processes in a changing environment. This has a conservation implication because, if the supposition is correct, the maintenance of species diversity becomes important. The concept of reduction in temporal variability is essentially the same as the concepts of reliability and predictability of ecosystem processes. Naeem and Li (1997) hypothesised that large numbers of species should enhance ecosystem reliability, which is the probability that a system will provide a consistent level of performance over a given unit of time. Their experiments with microbial microcosms supported the hypothesis, although it would be possible to erect alternative hypotheses that it might have supported. One such alternative hypothesis might be that there are a few 'keystone' species, at present unidentified among the large number of species, that have physiological traits that could individually promote stability.

Despite problems with experimental design and possible alternative interpretations of the results, there may be important implications for the conservation management of ecosystems. All of the soil functions – those listed by Bradford and Newington (2002) as well as the productivity of agroecosystems (Swift *et al.* 1996), coniferous forests (Huhta *et al.* 1998) and natural ecosystems (Neher 1999) – would become more predictable, and less temporally variable, with more species. Naeem and Li (1997) recognised that this is related to there being multiple species in each functional group. On the one hand, some argue that, provided each functional group is represented by one or a few species, then there is considerable redundancy in the species complement (Andrén *et al.* 1995; Liiri *et al.* 2002) because the ecosystem can still function effectively. On the other hand, the diversity of species in each functional group means that the ecosystem processes are more reliable or predictable, and hence that redundancy is a critical feature of ecosystems that needs to be conserved (Naeem 1998), and indeed that its conservation would be advocated by application of the 'precautionary principle' (Rosenfeld 2002). As Andrén and Balandreau (1999) said, 'the overwhelming diversity of organisms found [in the soil] makes a Linnean interpretation (every species counts and is necessary for the ecosystem) more and more illogical'. However, it becomes increasingly clear that, despite the truth of this statement, the soil ecosystem can only function reliably if there is an overwhelming diversity of species present in it, whether it be diversity per se or the presence of a few 'keystone' species that are encompassed by that diversity.

Principle 10 is concerned with balance, and often this can only be determined by case studies rather than experiments. For example, Ramakrishnan (2000) was concerned with determining a system of integrated land use management that conserved biodiversity on the basis of influencing farmers' choices of cropping combinations. The choices available in India were explored in relation to the multi-species complex agroecosystems of traditional human societies. Conservation of soil biodiversity has to involve both scientific knowledge and socio-economic considerations.

The difficult question is one of knowing when a balance has been achieved. One aspect is social, gaining consensus among users and conservationists that there is an appropriate balance. Another aspect is the provision of data that can be used for supporting the arguments leading to consensus. This is generally the result of collecting time series of values for indicators of the state of ecosystems (Eiswerth & Haney 2001). Although the pressure–state–response model is widely used in environmental audit (Birnie et al. 2002), it is indicators of the state of an ecosystem, as opposed to those of pressures and responses, which are most likely to be collected (Wilson et al. 2003). An example of the use of four of the state indicators of soil quality is quoted by Knoepp et al. (2000). More attention, however, needs to be given to the pressure and response indicators; within Europe there is still considerable discussion about what any of these indicators might be and how their measurement could be standardised (Black et al. 2003).

In the conservation of soil biota, it seems inappropriate to focus on rare species, endemic species, or other 'significant' species because of their charismatic character or keystone role in ecosystems, but pertinent to consider the functioning of the soil ecosystem. This has three particular facets. First, there is the multiplicity of ecological processes that occur in the soil, including decomposition, recycling, removal of pollutants, support for biological production, stabilisation of soil structure and storage of carbon. Second, there is a kind of 'insurance policy'; if all of these processes are to be temporally reliable, then the species richness, which can be viewed as the multiplicity of species available to perform each functional role, is important. We still do not understand this fully, and more research is needed to appreciate just how species rich the soil ecosystem needs to be to perform its functions reliably. Third, the soil biota can provide indicators telling us about the state of the soil ecosystem and its ability to function fully and reliably. Again, more research is needed to determine the best suite of indicators, which can be easily and reliably measured, that incorporate aspects of the soil's biodiversity.

Approaching sustainability?

The problem with addressing sustainability is that the associated time scale is very long. It is relatively easy to know that an activity is non-sustainable, but an activity today can, at this time, only be assessed as being potentially more or less sustainable than some alternative activity. We are unlikely ever to know if we have achieved sustainability. There is thus an element of probability; how can the probability that we are acting sustainably be increased? If an ecosystem process is reliable, then the probability that it is sustainable is likely to be greater. Alternatively, if an ecosystem process is not reliable, or has great temporal variability, then the probability that it is sustainable is likely to be less. Can sustainability ever be achieved, or is it a totally illusionary concept?

The context for approaching sustainability is, however, clear. As virtually all terrestrial ecosystems on the planet are to some extent managed by people, the aims of that management need to be considered, be they for single or multiple resources (Yaffee 1999). Management, including setting ecosystems aside for the conservation of their biodiversity, has to be a part of public policy (Ludwig et al. 2001). Science, based on the results of research, must be used in the public domain to support policy making by providing reliable evidence. Equally, the natural sciences need to be considered alongside socio-economic and ethical inputs. This is clearly seen in analyses of some of the multi-national research programmes, such as in the European Union, where Bennett and van Halen (2001) pointed to the continuing knowledge gap between the results of research and policy objectives. As the consideration of these issues in The Netherlands concluded, we must 'strive continuously for sustainability when negotiating international agreements on the management and qualitative protection of the global *res communis* [oceans, ozone layer, etc.] resources which support life-support functions' (Anonymous 1999, p. 29).

Against this developing policy background, what role does soil biodiversity play? Often it is forgotten, as for example in the book entitled *Conservation of Biodiversity for Sustainable Development* (Sandlund et al. 1992) in which the word 'soil' does not appear in the index! Where there is a consideration of soils and sustainable land management, it has generally used a proxy for soil biodiversity, which during the last century has been so little understood. For example, Carter (2002) focused on soil organic matter and soil aggregate stability in his assessments of the quality of Canadian soils and, similarly, Wander and Drinkwater (2000) considered that organic matter and organic matter-dependent properties of soils in the USA would be the most promising indicators of soil quality. Herrick (2000), however, defined soil quality as reflecting the capacity of soil to sustain plant and animal activity, to maintain and enhance water and air quality, and to promote plant and animal health – certainly a 'tall order' for soils to perform. He concluded that soil quality is a necessary, but not sufficient, indicator of sustainable land management. There is thus a substantial link between soil quality and sustainability (Taylor *et al.* 1996), but understanding that link and how it functions is largely still beyond our grasp, providing scope for further research.

How then do we conserve soils and their biodiversity? The traditional methods of protecting some species, and of designating some sites, may not be appropriate as this review has shown. Conservation must aim to protect the whole suite of ecosystem functions, both within and outwith the soil, and thereby to maintain and enhance the soil's ability to function reliably in the face of changing environmental conditions. If the soil functions reliably, then the probability of its management being sustainable is increased. National strategies for soil therefore need to focus more on sustainability than on protection, recognising that, despite the possible redundancy of many of the species, sustainability relies on reliability, which in turn relies on the biodiversity that exists in the soil. Reliability seems to be a key concept in the face of continual change, and it is finding a way of ensuring a greater probability of reliability and resilience in the face of change that requires a greater research effort and application in practice. Richter and Markewitz (2001) sum this up well when they say 'the management of soil at local, regional, and global scales must continue to improve'.

It is true that the soils of some designated areas, managed for the conservation of their habitats, might be useful for comparative research with soils supporting crops, grazing livestock, forestry, etc. However, conservation of soils must aim at ecosystem function, understanding by research how this ecosystem functioning is affected by the soil's biodiversity, getting closer to the sustainability of terrestrial ecosystems. This leaves much for us to understand about both soil biodiversity and the sustainable use of soils, and hence there is a continuing need for soil research. As Yaalon (2000) said 'why do we know more about distant celestial objects than we do about the ground beneath our feet?'

Acknowledgements

I should like to thank the referees of this chapter for their helpful comments. I also gratefully acknowledge the financial support of The Leverhulme Trust.

References

Adderley, W. P., Davidson, D. A., Grieve, I. C., Hopkins, D. W. & Salt, C. A. (2001). *Issues Associated with the Development of a Soil Protection Strategy for Scotland*. Unpublished report, University of Stirling.

André, H. M., Ducarme, X., Anderson, J. M., *et al.* (2001). Skilled eyes are needed to go on studying the richness of soil. *Nature*, **409**, 761.

André, H. M., Ducarme, X. & Lebrun, P. (2002). Soil biodiversity: myth, reality or conning? *Oikos*, **96**, 3–24.

Andrén, O. & Balandreau, J. (1999). Biodiversity and soil functioning: from black box to can of worms? *Applied Soil Ecology*, **13**, 105–108.

Andrén, O., Bengtsson, J. & Clarholm, M. (1995). Biodiversity and species redundancy among litter decomposers. *The Significance and Regulation of Soil Biodiversity* (Ed. by H. P. Collins, G. P. Robertson & M. J. Klug), pp. 141–151. Dordrecht: Kluwer Academic.

Anonymous (1989). *Guidelines for the Selection of Biological Sites of Special Scientific Interest.* Peterborough: Nature Conservancy Council.

Anonymous (1994a). *Guidelines for the Selection of Biological SSSIs: Bogs.* Peterborough: Joint Nature Conservation Committee.

Anonymous (1994b). *IUCN Red List Categories.* Gland: IUCN The World Conservation Union.

Anonymous (1995). *Biodiversity: the UK Steering Group Report*, Vol. 2, *Action Plans.* London: Her Majesty's Stationery Office.

Anonymous (1999). *Global Sustainability and the Ecological Footprint.* The Hague: VROM-Council.

Ball, D. F. & Stevens, P. A. (1981). The role of 'ancient' woodlands in conserving 'undisturbed' soils in Britain. *Biological Conservation*, **19**, 163–176.

Bennett, G. & van Halen, C. (2001). *Environmental Policy Priorities in Europe and EU Research Programming.* Rijswijk: RMNO.

Birkeland, P. W. (1984). *Soils and Geomorphology.* New York: Oxford University Press.

Birnie, R. V., Curran, J., MacDonald, J. A., *et al.* (2002). The land resources of Scotland: trends and prospects for the environment and natural heritage. *The State of Scotland's Environment and Natural Heritage* (Ed. by M. B. Usher, E. C. Mackey & J. C. Curran), pp. 41–81. Edinburgh: The Stationery Office.

Black, H. I. J., Hornung, M., Bruneau, P. M. C., *et al.* (2003). Soil biodiversity indicators for agricultural land: nature conservation perspectives. *Soil Erosion and Soil Biodiversity Indicators for Agricultural Land*, Rome, 25–28 March 2003. Unpublished paper, OECD.

Bouchet, P., Falkner, G. & Seddon, M. B. (1999). Lists of protected land and freshwater molluscs in the Bern Convention and European Habitats Directive: are they relevant to conservation? *Biological Conservation*, **90**, 21–31.

Bradford, M. A. & Newington, J. E. (2002). With the worms: soil biodiversity and ecosystem functioning. *Biologist*, **49**, 127–130.

Bundt, M., Widmer, F., Pesaro, M., Zeyer, J. & Blaser, P. (2001). Preferential flow paths: biological 'hot spots' in soils. *Soil Biology and Biochemistry*, **33**, 729–738.

Carter, M. R. (2002). Soil quality for sustainable land management: organic matter and aggregation interactions that maintain soil functions. *Agronomy Journal*, **94**, 38–47.

Copley, J. (2000). Ecology goes underground. *Nature*, **406**, 452–454.

Davison, A. D., Yeates, C., Gillings, M. R. & de Brabandere, J. (1999). Microorganisms, Australia and the Convention on Biological Diversity. *Biodiversity and Conservation*, **8**, 1399–1415.

Eijsackers, H. (2001). A future for soil ecology? Connecting the system levels: moving from genomes to ecosystems. *European Journal of Soil Biology*, **37**, 213–220.

Eiswerth, M. E. & Haney, J. C. (2001). Maximizing conserved biodiversity: why ecosystem indicators and thresholds matter. *Ecological Economics*, **38**, 259–274.

Ellis, N. V., Bowen, D. Q., Campbell, S., *et al.* (1996). *An Introduction to the Geological Conservation Review.* Peterborough: Joint Nature Conservation Committee.

Finlay, B. J. (2002). Global dispersal of free-living microbial eukaryote species. *Science*, **296**, 1061–1063.

Finlay, B. J. & Clarke, K. J. (1999). Ubiquitous dispersal of microbial species. *Nature*, **400**, 828.

Foster, S. & Smout, T. C. (Eds.) (1994). *The History of Soils and Field Systems.* Aberdeen: Scottish Cultural Press.

Gauld, J. H. & Bell, J. S. (1997). Soils and nature conservation in Scotland. *Scottish Natural Heritage Review*, **62**, 1–33.

Griffith, G. W., Easton, G. L. & Jones, A. W. (2002). Ecology and diversity of waxcap (*Hygrocybe* spp.) fungi. *Botanical Journal of Scotland*, **54**, 7–22.

Hadley, M. (Ed.) (2000). *Solving the Puzzle: the Ecosystem Approach and Biosphere Reserves*. Paris: UNESCO.

Heal, O. W. (1999). Looking north: current issues in Arctic soil ecology. *Applied Soil Ecology*, **11**, 107–109.

Heathwaite, A. L. & Göttlich, K. (Eds.) (1990). *Mires: Process, Exploitation and Conservation*. Chichester: Wiley.

Herrick, J. E. (2000). Soil quality: an indicator of sustainable land management? *Applied Soil Ecology*, **15**, 75–83.

Hopkins, J. (2003). Some aspects of geology and the British flora. *British Wildlife*, **14**, 186–194.

Huhta, V., Persson, T. & Setälä, H. (1998). Functional implications of soil fauna diversity in boreal forests. *Applied Soil Ecology*, **10**, 277–288.

Kapoor-Vijay, P. & Usher, M. B. (1993). *Identification of Key Species for Conservation and Socio-economic Development*. London: Commonwealth Secretariat.

Knoepp, J. D., Coleman, D.C., Crossley, D. A. & Clark, J. S. (2000). Biological indices of soil quality: an ecosystem case study of their use. *Forest Ecology and Management*, **138**, 357–368.

Liiri, M., Setälä, H., Haimi, J., Pennanen, T. & Fritze, H. (2002). Soil processes are not influenced by the functional complexity of soil decomposer food webs under disturbance. *Soil Biology and Biochemistry*, **34**, 1009–1020.

Loreau, M., Naeem, S., Inchausti, P., et al. (2001). Biodiversity and ecosystem functioning: current knowledge and future challenges. *Science*, **294**, 804–808.

Ludwig, D., Mangel, M. & Haddad, B. (2001). Ecology, conservation, and public policy.

Annual Review of Ecology and Systematics, **32**, 481–517.

Mermut, A. R. & Eswaran, H. (2001). Some major developments in soil science since the mid-1960s. *Geoderma*, **100**, 403–426.

Morris, M. G. (2000). The effects of structure and its dynamics on the ecology and conservation of arthropods in British grasslands. *Biological Conservation*, **95**, 129–142.

Naeem, S. (1998). Species redundancy and ecosystem reliability. *Conservation Biology*, **12**, 39–45.

Naeem, S. & Li, S. (1997). Biodiversity enhances ecosystem reliability. *Nature*, **390**, 507–509.

Neher, D. A. (1999). Soil community composition and ecosystem processes: comparing agricultural ecosystems with natural ecosystems. *Agroforestry Systems*, **45**, 159–185.

Puri, G. (2002). Soil restoration and nature conservation. *Scottish Natural Heritage Information and Advisory Note*, **150**, 1–4.

Ramakrishnan, P. S. (2000). An integrated approach to land use management for conserving agroecosystem biodiversity in the context of global change. *International Journal of Agricultural Resources, Governance and Ecology*, **1**, 56–67.

Ratcliffe, D. A. (Ed.) (1977). *A Nature Conservation Review*, Vol. 1. Cambridge: Cambridge University Press.

Richter, D. D. & Markewitz, D. (2001). *Understanding Soil Change: Soil Sustainability over Millennia, Centuries, and Decades*. Cambridge: Cambridge University Press.

Rosenfeld, J. S. (2002). Functional redundancy in ecology and conservation. *Oikos*, **98**, 156–162.

Sandlund, O. T., Hindar, K. & Brown, A. H. D. (Eds.) (1992). *Conservation of Biodiversity for Sustainable Development*. Oslo: Scandinavian University Press.

Simberloff, D. (1998). Flagships, umbrellas, and keystones: is single-species management passé in the landscape era? *Biological Conservation*, **83**, 247–257.

Swift, M. J., Vandermeer, J., Ramakrishnan, P. S., et al. (1996). Biodiversity and agroecosystem function. *Functional Roles of Biodiversity: a Global Perspective* (Ed by H. A. Mooney, J. H. Cushman, E. Medina, O. E. Sala & E.-D. Schulze), pp. 261–298. Chichester: Wiley.

Tamm, O. (1950). *Northern Coniferous Forest Soils*. Oxford: Scrivener Press.

Tansley, A. G. (1949). *The British Isles and their Vegetation*, Vol. I. Cambridge: Cambridge University Press.

Taylor, A. G., Usher, M. B., Gordon, J. E. & Gubbins, N. (1996). Epilogue: the way forward: soil sustainability in Scotland. *Soils, Sustainability and the Natural Heritage* (Ed. by A. G. Taylor, J. E. Gordon & M. B. Usher), pp. 295–305. Edinburgh: Her Majesty's Stationery Office.

Towers, W., Hester, A. J., Malcolm, A., Stone, D. & Gray, H. (2002). The use of soil data in natural heritage planning and management. *Soil Use and Management*, **18**, 26–33.

Trudgill, S. (1989). Soil types: a field identification guide. *Field Studies*, **7**, 337–363.

Usher, M. B. (Ed.) (1986a). *Wildlife Conservation Evaluation*. London: Chapman and Hall.

Usher, M. B. (1986b). Insect conservation: the relevance of population and community ecology and of biogeography. *Proceedings of the Third European Congress of Entomology*, Part 3 (Ed. by H. H. W. Velthuis), pp. 387–398. Amsterdam: Nederlandse Entomologische Vereniging.

Usher, M. B. (1996). The soil ecosystem and sustainability. *Soils, Sustainability and the Natural Heritage* (Ed. by A. G. Taylor, J. E. Gordon & M. B. Usher), pp. 22–43. Edinburgh: Her Majesty's Stationery Office.

Usher, M. B. (2001). Earth science and the natural heritage: a synthesis. *Earth Science and the Natural Heritage: Interactions and Integrated Management* (Ed. by J. E. Gordon & K. F. Leys), pp. 314–324. Edinburgh: The Stationery Office.

Wander, M. M. & Drinkwater, L. E. (2000). Fostering soil stewardship through soil quality assessment. *Applied Soil Ecology*, **15**, 61–73.

Wells, S. M., Pyle, R. M. & Collins, N. M. (1983). *The IUCN Red Data Book*. Gland: IUCN The World Conservation Union.

White, R. E. (1987). *Introduction to the Principles and Practice of Soil Science*, 2nd edition. Oxford: Blackwell.

Wilson, J., Mackey, E., Mathieson, S., et al. (2003). Towards a strategy for Scotland's biodiversity: developing candidate indicators of the state of Scotland's biodiversity. *Scottish Executive Environment and Rural Affairs Department Paper* 2003/6. Edinburgh: SEERAD.

Wolters, V. (2001). Biodiversity of soil animals and its function. *European Journal of Soil Biology*, **37**, 221–227.

Yaalon, D. H. (2000). Down to earth: why soil – and soil science – matters. *Nature*, **407**, 301.

Yaffee, S. L. (1999). Three faces of ecosystem management. *Conservation Biology*, **13**, 713–725.

PART VI

Conclusion

CHAPTER TWENTY

Underview: origins and consequences of below-ground biodiversity

KARL RITZ

Cranfield University

SUMMARY

1. The mechanistic origins and functional consequences of soil biodiversity, in terms of general principles and across a broad context, are reviewed.
2. The origins of below-ground biodiversity are discussed in terms of the spatial isolation that soil structure imparts, substrate diversity, competition and environmental fluctuation. Community structure is governed by many factors.
3. The consequences of soil biodiversity are explored in relation to the functional repertoire that the biota carries, the potential and realised interactions between components, and functional redundancy.
4. A wide variety of relationships are expressed between soil biodiversity and function. These are discussed in relation to resilience, the impact of biodiversity upon individual organisms, complexity, and above- and below-ground linkages.
5. Biodiversity per se, particularly in terms of species richness that prevails in most soils, is apparently of little functional consequence. The *functional repertoire* of the soil biota is considerably more pertinent.
6. Improved understanding of the relationships between soil community structure and function underpins the effective and sustainable management of ecosystems in an agricultural, forestry, conservation or restoration context. Knowledge is burgeoning and an improved understanding is following, but a unifying framework is currently elusive. Soil architecture may be the key.

Introduction

The aims here are to take an 'underview' of soil biodiversity within the broad context of the preceding 19 chapters, the presentations and discussions that ensued at the symposium on which this volume is based, and to discuss some

Biological Diversity and Function in Soils, eds. Richard D. Bardgett, Michael B. Usher and David W. Hopkins. Published by Cambridge University Press. © British Ecological Society 2005.

additional concepts and issues that received less emphasis. It is not intended to summarise the volume formally, since that is more appropriately a task for each reader to approach and synthesise from their widely different perspectives. From such diversity, perhaps new views will emerge.

Scientific study of life below ground has continued for over a century, but it is only recently that the true extent of soil biodiversity has become apparent. Darwin recognised the importance of earthworms in soil functioning, and others such as Winogradsky, Lawes and Lipman were early pioneers of the study of soil microbes. There has long been a realisation that life below ground is intimately linked to soil functioning and hence to the integrity of terrestrial systems, but the identification and mechanistic understanding of relationships between biological diversity and function in soils continues to fascinate and occupy scientists, as this volume testifies.

Biodiversity is a dominant contemporary topic in the environmental and life sciences. Citation indices show an exponential increase in the appearance of the term as a keyword, and the issue of biodiversity is climbing the political and public agenda. It is seen as important – the dogma is often that biodiversity is good; biodiversity is necessary; diverse systems are deemed more stable, more resilient, more productive, *more desirable*. The scientific evidence for the extent to which biodiversity underwrites the functional properties of ecosystems is growing, but it is apparent that such relationships are not straightforward and the search for simple or unifying frameworks remains elusive, and is perhaps unattainable.

Extensive usage of particular terms often leads to flexibility in their use, and biodiversity can mean many different things depending upon context. Gaston (1996) paraphrases it as *the biology of numbers and difference*. The standard definition is often seen as the base number and relative abundance of different species present within the confines of the system being considered, which is formally defined as richness and evenness. But it is increasingly being applied to other attributes and there can be consideration and quantification of genetic, phenotypic, functional and trophic biodiversity. The key intimation relates to expressions of difference, variety or complexity. The term is also often used as a synonym for population or community structure. No single parameter can adequately summarise what is essentially a frequency distribution of various properties, and there are hence many numerically based indices of biodiversity each of which embodies various subtleties (see, for example, Gaston 1996; Hill *et al.* 2003).

Soil biodiversity

An immense quantity and range of life resides in the upper layers of most soils. The statistics are now well rehearsed and always remarkable. For example, the total fresh weight mass of the biota below an old temperate grassland can exceed $45\,t\,ha^{-1}$, at least equal to the above-ground biomass, and equivalent to a stocking rate of about 900 sheep per hectare. A few grams of such soil contains billions of bacteria, hundreds of kilometres of fungal hyphae, tens of thousands of protozoa, thousands of nematodes, several hundred insects, arachnids and worms, and hundreds of metres of plant roots. These large numbers are matched by extreme levels of biodiversity, most particularly at the microbial scale. Nucleic acids were originally extracted from soils from the perspective of estimating biomass (Nannipieri et al. 1986), a theme which is still occasionally revisited (Blagodatskaya et al. 2003). However, early application of molecular genetic analysis to environmental DNA revealed astonishing levels of diversity in the prokaryotes, subsequently reinforced by many studies. The pioneering work of Torsvik (Torsvik et al. 1990, 1994), involving the broad-scale analysis of DNA by reassociation kinetics, was the first to demonstrate that upward of 10 000 genetically distinct prokaryotic types (operational taxonomic units, OTUs) can prevail in 100 g of soil. Subsequently, many high-resolution analyses of soil community DNA based on PCR amplification and sequencing of ribosomal DNA and RNA have confirmed such high levels of prokaryotic diversity. For example, in an agricultural soil from Wisconsin, 4% of 124 amplified sequences were duplicates (Borneman et al. 1996) and there were no duplicates among 100 sequences from Amazonian rainforest or pasture soils (Borneman & Triplett 1997). McCaig et al. (1999) found eight duplicate and two triplicate sets out of 275 eubacterial sequences in a UK upland grassland. Such data have led to the general calculation using species–abundance curves that there may be upward of 4×10^6 prokaryotic taxa in soil (Zak et al. 1996; Curtis et al. 2002).

Broad-scale molecular analyses of soil fungal diversity are currently infrequent, but the recent development of ostensibly universal fungal primers (Pennanen et al. 2001; Anderson et al. 2003; Kabir et al. 2003; Marshall et al. 2003) will advance such studies. Estimates of global species richness of soil fungi based on extrapolation from cultivable species range from 7 to 80×10^3 fungal taxa in soil (Bridge & Spooner 2001). Estimates of global diversity within freshwater protozoa are of the order of a few thousand species (Finlay & Esteban 1998) and within soil of several hundred (Finlay et al. 2001). Estimates of the diversity of other fauna are given by Bardgett et al. (this volume). For smaller multi-cellular eukaryotes, one to several hundred species for each group are likely; however, at the larger scales of earthworms only several species are apparent. There is thus a distinct general relationship between genetic diversity and body size, in that the larger the organisms the lower their diversity. Biodiversity below ground is considerably greater than that above ground or in aquatic systems, even at microbial scales. For example, reassociation kinetics of community DNA derived from marine bacterioplankton suggested that these were only as diverse as a cultivable subset of soil bacteria, an order of magnitude less diverse than the directly extractable bacterial fraction of the same soil, and three orders of magnitude less complex than whole-soil DNA (Ritz et al. 1997a). Extrapolation of a range of species abundance curves suggests aquatic bacterial diversity is between 40 and 250 times less than that in soil (Curtis et al. 2002).

Origins of diversity

Precisely why soil communities are so diverse is unknown, and there is little experimental work that formally explores such origins. Four basic mechanisms are postulated, all of which may interact in various ways to govern the generation of soil biodiversity.

1. *Spatial structure.* A primary driver relates to the extreme heterogeneity that exists in soils purely at a physical level (Bardgett et al. this volume; Young & Ritz this volume). The architecture of the soil, i.e. the spatial organisation of solids, pores, liquids, gases and solutes, defines the physical framework in which the soil biota lives and functions, and shows great heterogeneity across a very wide range of size scales (Fig. 20.1). The convoluted pore network regulates the movement and defines the relative location of all entities in soil including organisms and their substrates; the larder is a labyrinth. Due to the complex topology of pore networks, microorganisms in particular can be effectively separated by considerable distances in three dimensions at their scale. Such spatial separation in soils is akin to geographical isolation above ground, albeit on a considerably smaller scale, which leads to mechanisms of speciation. That microbial diversity in aquatic systems is much lower than that in soil is telling. A key difference is that in waters the spatial structure imparted by the solid phases of soil is absent. As demonstrated by Rainey et al. (this volume), even the most subtle spatial structure in systems can lead to adaptive radiation. In sand-based microcosms, Treves et al. (2003) imparted spatial isolation on model two-species systems by manipulating the matric potential. Where free mixing was enabled by a high matric potential, one species dominated, but at lower potentials co-existence was prevalent. Similar, but less-pronounced, trends were observed where strains of the same species were grown together. Such spatial mechanisms were hypothesised to operate in a range of soils under low substrate conditions (Zhou et al. 2002). Here, patterns of bacterial diversity in a range of low carbon-containing freely drained topsoils were found to be quite different to those in affiliated vadose zones or saturated subsoils. In the subsoils, the communities showed low diversity and greater dominance by fewer OTUs, while in the topsoils there was high OTU diversity and a total lack of dominance (Fig. 20.2a). Such patterns were attributed to competitive interactions that occur where substrate is limiting but free interaction is attainable; some species are eliminated by competitive exclusion and a few species better adapted to conditions, dominate. In the topsoils, spatial isolation prevented

Figure 20.1. The subterranean habitat: soil structure across five orders of magnitude. (a) field scale; (b) profile scale (bar = 1 cm); (c) root scale (bar = 3 mm); (d) pore scale (false-colour image of soil thin-section, bar = 1 mm); (e) microscale (computer-aided tomographic slices, bar = 500 μm). Note that the physical size of the soil biota spans five similar orders of magnitude. Sources: (a–d) Karl Ritz; (e) Iain Young, Karl Ritz and Naoise Nunan.

such extensive interaction, competition was modulated and abundance was unified.

2. *Substrate diversity.* There is considerable chemical diversity in soils, particularly associated with organic matter, which is ultimately the substrate for the majority of soil organisms (Hopkins & Gregororich this volume). Compounds range from the extremely simple through to vast random polymers with turn-over times ranging from minutes to centuries. It is intuitively reasonable that such diversity in substrate will lead to extensive variety in the organisms that consume such material; however, there is virtually no formal experimental exploration of such issues. Maire *et al.* (1999) proposed that greater diversity in food sources in alpine pastures in spring led to diversity in the decomposer communities that feed upon them. Further components in the work of Zhou *et al.* (2002) also point towards this. In the suite of soils they studied, those containing greater amounts of carbon, supported

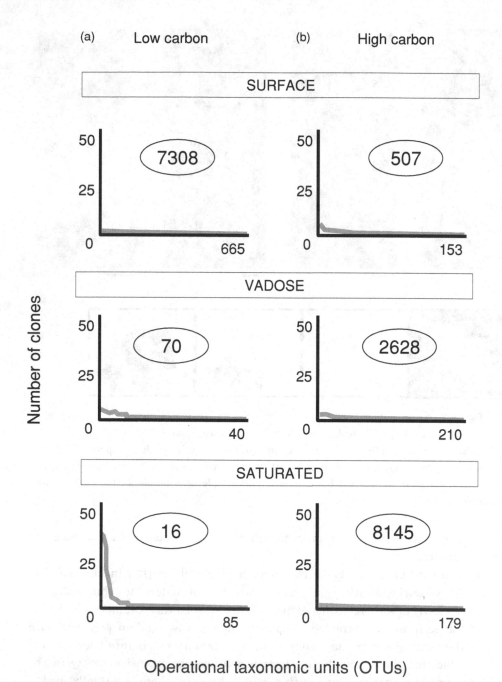

Figure 20.2. Representative bacterial community diversity patterns in three soil zones, expressed as rank-order curves (graphs) and reciprocal of Simpson's index (values in ellipses), based on PCR amplification and sequencing of community DNA. A total of 1564 clones was analysed in these instances. (a) low carbon soil; (b) high carbon soil. Redrawn from Zhou *et al.* (2002).

bacterial-diversity patterns that were uniform with no dominance, even in saturated soils (Fig. 20.2b). They hypothesised that if resources are available in many different forms, populations can avoid competition (leading to a reduction in dominant types) by specialisation in feeding preferences. A number of experiments have demonstrated that mixtures of litters tend to support a greater diversity of mites than the individual components (Liddell & Hansen 1993; Kaneko & Salamanca 1999; Hansen 2000). It is apparent that the more complex a compound, especially in tertiary structural terms, or the more energy required to cleave bonds therein, the narrower the range of individual organisms that can utilise such sources. For example, glucose, which is a universal component of catabolism, can be utilised by the majority of soil organisms (hence the efficacy of the glucose-induced respiration assay as a measure of active biomass). But the more complex polymers require interactions between consortia of organisms with complementary properties in order to be degraded. This will act as a driver for diversification, including across trophic levels where biochemical action is supplemented by physical comminution by fauna.

3. *Competition.* Many soils are oligotrophic environments, and it follows that competition between organisms for substrate is severe. Whether the predominant limiting element in soils is carbon, nitrogen or otherwise is open to debate (Schimel *et al.* this volume). Even where nutrients are relatively abundant, physical protection mechanisms further serve to render much substrate unavailable to organisms. This is demonstrated by the flush of decomposition that invariably follows any physical disturbance of soil (e.g. Ritz *et al.* 1997b; Watts *et al.* 2000). Competition for resources generally is known to be a driver for adaptive radiation, by forcing evolutionary exploration of potential niches. In soil the variety of niches is huge and at smaller spatial scales borders on the limitless. This explains why the greatest diversity prevails at smaller spatial scales. While the lack of dominance in OTU profiles found for many soil bacterial communities can at one level be interpreted as 'non-competitive' (Zhou *et al.* 2002), this does not account for the fact that addition of glucose (Anderson & Domsch 1978), even at trace concentrations (de Nobili *et al.* 2001), invariably leads to a rapid respiratory response by the soil biomass. Despite the likely importance of competition being a driver of microbial diversity at a local scale, for example on substrate hot spots, there is no evidence available to indicate that it is an important driver of soil food web diversity *at larger spatial scales* (Bardgett *et al.* this volume).

4. *Environmental fluctuation.* Temporal variation in environmental conditions in soils can be highly dynamic, especially in near-surface horizons, in terms of hourly, diurnal, through to seasonal variation in temperature, moisture, substrate deposition, solute concentration, physical disturbance, etc. Such variation adds to the variety of niches that prevails in soil, across a range of time scales, and hence the range of opportunities for organisms to adapt to.

Drivers of community structure

The above discussion has largely focused upon the origins of diversity per se in the soil biota. Factors that regulate such diversity also impact markedly upon the higher-order organisation, or compositional structure, of soil communities. The evidence is that such interactions are varied and complex, and there are a myriad of studies on the relationships between environmental and biotic factors and soil community structure. Consider substrate as an example. Under extremely controlled conditions, substrate loading (i.e. the rate of delivery of a defined, complex substrate to a soil) can have a very coherent and consistent effect upon the genetic and phenotypic composition of soil communities (Fig. 20.3). Most carbon substrate in soils is derived from plants; different plants will deliver different substrate to soil, and there will be some variety and consistency in the quality and quantity of such substrate between plant species. Hence specific community-level associations between individual plant species and soil communities might be expected, and there is experimental evidence to support this (e.g. Lemanceau *et al.* 1995; Mahaffee & Kloepper 1997; Marilley *et al.* 1998; Siciliano *et al.* 1998). Such associations may be more pronounced at the smaller (rhizosphere) scale (Kowalchuk *et al.* 2002). Changes in plant community composition may act as a driver of change in soil biotic communities (Wardle this volume) and there is some evidence for whole community-level coupling between vegetation and microbial assemblages (e.g. Donnison *et al.* 2000; Grayston *et al.* 2004), but this is generally less coherent, implying that factors other than purely substrate quality and quantity drive community structure. Top–down forces of predation may act as a driver of food web architecture and ecosystem processes (Setälä *et al.* this volume) and there is evidence that selective grazing of microbes can lead to changes in microbial community structure in the field and under laboratory conditions (Bardgett & Griffiths 1997).

Consequences of soil biodiversity

To soil biologists, one of the primary consequences of soil biodiversity is that it is innately intriguing: *biodiversity is beguiling*. But it is important to stress that an improved understanding of such consequences underwrites the management- and policy-related applications of the discipline, as explored in a number of the preceding chapters. The inherent complexity means that it is very challenging to study, and advances in techniques that enable increasing resolution lead to the possibility of increasing reductionism and a virtually infinite opening of the 'black box'. However, results from the genetic analyses alluded to above make the concept sometimes alluded to as 'all taxa biotic inventories' (ATBI) for soils both untenable and unattainable. A key task in the quantification and description of soil communities must therefore be to establish the appropriate levels of resolution of analyses pertinent to the questions being asked. This is particularly important in relation to considerations of the practical application

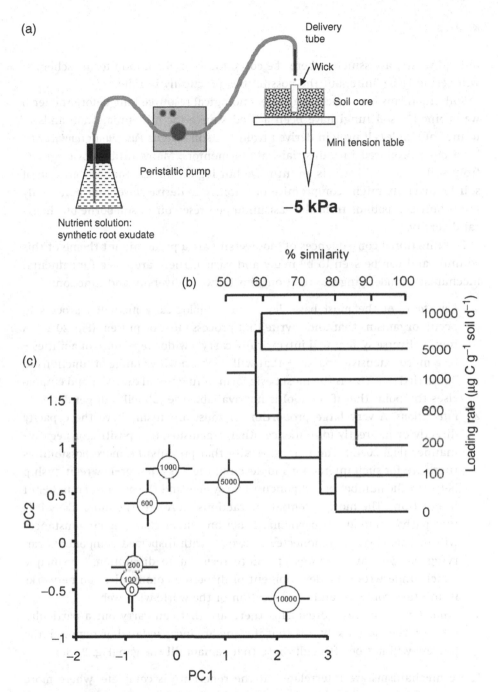

Figure 20.3. Effect of substrate loading rate upon microbial community structure.
(a) Experimental system ensuring highly controlled delivery rates and consistent
incubation conditions. (b) Genetic similarity between communities, based on community
DNA cross-hybridisation. Note the consistent decline in similarity with increasing
loading rate. (c) Phenotypic profiles (based on phospholipid fatty acid analyses) of soil
communities shown as principal component (PC) plot. Values in points are loading rates
in $\mu g\ C\ g^{-1}\ d^{-1}$. Note consistent trajectory in PC space with increasing loading rate.
Derived from Griffiths *et al.* (1999).

of biodiversity assessment where the costs, for example of monitoring schemes, will certainly be finite, but the possibilities practically infinite.

Soil organisms have acted as a biotechnological resource long before the term was coined – soil fungi have been a food source for millennia. Thousands of tonnes of single-cell protein derived from the soil fungus *Fusarium graminearum* have been produced in industrial-scale fermentors. Many antibiotics originate from soil microbes. There is an intuitive but logical unease that any erosion of soil biodiversity might compromise our ability to derive novel and potentially important compounds from the vast untapped reservoir of soil-borne biochemical diversity.

The functional consequences of biodiversity are a predominant theme of this volume, and can be seen to be many and varied. There are three fundamental mechanisms underlying the relationships between diversity and function:

1. *Repertoire.* At the most basic level, for a biologically mediated process to occur, organisms that underwrite that process must be present (Fig. 20.4a). A highly diverse system will intrinsically carry a wider repertoire of abilities – or a more extensive 'toolkit' – that will enable a wider range of functions to be carried out. This is the most basic form of functional diversity and emphasises the point that if the tool is not available the job will not get done.

2. *Interactions.* A very large proportion of most organisms have the capacity directly or indirectly to influence other organisms, in a positive or negative manner (Fig. 20.4b). The more diversity that prevails the more possibilities there are for such interactions to develop; there is a power–law relationship between the number of components in a system and the potential number of interactions. The more potential interactions there are, the more likely it is that pathways to attain particular functions under changing circumstances will be manifest. Interconnected networks with dispersed components carrying out similar functions are more resistant to disruption, a principle which underwrote the development of dispersed communication networks in military contexts, and the evolution of the world-wide web.

3. *Redundancy.* The more organisms there are that can carry out a particular process, the more likely it is that if some are incapacitated or removed, the process will not be affected; those that remain fill the gap (Fig. 20.4c).

These mechanisms are interrelated: if the repertoire is complete, where more components are present, there is a greater potential for 'rewiring' (interaction) to occur to enable a process to prevail.

There are a variety of potential relationships between biodiversity and function including the so-called rivet, predictable, redundant, idiosyncratic, hump-shaped, U-shaped and keystone forms (see Waide *et al.* 1999; Ritz & Griffiths 2000; Naeem *et al.* 2002; Robinson *et al.* this volume). These have been determined using four contrasting but complementary approaches based upon construction

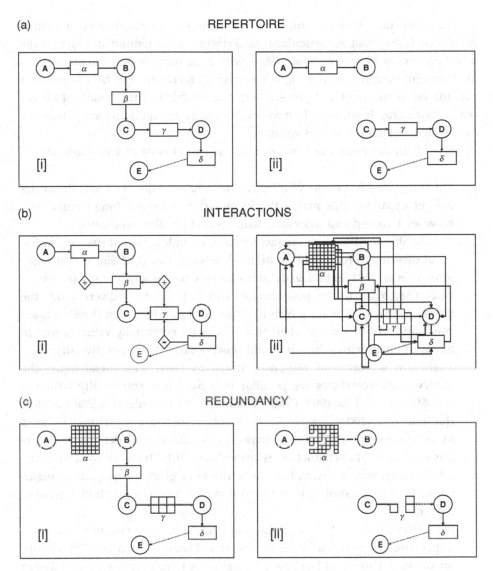

Figure 20.4. Schematic representation of basis for consequences of biodiversity in ecosystems. The scheme denotes function in terms of transformation of ecosystem components A–E by biota α–δ. (a) If the repertoire is complete, all transformations occur and the system is coherent and fully functional; if a component is missing, in this case β, then there is incoherence and a failure of function [ii]. (b) Both positive and negative interactions between organisms can occur, affecting overall function; the greater the diversity within functional groups, the greater the potential for interactions, for complex interactions, and potential for non-linearity in function [ii]. (c) If the system is perturbed such that a proportion of all biota is removed, greater diversity results in less impairment of the function. This principle also applies where a large number of interactions exist, as in (b)[ii]; removing a proportion of the interaction arrows may still result in a functionally intact system.

or deconstruction of communities, study of natural gradients and modelling (Ritz & Griffiths 2000). A particular manifestation of the redundancy form is the so-called gas-box of Andrén *et al.* (1999), where the tenet is that the biota act as an 'averaging engine', by-passing the ecological hierarchy and that function is considered as an integrated process with the omission of the details of all the players. At a gross level, this often works, but important subtleties are potentially overlooked (Schimel *et al.* this volume).

Some of the key points to emerge from the extant body of knowledge are:

1. There are manifestations of all forms of relationships. No overriding rules are yet apparent. This makes the formulation of over-arching frameworks, however favoured and desirable, arduous and possibly untenable.
2. 'Predictable' relationships characterised by a combination of rivet at low levels of diversity to redundant at higher levels (e.g. van der Heijden *et al.* 1998; Jonsson *et al.* 2001; Setälä & McLean 2004) are potentially of great relevance, since they point to the possibility of threshold levels of diversity (i.e. the point of inflexion), below which function is impaired. If such thresholds are consistent within systems then this provides a potentially valuable tool in terms of monitoring, since it would provide target values of diversity.
3. There are instances of responses stated as being idiosyncratic, i.e. the observed characteristics are peculiar to a particular community structure (e.g. Mikola & Setälä 1998; Cragg & Bardgett 2001; Bardgett & Wardle 2003; Hedlund *et al.* 2003; Porazinska *et al.* 2003). These are sometimes referred to as 'context dependent' (e.g. Jonsson *et al.* 2001), and demonstrate non-linearity in diversity:function relationships. To be truly idiosyncratic, the relationships would always have to be the same given a particular configuration. To date, no studies have tested this with a sufficiently high degree of rigour.
4. In all circumstances, but particularly in relation to 'context dependency', it is imperative that repeatability is demonstrated before particular relationships are invoked. This should go beyond statistically robust replication and design within experiments (Schmid *et al.* 2002), but via replication over time as well. This is rarely done (but see, for example, Griffiths *et al.* 2000). A further complication arises in the light of the recent demonstration that in aquatic systems at least, the *sequence of assembly* of communities can significantly affect productivity–biodiversity relationships (Fukami & Morin 2003).
5. Context dependency also relates to the relative importance of abiotic versus biotic controls on processes (e.g. Gonazalez & Seastedt 2001). Biota may be of limited importance in some sites, whereas of greater importance in others. Hence relationships between biodiversity and function can vary between ecosystem types.
6. Base richness (in terms of the number of species present within a soil) may show little relationship to function because, as discussed above, it is the

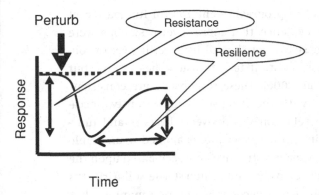

Figure 20.5. Graphical representation of trajectory of resistance and resilience in perturbed systems. Broken line shows time course of response variable in unperturbed (control) systems, solid line shows response following perturbation. Resistance is measured as the degree of impairment of response relative to control; resilience as the rate and extent of recovery. Recovery may be incomplete within the measured time scale.

functional repertoire of the community that is of more relevance. For a great many soil organisms, the relationships between taxonomy and function are not necessarily robust. For example, denitrification as a property is scattered throughout the eubacterial phylogenetic tree; bacterial-feeding nematodes belong to a wide range of genera. Furthermore, a single species may be multi-functional. For example, a species of *Rhizobium* bacteria may fix N_2 when in a symbiotic state in the host legume, but denitrify when in a free-living state (e.g. Garciaplazaola *et al.* 1993), decompose organic matter during heterotrophic growth, enhance soil aggregation by the production of extra-cellular polysaccharides, and carry out a number of other biochemical transformations that are not yet apparent due to the difficulty of studying microbial physiology *in situ* in the soil. If this bacterium were alone in a soil, there would be a richness value of unity, but functionally the system would be quite diverse! The lack of a connection between taxonomy and function is also manifest in the fauna.

Relationships between biodiversity and resilience

Soil resilience is another term open to a variety of interpretations. A simplified definition is that it is the 'soil's ability to recover after disturbance' (Greenland & Szabolcs 1994). Responses of soils to disturbance have two components: *resistance*, i.e. the inherent capacity of the system to withstand the perturbation; and *resilience*, the capacity to recover following such disturbance (McNaughton 1994). These properties can be formalised and quantified by measurement of the dynamics of changes in system properties following perturbation (Fig. 20.5). It is hypothesised that systems that are more biologically diverse have greater resilience (Wardle & Giller 1996). A mechanistic basis for this could be the repertoire–interactions–redundancy concepts discussed above. In studies utilising constructive and deconstructive approaches to manipulate soil biodiversity, the relationships between diversity and functional resilience have been shown to be variable in relation to the short-term decomposition of grass residues. In a deconstructive approach based upon the progressive chloroform fumigation of an arable grassland soil, there was evidence that more diverse systems were more resistant to a persistent copper stress but showed little resilience; in contrast, more diverse systems were less resistant to a transient heat stress, but showed clear resilience (Griffiths *et al.* 2000). These responses were consistent between two experiments conducted with the same soil sampled on two occasions one year apart. However, such relationships between diversity and functional resilience were not apparent in similar experiments applied to an arable soil where diversity was manipulated via a constructive approach based upon the reinoculation of sterile soil with progressively diluted non-sterile soil (Griffiths *et al.* 2001). Here, resistance and resilience of decomposition were manifest, but there was no interaction with diversity. Furthermore, when this constructive approach was applied to a permanent upland pasture soil there was no effect of diversity upon resistance to heat stress, and some indication that soils containing greater levels of diversity were more resilient, showing a greater rate of recovery. However, there was a strong and consistent trend of decreasing resistance to copper stress with decreasing diversity, but no resilience at all over 28 days (Fig. 20.6).

Impact of biodiversity upon individual organisms

The prevailing diversity within soils has a strong impact upon the growth and function of individual organisms. This is manifest as community-level controls upon organisms and processes (e.g. Janzen *et al.* 1995; Janzen & McGill 1995; de Nobili *et al.* 2001; Wheatley *et al.* 2001). This is important in relation to the development of effective biocontrol strategies, the persistence of undesirable organisms in soils, and the way in which soil diversity might develop in pristine environments. In a microcosm study, Toyota *et al.* (1996) demonstrated that the ability of the fungus *Fusarium oxysporum* to colonise discrete aggregates of soil was impaired by increasing levels of biodiversity within such aggregates (Fig. 20.7). This may be related to the availability of niches; in diverse systems, fewer vacant niches are likely. Alternatively allelopathic effects mediated by inhibitory compounds may be a mechanism. Greater levels of diversity may increase the propensity for allelopathy due to the number of potential interactions developing between organisms, as discussed above. However, some single species could be as effective at preventing colonisation by the fungus as fully diverse communities (Fig. 20.7b). This again demonstrates how diversity per se is not necessarily a key determinant of function, rather it is the properties that are manifest by the organisms that is pertinent.

Complexity

It is well established that a characteristic property of complex systems (i.e. those with a large number of components that may interact in a non-linear fashion)

(a) Heat perturbation

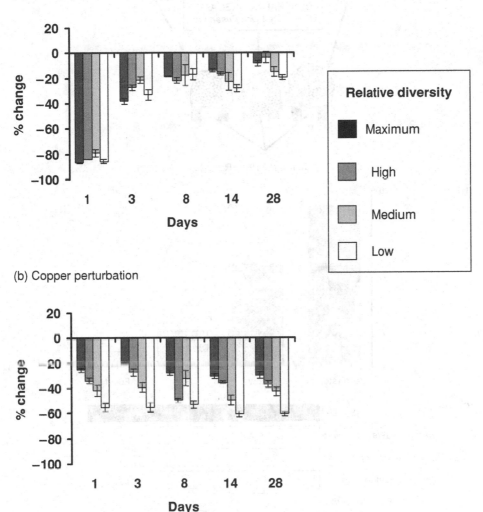

Figure 20.6. The percentage change in respiration (CO_2 evolution over 24 h), from powdered grass added to upland grassland soils, 1, 3, 8, 14 or 28 days after perturbation by (a) heat or (b) copper, compared to an unperturbed control. The soils were manipulated by irradiation and reinoculation with a 10^{-2} (■), 10^{-4} (■), 10^{-6} (■) or 10^{-8} (□) dilution of a non-sterile soil suspension, resulting in a decreasing net biodiversity within them. Data represent mean ± s.e., $n = 6$. Derived from Griffiths et al. (2004).

is the sensitivity to initial state. Identical starting configurations may result in quite different outcomes, a property sometimes ascribed to so-called chaotic systems. Another feature of complex systems is that of self-organisation, where an emergent property is organisationally impelled to arise under certain conditions. If soil structure is destroyed by physical force, then such structure typically

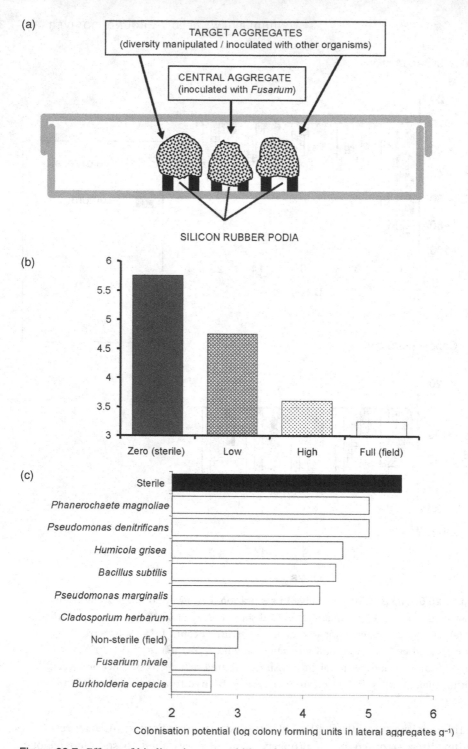

Figure 20.7. Effects of biodiversity upon ability of the fungus *Fusarium oxysporum* f. sp. *raphani* to colonise aggregates of soil. (a) Experimental system. (b) Effect of general reduction of diversity in target aggregates by deconstructive approach. (c) Effect of pre-colonisation of target aggregate by single species of fungi (stippled bars) and bacteria (spotted bars). Derived from Toyota *et al.* (1996).

reforms over time, with the soil biota strongly implicated in such processes (Young & Ritz this volume). There is almost no experimental work concerning the impact of soil biodiversity upon soil structural dynamics. De Ruiter *et al.* (this volume) show how weak links always apparently form in long loops in food webs; in other words, food web organisation is apparently driven to form stable systems. Given the close relationship between biodiversity and complexity, these concepts warrant further serious attention.

Upstairs, downstairs

In seeking over-arching principles, there is often discussion by soil ecologists as to whether and how basic ecological principles derived from the large body of knowledge relating to above-ground ecosystems can be mapped below ground (e.g. Wardle & Giller 1996), and this issue is frequently discussed within this volume. The evidence is equivocal: diversity:productivity curves and species abundance curves (e.g. Fig. 20.2) for soil systems sometimes, but not always, follow above-ground patterns and, in any case, the form of the relationship varies widely above ground (Waide *et al.* 1999). There little evidence that productivity–diversity relations exist in soil as they do above ground, other than at extremely unproductive sites where added resources increase productivity (Wardle 2002; Bardgett *et al.* this volume). It is stressed again that the critical difference between above-ground and below-ground systems is, obviously, *soil* and its attendant structural heterogeneity. Perhaps then soil ecologists need to think anew, and derive concepts from first principles, crucially involving soil structural frameworks as a fundamental basis in theories and models.

Does soil biodiversity matter?

To conclude with a simple question: *Does soil biodiversity matter?* The preceding chapters suggest the answer is *possibly*. If biodiversity is strictly defined as conventional richness, i.e. the base number of different species present, then for below-ground systems the answer is apparently no. Even at a planetary level, given that there have been five mass extinctions to date, with the Permian estimated to have involved the disappearance of over 90% of species, the general answer seems to be no. However, by expanding the question to '*Does soil biodiversity matter for the future of humankind in its present form?*', and accepting the looser definition of the term, then we have a more complicated question to which the simple answer is an unequivocal yes. Reasons abound in these pages and the literature. Monetary estimates of the value of soil biodiversity, crude though they must necessarily be, are substantial. But is current knowledge about the structure and function of soil communities adequate? The exciting work reported and discussed in this volume and the further questions it raises suggests that it is not, but the breadth demonstrates that substantial and significant progress is being made. A consequence of diversity is complexity, and all that entails for determining a mechanistic understanding.

Ten years ago, a similar symposium was held with the theme of opening the black box of the soil biomass (Ritz *et al.* 1994). Over the decade since that meeting, knowledge about the soil biota and its attendant functioning has vastly increased – we are now truly way *Beyond the Biomass*. The key task for the next decade is to integrate our increasing knowledge of what appears to be the most diverse and complex system on the planet and to attempt to formulate unifying theories. One more consequence of soil biodiversity is that this will require substantial and coordinated research on a global scale and some particularly clear thinking. The need is to revisit the ten tenets and six challenges of Wall *et al.* in the seminal first chapter of this volume. Clearly, *we need to establish a dialogue to make this happen.*

References

Anderson, I. C., Campbell, C. D. & Prosser, J. I. (2003). Potential bias of fungal 18S rDNA and internal transcribed spacer polymerase chain reaction primers for estimating fungal biodiversity in soil. *Environmental Microbiology,* 5, 36–47.

Anderson, T.-H. & Domsch, K. H. (1978). A physiological method for the quantitative measurement of microbial biomass in soils. *Soil Biology and Biochemistry,* 10, 215–221.

Andrén, O., Brussaard, L. & Clarholm, M. (1999). Soil organism influence on ecosystem-level processes bypassing the ecological hierarchy? *Applied Soil Ecology,* 11, 177–188.

Bardgett, R. D. & Griffiths, B. (1997). Ecology and biology of soil protozoa, nematodes and microarthropods. *Modern Soil Microbiology* (Ed. by J. D. van Elasas, E. Wellington and J. T. Trevors), pp. 129–163. New York: Marcel Dekker.

Bardgett, R. D. & Wardle, D. A. (2003). Herbivore-mediated linkages between aboveground and belowground communities. *Ecology,* 84, 2258–2268.

Blagodatskaya, E. V., Blagodatskii, S. A. & Anderson, T. H. (2003). Quantitative isolation of microbial DNA from different types of soils of natural and agricultural ecosystems. *Microbiology,* 72, 744–749.

Borneman, J., Skroch, P. W., O'Sullivan, K. M., *et al.* (1996). Molecular microbial diversity of an agricultural soil in Wisconsin. *Applied and Environmental Microbiology,* 62, 1935–1943.

Borneman, J. & Triplett, E. W. (1997). Molecular microbial diversity in soils from eastern Amazonia: evidence for unusual microorganisms and microbial population shifts associated with deforestation. *Applied and Environmental Microbiology,* 63, 2647–2653.

Bridge, P. & Spooner, B. (2001). Soil fungi: diversity and detection. *Plant and Soil,* 232, 147–154.

Cragg, R. G. & Bardgett, R. D. (2001). How changes in soil faunal diversity and composition within a trophic group influence decomposition processes. *Soil Biology and Biochemistry,* 33, 2073–2081.

Curtis, T. P., Sloan, W. T. & Scannell, J. W. (2002). Estimating prokaryotic diversity and its limits. *Proceedings of the National Academy of Sciences, USA,* 99, 10 494–10 499.

De Nobili, M., Contin, M., Mondini, C. & Brookes, P. C. (2001). Soil microbial biomass is triggered into activity by trace amounts of substrate. *Soil Biology and Biochemistry,* 33, 1163–1170.

Donnison, L., Griffith, G. S., Hedger, J., Hobbs, P. J. & Bardgett, R. D. (2000). Management influences on soil microbial communities and their function in botanically diverse haymeadows of northern England and Wales. *Soil Biology and Biochemistry*, 32, 253–263.

Finlay, B. J. & Esteban, G. F. (1998). Freshwater protozoa: biodiversity and ecological function. *Biodiversity and Conservation*, 7, 1163–1186.

Finlay, B. J., Esteban, G. F., Clarke, K. J. & Olmo, J. L. (2001). Biodiversity of terrestrial protozoa appears homogeneous across local and global spatial scales. *Protist*, 152, 355–366.

Fukami, T. & Morin, P. J. (2003). Productivity-biodiversity relationships depend on the history of community assembly. *Nature*, 424, 423–426.

Garciaplazaola, J. I., Becerril, J. M., Arreseigor, C., Gonzalezmurua, C. & Apariciotejo, P. M. (1993). *Rhizobium meliloti* denitrification in soils. *Plant Physiology*, 102, 177.

Gaston, K. J. (1996). *Biodiversity: A Biology of Numbers and Difference*. Oxford: Blackwell Science.

Gonzalez, G. & Seastedt, T. R. (2001). Soil fauna and plant litter decomposition in tropical and subalpine forests. *Ecology*, 82, 955–964.

Grayston, S. J., Campbell, C. D., Bardgett, R. D., *et al.* (2004). Assessing shifts in microbial community structure across a range of grasslands of differing management intensity using CLPP, PLFA and community DNA techniques. *Applied Soil Ecology*, 25, 63–84.

Greenland, D. J. & Szabolcs, I. (1994). *Soil Resilience and Sustainable Land Use*. Wallingford: CAB International.

Griffiths, B. S., Kuan, H. L., Ritz, K., *et al.* (2004). The relationship between microbial community structure and functional stability, tested experimentally in an upland pasture soil. *Microbial Ecology*, 47, 104–113.

Griffiths, B. S., Ritz, K., Bardgett, R. D., *et al.* (2000). Ecosystem response of pasture soil communities to fumigation-induced microbial diversity reductions: an examination of the biodiversity–ecosystem function relationship. *Oikos*, 90, 279–294.

Griffiths, B. S., Ritz, K., Ebblewhite, N. & Dobson, G. (1999). Soil microbial community structure: effects of substrate loading rates. *Soil Biology and Biochemistry*, 31, 145–153.

Griffiths, B. S., Ritz, K., Wheatley, R. E., *et al.* (2001). An examination of the biodiversity-ecosystem function relationship in arable soil microbial communities. *Soil Biology and Biochemistry*, 33, 1713–1722.

Hansen, R. A. (2000). Effects of habitat complexity and composition on a diverse litter microarthropod assemblage. *Ecology*, 81, 1120–1132.

Hedlund, K., Regina, I. S., van der Putten, W. H., *et al.* (2003). Plant species diversity, plant biomass and responses of the soil community on abandoned land across Europe: idiosyncracy or above–belowground time lags. *Oikos*, 103, 45–58.

Hill, T. C. J., Walsh, K. A., Harris, J. A. & Moffett, B. F. (2003). Using ecological diversity measures with bacterial communities. *FEMS Microbiology Ecology*, 43, 1–11.

Janzen, R. A., Dormaar, J. F. & McGill, W. B. (1995). A community-level concept of controls on decomposition processes: decomposition of barley straw by *Phanerochaete chrysoporium* or *Phlebia radiata* in pure or mixed culture. *Soil Biology and Biochemistry*, 27, 173–179.

Janzen, R. A. & McGill, W. B. (1995). Community-level interactions control the proliferation of *Azospirillium brasilense* Cd in microcosms. *Soil Biology and Biochemistry*, 27, 189–196.

Jonsson, L. M., Nilsson, M. C., Wardle, D. A. & Zackrisson, O. (2001). Context dependent effects of ectomycorrhizal species richness on tree seedling productivity. *Oikos*, 93, 353–364.

Kabir, S., Rajendran, N., Amemiya, T. & Itoh, K. (2003). Quantitative measurement of fungal DNA extracted by three different methods using real-time polymerase chain reaction. *Journal of Bioscience and Bioengineering*, 96, 337–343.

Kaneko, N. & Salamanca, E. F. (1999). Mixed leaf litter effects on decomposition rates and soil microarthropod communities in an oak–pine stand in Japan. *Ecological Research*, **14**, 131–138.

Kowalchuk, G. A., Buma, D. S., de Boer, W., Klinkhamer, P. G. L. & van Veen, J. A. (2002). Effects of above-ground plant species composition and diversity on the diversity of soil-borne microorganisms. *Antonie Van Leeuwenhoek International Journal of General and Molecular Microbiology*, **81**, 509–520.

Lemanceau, P., Corberand, T., Gardan, L., *et al.* (1995). Effect of two plant species, flax (*Linum usitatissinum* L.) and tomato (*Lycopersicon esculentum* Mill.), on the diversity of soilborne populations of fluorescent pseudomonads. *Applied and Environmental Microbiology*, **61**, 1004–1012.

Liddell, C. M. & Hansen, D. (1993). Visualising complex biological interactions in the soil ecosystem. *Journal of Visualisation and Computer Animation*, **4**, 3–12.

Mahaffee, W. F. & Kloepper, J. W. (1997). Temporal changes in the bacterial communities of soil, rhizosphere, and endorhiza associated with field-grown cucumber (*Cucumis sativus* L.). *Microbiology Ecology*, **34**, 210–223.

Maire, N., Borcard, D., Laczko, E. & Matthey, W. (1999). Organic matter cycling in grassland soils of the Swiss Jura mountains: biodiversity and strategies of the living communities. *Soil Biology and Biochemistry*, **31**, 1281–1293.

Marilley, L., Vogt, G., Blanc, M. & Aragno, M. (1998). Bacterial diversity in the bulk soil and rhizosphere fractions of *Lolium perenne* and *Trifolium repens* as revealed by PCR restriction analysis of 16S rDNA. *Plant and Soil*, **198**, 219–224.

Marshall, M. N., Cocolin, L., Mills, D. A. & VanderGheynst, J. S. (2003). Evaluation of PCR primers for denaturing gradient gel electrophoresis analysis of fungal communities in compost. *Journal of Applied Microbiology*, **95**, 934–948.

McCaig, A. E., Glover, E. A. & Prosser, J. I. (1999). Molecular analysis of bacterial community structure and diversity in unimproved and improved grass pastures. *Applied and Environmental Microbiology*, **65**, 1721–1730.

McNaughton, S. J. (1994). Biodiversity and function of grazing ecosystems. *Biodiversity and Ecosystem Function* (Ed. by E. D. Schulze & H. A. Mooney). Berlin: Springer-Verlag.

Mikola, J. & Setälä, H. (1998). Relating species diversity to ecosystem functioning: mechanistic backgrounds and experimental approach with a decomposer food web. *Oikos*, **83**, 180–194.

Naeem, S., Loreau, M. & Inchausti, P. (2002). Biodiversity and ecosystem functioning: the emergence of a synthetic ecological framework. *Biodiversity and Ecosystem Functioning* (Ed. by M. Loreau, S. Naeem & P. Inchausti), pp. 3–11. Oxford: Oxford University Press.

Nannipieri, P., Ciardi, C. & Badalucco, L. (1986). A method to determine soil DNA and RNA. *Soil Biology and Biochemistry*, **18**, 275–281.

Pennanen, T., Paavolainen, L. & Hantula, J. (2001). Rapid PCR-based method for the direct analysis of fungal communities in complex environmental samples. *Soil Biology and Biochemistry*, **33**, 697–699.

Porazinska, D. L., Bardgett, R. D., Blaauw, M. B., *et al.* (2003). Relationships at the aboveground–belowground interface: plants, soil biota, and soil processes. *Ecological Monographs*, **73**, 377–395.

Ritz, K., Dighton, J. & Giller, K. E. (1994). *Beyond the Biomass: Compositional and Functional Analysis of Soil Microbial Communities.* Chichester: Wiley.

Ritz, K. & Griffiths, B. S. (2000). Implications of soil biodiversity for sustainable organic matter management. *Sustainable Management of Soil Organic Matter* (Ed. by R. M. Rees, C. A. Watson, B. C. Ball & C. D. Campbell), pp. 343–356. Wallingford: CAB International.

Ritz, K., Griffiths, B. S., Torsvik, V. L. & Hendriksen, N. B. (1997a). Analysis of soil and bacterioplankton community DNA by melting profiles and reassociation kinetics. *FEMS Microbiology Ecology*, **149**, 151–156.

Ritz, K., Wheatley, R. E. & Griffiths, B. S. (1997b). Effects of animal manure application and crop plants upon size, activity and survival of soil microbial biomass under organically grown spring barley. *Biology and Fertility of Soils*, **24**, 378–383.

Setälä, H. & McLean, M. A. (2004). Decomposition rate of organic substrates in relation to the species diversity of soil saprophytic fungi. *Oecologia*, **139**, 98–107.

Siciliano, S. D., Theoret, J. R., de Freitas, J. R., Huci, P. J. & Germida, J. J. (1998). Differences in the microbial communities associated with the roots of different cultivars of canola and wheat. *Canadian Journal of Microbiology*, **44**, 844–851.

Torsvik, V. L., Goksøyr, J. & Daae, F. L. (1990). High diversity in DNA of soil bacteria. *Applied and Environmental Microbiology*, **56**, 782–787.

Torsvik, V. L., Goksøyr, J., Daae, F. L., et al. (1994). Use of DNA analysis to determine the diversity of microbial communities. *Beyond the Biomass: Compositional and Functional Analysis of Soil Microbial Communities* (Ed. by K. Ritz, J. Dighton & K. E. Giller), pp. 39–48. Chichester: Wiley.

Toyota, K., Ritz, K. & Young, I. M. (1996). Microbiological factors affecting the colonisation of soil aggregates by *Fusarium oxysporum* F. sp. *raphani*. *Soil Biology and Biochemistry*, **28**, 1513–1521.

Treves, D. S., Xia, B., Zhou, J. & Tiedje, J. M. (2003). A two-species test of the hypothesis that spatial isolation influences microbial diversity in soil. *Microbial Ecology*, **45**, 20–28.

Van der Heijden, M. G. A., Klironomos, J. N., Ursic, M., et al. (1998). Mycorrhizal fungal diversity determines plant biodiversity, ecosystem variability and productivity. *Nature*, **396**, 69–72.

Waide, R. B., Willig, M. R., Steiner, C. F., et al. (1999). The relationship between productivity and species richness. *Annual Review of Ecology and Systematics*, **30**, 257–300.

Wardle, D. A. (2002). *Communities and Ecosystems: Linking the Aboveground and Belowground Components*. Princeton, NJ: Princeton University Press.

Wardle, D. A. & Giller, K. E. (1996). The quest for a contemporary ecological dimension to soil biology. *Soil Biology and Biochemistry*, **28**, 1549–1554.

Watts, C. W., Eich, S. & Dexter, A. R. (2000). Effects of mechanical energy inputs on soil respiration at the aggregate and field scales. *Soil and Tillage Research*, **53**, 231–243.

Wheatley, R. E., Ritz, K., Crabb, D. & Caul, S. (2001). Temporal variations in potential nitrification dynamics in soil related to differences in rates and types of carbon and nitrogen inputs. *Soil Biology and Biochemistry*, **33**, 2135–2144.

Zak, D. R., Ringelberg, D. B., Pregitzer, K. S., et al. (1996). Soil microbial communities beneath *Populus grandidentata* grown under elevated atmospheric CO_2. *Ecological Applications*, **6**, 257–262.

Zhou, J. Z., Xia, B. C., Treves, D. S., et al. (2002). Spatial and resource factors influencing high microbial diversity in soil. *Applied and Environmental Microbiology*, **68**, 326–334.

Index

Printed in the United States
By Bookmasters